Obesity and Diabetes Mellitus

Obesity and Diabetes Mellitus

E.I. Sokolov

///VSP///
Utrecht, The Netherlands, 1997

VSP BV
P.O. Box 346
3700 AH Zeist
The Netherlands

First published in 1997

ISBN 90-6764-212-6

Printed in the Netherlands by Ridderprint bv, Ridderkerk.

CONTENTS

LIST OF ABBREVIATIONS

ACTH	Adrenocorticotropic hormone
ADP	Adenosine diphosphate
AI	Atherogenic index
AP	Arterial pressure
Apo-A (B, C)	Apolipoprotein A (B, C)
AT	Antithrombin
ATP	Adenosine triphosphate
BWI	Body weight index
BPM	Beats per minute
CA	Catecholamines
cAMP	Cyclic adenosine monophosphate
cGMP	Cyclic guanosine monophosphate
Cl	Clearance
CL	Cardiolipin
CM	Chylomicrons
CNS	Central nervous system
CRLV	Contraction rate of left ventricle
CS	Cholesterol
CSE	Cholesterol esters
DIC	Disseminated intravascular coagulation (syndrome)
DLP	Dyslipoproteinemia
DM	Diabetes mellitus
DNA	Desoxyribonucleic acid
DOPA	Dihydroxyphenylalanine
DPR	Duration of plasma recalcification
DT	Diet-dependent thermogenesis
EDD	End diastolic distance
EDV	End diastolic volume
EEI	Energy expenditure index
EF	Ejection factor
ESD	End systolic distance
ESV	End systolic volume
F	Fibrinogen
FAB	Fibrinolytic activity of blood

FC	Free cholesterol
FFA	Free fatty acids
FPOL	Free-radical peroxide oxidation of lipids
GC	Glucocorticoids
GHDLP	Glycosylized high density lipoproteins
HDLP	High density lipoproteins
HL	Hyperlipacidemia
HMG	3-hydroxy-3-methyl glutaryl
HSL	Hormonosensitive lipase
Ht	Hematocrit
HTG	Hypertriglyceridemia
IDLP	Intermediate density lipoproteins
IHD	Ischemic heart disease
IIDM	Insulin-independent diabetes mellitus
IRI	Immunoreactive insulin
LA	Lactic acid
LAT	Lecithin acyl transferase
LCAT	Lecithin cholesterol-acyl transferase
LDLP	Low density lipoproteins
LP	Lipoprotein
LPC	Lysophosphatidylcholine
LPL	Lipoproteinlipase
MOC	Maximum oxygen concentration
NA	Noradrenaline
NADP, $NADPH_2$	Oxidized and reduced forms of nicotinamide adenine dinucleotide phosphate (coenzyme II)
NEFA	Nonesterified fatty acids
NMOC	Normal maximum oxygen consumption
PC	Phosphatidylcholine
pct	Pulse coefficient
PEA	Phosphatidylethanolamine
PG	Prostaglandin
PL	Phospholipids
PPh	Plasmapheresis
PRA	Pyruvic acid
PrC	Prostacyclin

PS	Phosphatidylserine
PUFA	Polyunsaturated fatty acids
RNA	Ribonucleic acid
RRLV	Relaxation rate of left ventricle
SFMC	Soluble fibrin-monomer complexes
SM	Sphingomyelin
SNS	Sympathetic nervous system
STH	Somatotropic hormone
TAF	Thrombocyte activation factor
TF	Thrombocyte factor
TG	Triglycerides
TP	Total phospholipides
TPH	Tolerance of plasma to heparin
TPR	Total peripheral resistance
TSH	Thyroid-stimulating hormone
TX	Thromboxan
V_{cf}	Rate of circular shortening of the myocardium fiber
VLDLP	Very low density lipoproteins

INTRODUCTION

Obesity develops when the energy received with food exceeds its expenditure. There is no doubt that obesity is a genetically determined process and may be inherited. Lipid tissue is a highly active link in metabolism and functions under conditions of a constant exchange with its other participants. In high caloric diet lipogenesis exceeds lipolysis resulting in an increased triglyceride accumulation in adipocytes. Obesity leads to metabolic and humoral changes that may be completely reversible when the body weight returns to its previous "ideal" level. Clinicians noted long ago that diabetes mellitus (DM) develops more often in obese persons with disturbed carbohydrate and lipid metabolism.

Patients with obesity have been revealed to have an increased appetite, hyperinsulinism, and high activity of the cortical suprarenal hormones. Metabolic disturbances appear when obesity is combined with hyperinsulinism and a disturbed tolerance to glucose and hyperglycemia. These metabolic disturbances cause DM to develop. Lipid tissue metabolism and insulin secretion are closely related to the activity of the autonomous nervous system. Catecholamines are its most important link. Lipolysis in the lipid tissue and the level of free lipid acids are controlled by adrenaline and noradrenaline.

Hypertriglyceridemia and insulin resistance are close companions. Persons with hypertriglyceridemia have a low level of glucose utilization by tissues. It was also shown experimentally that VLDLP decrease glucose utilization by tissues and stimulate insulin secretion provoking hyperinsulinemia. The high level of latter stimulates the synthesis of triglycerides that are part of VLDLP and thus circulus virtiosus is closed. The intensification of glucogenesis in obesity stimulates hyperglycemia and hyperinsulinemia. In DM the transformation of carbohydrates into fats is an insulin-dependent process and takes place in lipid tissue. In these patients, lipoproteinlipase plays an important role in all lipid metabolic transformations, as the lipid acids supply of cells depends on this substance. The circulating lipid acids perform fat synthesis from glucose in DM patients. Lipid acids reach the lipid depots as a triacylglycerol component of lipoproteins and chylomicrons. The hyperlipidemia in DM depends on the severity of the disease, the nature of its decompensation, and the level of hypertriglyceridemia. The blood level of free lipid acids generally increases in all DM patients, especially in the ketose state.

Insulin promotes glucose penetration into the lipocytes, enhances its transformation into glycerophosphate and lipid acids. It not only augments the synthesis, but also diminishes fat dissociation and lipid acid liberation from the lipid tissue. The disturbed carbohydrate metabolism in DM patients involves suppression of liver carbohydrate oxidation and intensification of lipid oxidation. The increased synthesis of hepatic carnitine in DM patients intensifies lipid acid oxidation.

CHAPTER 1. FEATURES OF LIPID METABOLISM IN DIABETES MELLITUS AND ISCHEMIC HEART DISEASE

1.1. Metabolic Reaction of Lipoproteins and Hormones in a Healthy Person's Organism

Clinical studies of recent years revealed that the main risk factors of IHD and atherosclerosis are disorders of LP metabolism – dyslipoproteinemia (DLP), arterial hypertension, smoking, obesity, DM. Disorder of LP metabolism is almost always a component of IHD and atherosclerosis The most severe manifestation of hyperlipidemia – a growth in the level of total cholesterol in the blood plasma, i.e. hypercholesterolemia – is widely known. A growth in the content of total cholesterol to above 225 mg/dl drastically increases the probability of the development of atherosclerosis. If the level of total cholesterol is 275 mg/dl and more, atherosclerosis appears still more rapidly [30, 51, 77, 189, 267, 301, 537].

In clinical observations, the pathogenesis of IHD is considered from the viewpoints of the atherogenic effect of DLP, i.e. the qualitative and quantitative disturbances of the level of blood components such as cholesterol, triglycerides, apoproteins, that are transported in the blood in the composition of LP of a very low density (VLDLP), low density (LDLP), and high density (HDLP). Lipoproteins form the structure basis of all the biological membranes of the cells of the blood and endothelium. A considerable part of them forms in the liver and in the walls of the small intestine. The LP are the main transportation form of lipids in an organism.

Structurally, LP are an association of lipids and proteins in various proportions. In the rate of settling in ultracentrifuging, they are divided into five main classes: (i) chylomicrons (CM) – they contain 2% of protein and 98% of lipids (mainly TG up to 90%); under normal conditions no CM are contained in the blood plasma when food is not being taken, (ii) VLDLP – 10% of protein and 90% of lipids (mainly TG); (iii) lipoproteins of an intermediate density (IDLP) – they are intermediate between VLDLP and LDLP in their physicochemical composition, (iv) LDLP contain 25% of protein and 75% of lipids (mainly cholesterol esters), and (v) HDLP – about 50% of protein and 50% of lipids (phospholipids form about 30%, TG and free cholesterol – 70%). Apoproteins are the protein component of LP.

The main functional role of chylomicrons is the transportation of triglycerides. They form in the walls of the small intestine and then along the lymph tracts get into a lymphatic duct and the blood. They are not atherogenic because they cannot penetrate into an artery wall owing to the large particle size [107, 167, 176, 292].

The VLDLP mainly synthesized in the liver and to a smaller extent in the wall of the small intestine are a significant atherogenic factor. They are the main transportation form of endogenic triglycerides. They are atherogenic, because owing to the smaller particle size (in comparison with that of the chylomicrons), they penetrate more readily into an artery wall and may remain in it. Moreover, in the hydrolysis of triglycerides they first transform into IDLP – remnant fine particles that can also penetrate with the blood into an arterial wall [33, 74, 302, 347, 351, 377].

The LDLP, whose main function is the transportation of cholesterol into the peripheral cells, are the richest in it; their particles are still smaller, owing to which the LDLP readily penetrate into a cell and can be deposited in it. They form mainly in the hydrolysis of TG-containing VLDLP with the participation of the enzyme lipoproteinlipase (LPL). In healthy persons, two thirds of the plasma cholesterol is contained in the LDLP. Of all the classes of LP, the most atherogenic are LDLP.

In recent years, great significance is being attached to HDLP (in particular, alpha-LP), which are divided in turn into two subclasses: $HDLP_2$ and $HDLP_3$ differing in their average hydrated density, magnitude of the particles, physicochemical composition, and functions. The main function of the LP of this class is the transport of cholesterol and phospholipids (PL) from a cell. The HDLP are synthesized in the liver and small intestine [116, 148, 270, 333, 423]

The work of the LP transportation system is determined in many ways by the function of the apolipoproteins, which create the pathophysiological basis of development (or inhibition) of atherosclerosis of the coronary arteries. By forming soluble complexes with the lipids owing to which the latter are readily transported by the blood, they activate individual reactions of lipid metabolism and participate in a reaction with the specific receptors of a cell, ensuring the capture of LP particles from the cells.

The apolipoproteins B (apo-B) are the main protein component of LDLP, VLDLP, and chylomicrons. The catabolism of apo-B-LP occurs in the peripheral tissues. By combining with receptors on the surface membrane of the cells, the apo-B-LP are transported into the cells, where the hydrolysis reactions proceed.

A growth in the level of apo-B-LP is generally attended by the transportation of more cholesterol into the tissues and membranes of the cells.

The main part of apo-A-1 is in the HDLP; an important role in the catabolism of apo-A-1 is played by parenchymal cells of the liver. Some authors discovered a relation between the LPL level in the blood and the apo-A-1 concentration. Forming the main part of the protein component of the HDLP, the apo-A-1 accepts and transports cholesterol into the liver and accordingly ensures its catabolism.

As already indicated, metabolism of the plasma LP regulates the content of the specific apo-LP, the main ones of which are apo-A-1, apo-A-2, apo-A-4, and apo-B. Moreover, there are also minor apo, namely apo-E, apo-C-1, apo-C-2, and apo-C-3. From the standpoints of functional biochemistry, the apo-LP play the role of specific enzymes or cofactors [58, 88, 116, 234, 453, 567].

The apo-E/apo-A-1 ratio is an important criterion of the atherogenic properties of healthy persons and of patients with IHD. The atherogenic index (AI) determines the relation between the intensity of transporting cholesterol to the tissues and the rate of its removal from the cells. The AI is calculated from the content of the total cholesterol C_{tot} and that of the HDLP C_{hdlp}:

$$\frac{C_{tot} - C_{hdlp}}{C_{hdlp}}$$

The functional reaction of the apolipoproteins with the tissue elements ensures a system of transportation of the lipids in an organism and facilitates the formation of atherosclerotic changes in the vessels. The directed transfers of apolipoproteins between the lipoproteins of different classes during fasting or taking food, performing physical exercise, and taking various metabolic drugs determine in every person the individual nature of the physiological relations of lipid metabolism.

When speaking of apo-A-1 (the main protein of HDLP), we cannot fail to consider briefly the antiatherogenic role of HDLP.

It was shown in clinical and experimental studies that the antiatherogenic function of HDLP is achieved as a result of the acception by them of unesterified cholesterol from the membranes of the smooth muscle and endothelial cells of the vessels, and also from the VLDLP and chylomicrons in the hydrolysis of their triglycerides in the blood plasma. The biochemical transformation of unesterified cholesterol into the esterified material proceeds with the participation of the enzyme lecithin acyl transferase (LAT). The product of a reaction catalyzed by LAT – esterified cholesterol – forms only on the surface of HDLP particles containing apo-A-1 in a complex with apo-D and LAT [490, 515, 617].

The particles of $HDLP_2$ transfer cholesterol esters into the liver, where the cholesterol transforms into bile acids that are excreted from the organism. The system of withdrawing cholesterol from tissues in a healthy organism achieved by HDLP consists of several units, each of which performs a strictly definite function: (i) the protein-lipid structure of the surface monolayer ensures the building into it of unesterified cholesterol, (ii) the enzyme system conducts a reaction catalyzed by LAT with its activator apo-A-1 and inhibitor apo-A-2, (iii) the system of nonenzymatic directed transportation of cholesterol esters includes lipid-carrying proteins.

This self-regulating system under physiological conditions performs an antiatherogenic function autonomously. The degree of activity of the entire system is determined in many ways by the activity of the LAT and the intensity of removing the product of the reaction it catalyzes (esterified cholesterol) from the surface of the HDLP particles. When the rate of this reaction is higher, the inflow to the fine spherical particles of HDLP of the unesterified cholesterol from both the LP enriched in TG (chylomicrons and VLDLP) and the cell membranes is more active.

The development of the atherosclerotic process in IHD patients is affected by the ratio of the intensities of two nonenzymatic processes of transportation of cholesterol esters, namely, in the composition of $HDLP_2$ into the liver (for the removal of cholesterol from an organism), and in the composition of VLDLP and LDLP into the smooth muscle and endothelial cells of a vessel wall [115, 203, 282, 288].

In view of the fact that the nature of lipid metabolism in every person is an integrative quantity depending both on genetic and environmental factors (the nature of feeding, physical activity, level of hormonal regulation, age, etc.), we studied 37 healthy men from 20 to 51 years old (23 from 20 to 29 years old, 6 from 30 to 39, and of 8 from 40 to 51). All of them were mental workers: students and instructors of a medical institute (22), scientific workers (7), physicians and engineers (8). None of them had arterial hypertension, endocrine disorders, obesity, diseases of the liver or kidneys, or took drugs during the investigation period, To exclude IHD, all of them were examined with the aid of instrumental and laboratory methods. No indications of disturbance of coronary blood circulation or other lesions of the myocardium were discovered on the ECG at rest. When making tests with a physical load according to the conventional scheme, all of the examined persons revealed the retaining of a high physical workability. The individual power of the threshold load varied from 900 to 1320 kgm/min (on an average 1040±42 kgm/min). No indications of ischemia of the myocardium, disorders of the rhythm and conductance (according to the results of the ECG), appearance of other symptoms requiring the stopping of the physical load in veloergometry was discovered in any of the persons tested. When collecting anamnesis for excluding IHD, we used J.Rose's standard questionnaire.

Table 1. Distribution of Healthy Persons with Consideration of the IHD Risk Factors.

Group	Age, years	BWI	Smokers, % of total number in group	Inherited predisposition, % of total number in group
1-я	26	23	75	25
2-я	34	26	71	71

The content of cholesterol, triglycerides, and HDLP cholesterol was determined by the chemical method in an analyzer AA-11 of the Tekhnikon firm, and the content of apo-A-1 and apo-B by the method of quantitative rocket immunoelectrophoresis with the use of antiserums at the Center of prophylactic medicine.

To reveal the group with atherogenic DLP, we used the criteria proposed by the European society for atherosclerosis. We included in this group persons in whom the content of cholesterol on an empty stomach was over 200 mg/dl, of TG over 200 mg/dl, and of HDLP cholesterol under 35 mg/dl. Simultaneously with determination of the lipid parameters, we measured the level of the hormones insulin and cortisole.

Group 1 (persons with normal lipid parameters of the LP spectrum, i.e. a content of cholesterol under 200 mg/dl, of TG under 200 mg/dl, and of HDLP cholesterol over 35 mg/dl) included 20 persons (54%), group 2 (persons with atherogenic disorders, i.e. with an increased content of cholesterol or triglycerides, or with a reduced level of alpha cholesterol) included 17 persons (46%). Apart from the differences in the background parameters (cholesterol, TG, HDLP cholesterol), in group 2 persons the content of LDLP cholesterol, the level of apo-B-1, and the ratio apo-B/apo-A-1 were higher. The differences in the content of HDLP cholesterol and apo-A-1 in these two groups are not authentic.

In connection with the fact that in the anamnesis of the healthy persons of both groups there were no diseases capable of causing secondary DLP, we singled out the factors that cause dyslipidemia in them. They were as follows:

(1) an excess body mass with a body weight index (BWI) over 25;

(2) regular smoking;

(3) hypodynamia (when working sitting down over 50% of the working time);

(4) an inherited predisposition (in persons whose parents suffered from cardiovascular diseases such as IHD, hypertension, insultus).

When analyzing these factors, we found that men of group 2 (with DLP) were older, had a higher body weight index, and most of them had an inherited predisposition to

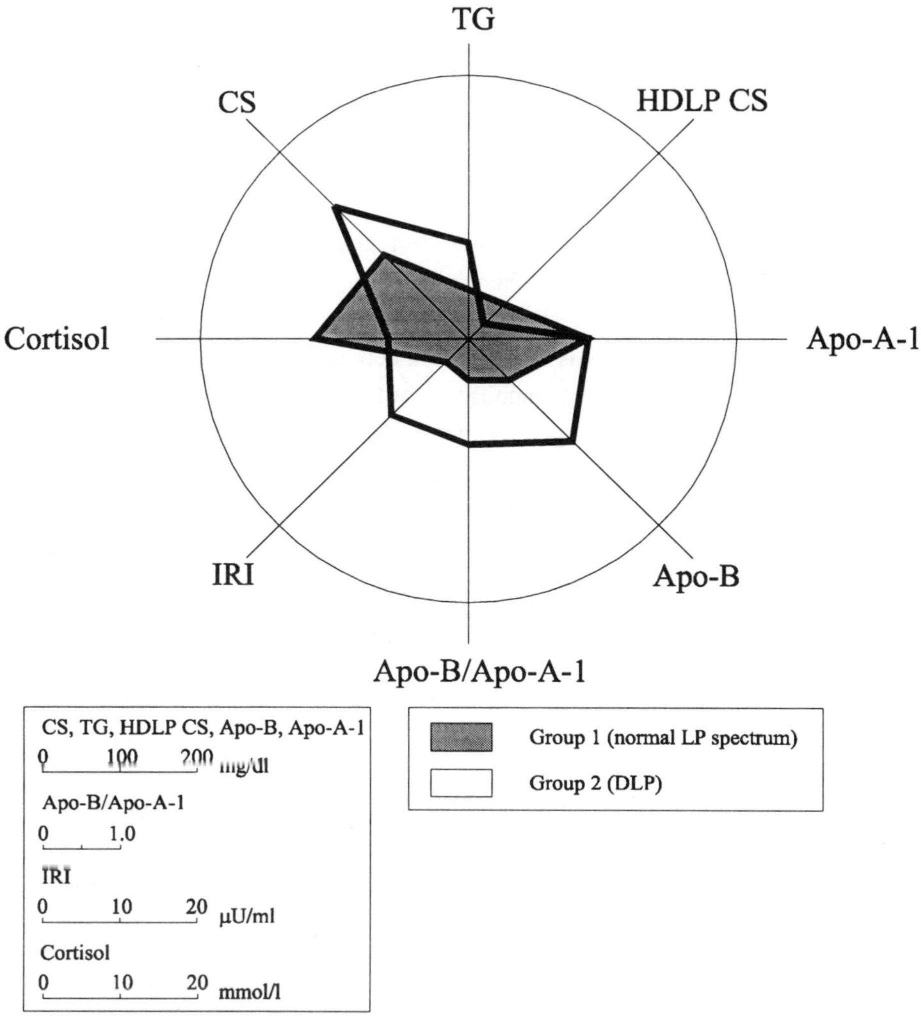

Fig. 1. Content of Lipids, Apoproteins, and Hormones in Blood of Healthy Persons.

IHD and hypodynamia. The number of smokers was about the same in both groups (Table 1).

Consequently, the differences in the lipid parameters in the two groups may be due either to hereditary factors or to a difference in age, body weight, mode of life, where the main role is played by improper feeding and hypodynamia.

When comparing the parameters of lipid metabolism in healthy persons of groups 1 and 2, we noted clear distinctions, namely, the cholesterol level in group 1 was 166 mg/dl, and in group 2 it was 260 mg/dl (Fig. 1).

What is such a difference in the cholesterol level due to?

Whether or not an excess of cholesterol in the food causes the appearance of "atherogenic factors" still remains a puzzle. In experiments involving the creation of hypercholesterolemia, we can increase the cholesterol content in the cells, but must remember that it is distributed unevenly in a human organism: 7% of the cholesterol is in the transportation form in the composition of the LP circulating in the blood, while 93% of it is in the cells. Normally homeostasis is maintained between these two forms of cholesterol. The latter gets into a liver cell in the composition of LDLP owing to the presence of LDLP receptors. Confirmation of the fact that the amount and activity of these LDLP receptors is regulated automatically by the level of the cholesterol in the blood is very interesting. Atherosclerosis is incurred most frequently when mismatching of the regulatory processes of exchange between the LP in the plasma and the cholesterol in the cells occurs in an organism. The function of the key regulators in matching is played by the LDLP receptors whose molecular defects in atherosclerosis associated with hereditary factors play the leading role. A strategic direction in studying atherosclerosis is the studying of the bioregulators responsible for the redistribution of the cholesterol flows between the blood and wall of a vessel. This is a promising direction when developing methods of therapeutic action on atherosclerosis.

With a growth in the total cholesterol content in the blood and a change in the rate of the forward and reverse transportation thereof into the tissues of the organs of the cardiovascular system, a multitude of adaptive changes appears. An important role here is played by activation of the peroxide oxidation of the lipids in the cells and tissues, consequently, the addition of antioxidants for diminishing side effects on various enzyme systems, including membrane ATPases, lowers the damaging effect in an atherosclerotic process [77, 99, 126, 519].

An important matter in cholesterol metabolism is the presence of specific protein receptors for LDLP on the surface of membranes of endothelial and smooth muscle cells. With the aid of these receptors, the lipoprotein particles enter the lysosomes for decomposition. A biochemical reaction in the lysosomes yields unesterified cholesterol that has a multilateral effect on a cell. In particular, it suppresses the key enzyme hydroxymethylglutaryl-CoA-reductase regulating the rate of cholesterol biosynthesis. The functional activity of the receptors on the membranes of the endothelial and smooth muscle cells ensures the capture of LDLP by a cell, thus retaining the normal level of the LP. It is exactly the functional activity of an endothelial cell with an increased receptor-dependent capture of LDLP that creates conditions for the prevention of atherosclerosis. Prolonged hyperlipidemia disturbs the physicochemical characteristics of the endothelial membranes with the subsequent change of both the receptor apparatus and the enzymatic properties of the cells. After this, the excess of cholesterol in the blood causes damaging of the endothelial cells and the rapid development of atheromatous plaques [169, 347, 444, 454, 515].

When analyzing such an important parameter of lipid metabolism as the triglyceride content in the blood of persons of both groups, we noted the following distinctions: in group 1 it is 70 mg/dl, and in group 2 – 109 mg/dl. In the physiological studies it was found that the level of the triglycerides is closely related to the activity of LPL in the blood.

Triglycerides are a form of accumulating and carrying energy in an organism, and a large amount of them is constantly carried between individual organs, This exchange is carried out with the aid of such specific transportation forms as chylomicrons and VLDLP, and also in the form of the products of hydrolysis of TG, FFA, and glycerol [315, 316, 376, 425, 489].

The energy needed for an organism (90%) is ensured by the free fatty acids (FFA), which have a very high rate of metabolism. Notwithstanding intense metabolism, the content of the FFA in the serum is very insignificant. After a fat food load, the major part of the FFA forms from the food fats along the way of their transformation into chylomicrons. Under the influence of the lipase of the fat tissue, the TG of a fat depot decomposes into glycerol and fatty acids (FA) that are used for the resynthesis of the triglycerides or enter the blood.

Many clinicians give special significance to the level of the HDLP cholesterol. A drop in this parameter is an unfavorable prognostic feature. We can see from Fig. 1 that if in group 1 the HDLP level was 52.0 mg/dl, in group 2 it was 45.0 mg/dl. The apo-A-1 content is almost the same: 139.0 mg/dl in group 1 and 142.0 in group 2.

A higher apo-B level is a very important indication of dyslipoproteinemia in healthy persons of group 2. If in group 1 it is 100.0 mg/dl, in group 2 it is 163.0. Accordingly, the parameter apo-B/apo-A-1 in group 1 is 0.72 and in group 2 it is 1.17. Generally in IHD, this parameter is greater than unity.

Our investigations showed that in persons of group 2 the level of HDLP lowers, and the content of apo-B grows. This suggests the following conclusion: first of all, in persons of group 2, the removal of cholesterol from the tissues into the liver (performed by the HDLP) is disturbed. The increased level of apo-B in the studied persons of group 2 also facilitates the development of atherosclerotic changes [234, 302, 310, 342, 347, 425, 625].

The apo-B/apo-A-1 ratio determines the relation between the intensity of transportation of cholesterol into the tissues and the rate of its removal from the cells. A growth in the level of apo-B-LP is generally attended by the increased transportation of cholesterol into the tissues and membranes of the cells.

Major significance in regulating lipid metabolism in an organism is attached to the state of hormonal regulation, especially to the activity of insulin, cortisol, the adrenocorticotropic hormone (ACTH), and the somatotropic hormone (STH). Most investigators consider that insulin plays a special role here.

Many clinicians note the relation between the insulin content in the blood and the level of triglycerides. In some patients, this can be explained by obesity, which is attended by hypertriglyceridemia, in others obesity is absent, but they have resistance to insulin.

LPL is discovered chiefly in adipose tissue and the skeletal muscles, where it is bound to glycosaminoglycanes. There is also a liver lipase that is not sensitive to inhibitors and activators. Both lipases participate in the catabolism of VLDLP and chylomicrons, i.e. lipoproteins rich in triglycerides.

It was proved that LPL is synthesized under the influence of insulin, while when there is a lack of the latter there is a deficiency of this enzyme. In DM, when there is a lack of insulin, the level of LPL drops, and a large amount of lipoproteins rich in triglycerides accumulate in the blood (diabetic lipemia) [420, 423, 424, 469, 474, 547].

When comparing the background level of hormones in persons of the two groups, we revealed a difference in the insulin level. If in persons of group 1 its value was 5.2 μU/ml, in these of group 2 it was 10.9 μU/ml. The content of cortisol and ACTH differed insignificantly: in group 1 persons the cortisol level was 18.6 mg/dl and of ACTH was 15.0 pg/ml, in group 2 the relevant figures were 16.1 mg/dl and 25.8 pg/ml.

How ACTH affects lipid metabolism (in particular cholesterol) remains to be clarified. When ACTH is administered, some authors noted a growth in the cholesterol level, while others did not. When ACTH was administered to animals, the content of free fatty acids and total fatty acids increased. This allows us to conclude that the level of triglycerides associated with the metabolism of the FFA grows. Klimov [291] considers that a growth in the level of FFA accelerates the synthesis of atherogenic lipoproteins.

Apart from comparing the numerical parameters, of great importance is an analysis of the relations between the parameters of the lipids and hormones. A glance at Table 2 reveals that in healthy persons of group 1 (without DLP), there is a positive correlation between the content of cholesterol and triglycerides, cholesterol and apo-B, cholesterol

Table 2. Coefficient of Correlation between Content of Lipids and Hormones in the Blood of Healthy Persons of Groups 1 and 2.

Parameters	Coefficient of correlation			
	CS	TG	HDLP CS	Apo-B
TG	0.7720* / 0.6152*			
Apo-B	0.7244* / 0.7029*	0.6448 / 0.9427		
Apo-B/Apo-A-1	0.7814* / 0.7029*	0.5708 / 0.7790	-0.2415 / -0.8311	
Cortisol	0.5361 / 0.8726*	0.8209* / 0.7545*	-	0.4948 / 0.7307*

* The results are authentic.

The numerator gives the coefficient for group 1, the denominator – for group 2.

and the apo-B/apo-A-1 ratio. The correlation of the level of cholesterol and apo-B, cholesterol and apo-B/apo-A-1 also remained in force for persons of group 2. However, correlations were established in group 2 that are absent in group 1, namely, a positive correlation of the level of triglycerides and apo-B, and of triglycerides and apo-B/apo-A-1, a negative correlation of the HDLP cholesterol and apo-B/apo-A-1, and a positive correlation of the content of cholesterol and cortisole, apo-B and cortisole, apo-B and ACTH. Some of these relations are also typical of IHD. This indicates changes in persons of group 2 in the normally proceeding physiological processes and the appearance of disorders of lipoprotein and hormone metabolism that in their direction are identical to the changes discovered in patients with IHD.

Our results confirm the opinion of many clinicians that DLP is a significant risk factor in the development of IHD. Here great significance is attached to functional disintegration of the hormonal systems, which is an important link in the pathogenesis of IHD. In most cases, however, samples for analysis were taken in the persons being investigated on an empty stomach. However, after each taking of food (especially with a high content of saturated fats), temporary lipemia develops that may cause atherogenic shifts in the lipid transportation system of the plasma. This is why clinical studies appeared in recent years in which a fat load was used to study the metabolism of lipoproteins [68, 89, 116, 186, 237, 270, 385, 449].

A lipid load is the best physiological test for determining the functional possibilities of the metabolic activities of an organism. Fat loads are quite popular under clinical conditions. They vary in composition, but do not differ noticeably in their energy value and average about 1300 kcal (or 5443 kJ). All investigators noted an increase in the triglyceride level in the blood plasma in from two to six hours after a fat load. After eight or nine hours, the level returned to normal. The lipolytic activity of the blood plasma also increased in from three to six hours, while in five or six hours the content of nonesterified fatty acids (NEFA) in the plasma increased. Most clinicians find that with a lipid load, the level of triglycerides in the plasma changes the most obviously. It depends both on the amount of fat eaten and on the capture of triglycerides by extraliver tissues. Very contradictory data were obtained when determining the cholesterol level after a fat load: some authors noted only an insignificant increase of it, while others found no differences in the cholesterol amount.

When investigating 28 healthy volunteers after a lipid load, we found a negative relation between the initial triglyceride level and the content of $HDLP_2$ particles in the plasma. Accordingly, after a single fat load (sweet cream) in persons with a low $HDLP_2$ level in the plasma, a more pronounced triglyceridemic response was registered in 2, 4, 6, and 8 hours, which may be one of the factors affecting the development of atherosclerosis.

However, when using a fat load, we studied only the biochemical parameters of the lipoprotein spectrum and how pronounced the changes in the lipoproteins are in response to a fat load. A number of questions remain a puzzle: how does the atherogenic effect of food manifest itself, how can one explain the differences in the reaction of the parameters of the lipid transportation system of the plasma to a fat load in healthy persons and in patients with IHD, and is it possible to detect early atherogenic disorders?

Studying of the pathogenetic mechanisms of development of atherogenic shifts in the lipid transportation system of the plasma after taking fat food makes it possible to substantiate clinically the principles of correcting the metabolic processes of the lipid and hormonal units for the prophylaxis of IHD.

A fat load was applied according to the procedure used by J.R.Patsch (1983) [the taking during five minutes of 20% sweet cream on the basis of 130 g of fat per 2 m^2 of the body's surface area (660 g on an average) and 50 g of white bread]. The total energy value of the proposed breakfast was 1300 kcal, and the cholesterol content was 420 mg.

From clinical standpoints, we posed the following tasks:

(1) studying the dynamics of the parameters of the plasma lipid transportation system in healthy persons during a day after a fat load;

(2) studying the reactions of the hypophyseal-adrenal and vagoinsular systems in healthy persons at different times after a lipid load.

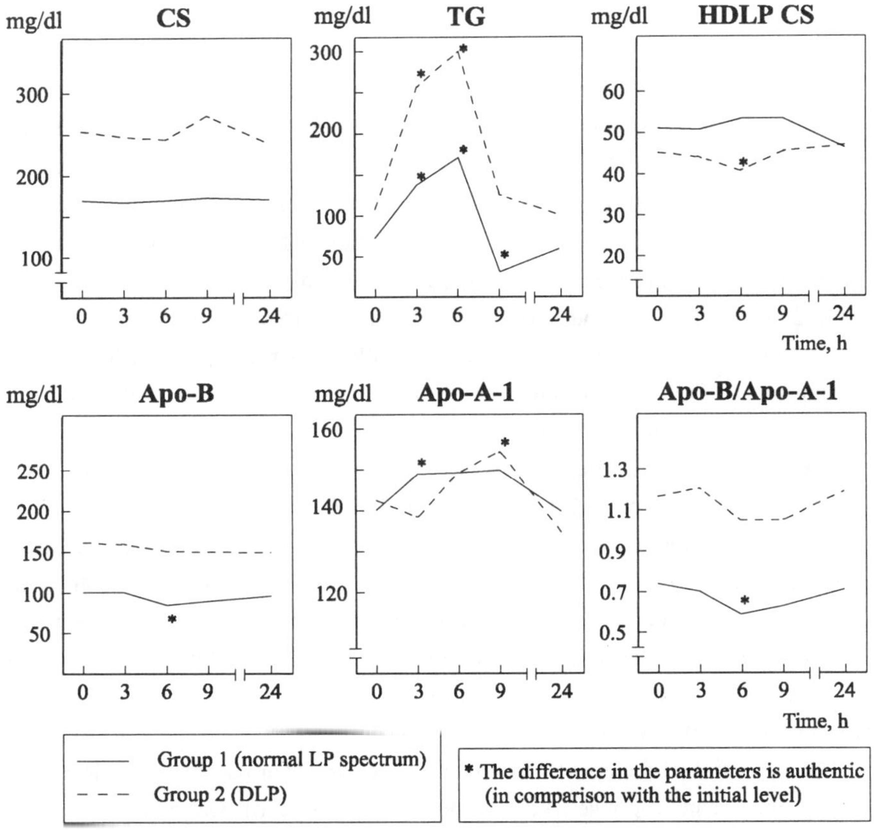

Fig. 2. Influence of Lipid Load on Content of Lipids and Apoproteins in Blood of Healthy Persons.

To solve these problems, we conducted the following biochemical investigations: determination of the content of total cholesterol, triglycerides, HDLP cholesterol, LDLP cholesterol, the concentration of apo-A-1 and apo-B, the part of the subfractions $HDLP_2$ in the total HDLP fraction and the part of the subfractions $HDLP_3$ in the HDLP fraction, and also the dynamics of the hormonal profile of the plasma (content of insulin, cortisol, ACTH, and STH).

The results indicate (Fig. 2) that the cholesterol level with a lipid load does not virtually change in healthy persons of both group 1 and group 2. The background value of the cholesterol in healthy persons of group 1 was 167 mg/dl, in three hours after the load it was 161 mg/dl, in six hours it was 164 mg/dl, in nine hours it was 171 mg/dl, and in 24 hours it was 160 mg/dl. There was also no dynamics in group 2: 259, 249, 245, 274, and 233 mg/dl.

The dynamics of the triglyceride level in response to a lipid load is more pronounced. In group 1 persons, the background value of the TG was 70 mg/dl, in three hours it grew to 138 mg/dl, in six hours to 172 mg/dl, in nine hours to 41 mg/dl, and in 24 h to 62 mg/dl. It is very important to note the obvious drop in the TG level after nine hours.

In analyzing the TG content after a lipid load in healthy persons of group 2, we found considerable differences: the background level was 109 mg/dl, in three hours it increased to 258 mg/dl, in six hours to 299 mg/dl, in nine hours it was 133 mg/dl, and in 24 h it was 101 mg/dl. These results show that in healthy persons of group 2 there is no phase of lowering of the TG level in nine hours to below the background levels, and a sharper growth in three and six hours is noted.

How can we explain these changes in persons of group 2? Doubtlessly, by the activity of the lipolytic enzymes [126, 343, 376, 424, 519].

VLDLP form mainly in the liver and ensure the transportation of endogenic triglycerides. The rate of formation of the VLDLP depends on the flow of FFA entering the liver and grows with a carbohydrate diet, when the synthesis of the FFA in the liver increases. As a result of lipolysis, the VLDLP transform into HDLP. The latter are the lipoproteins that transport the cholesterol. Unlike the VLDLP, the LDLP contain less triglycerides, but apo-B is a component of them.

The dynamics of the apo-B level is the next important parameter of development of dyslipoproteinemia in response to a lipid load. In group 1 of healthy persons, the apo-B content is insignificant (100 mg/dl), in three hours after a lipid load it remained at the same level, in six hours it dropped to 86 mg/dl, in nine hours it was 96 mg/dl and in 24 h it was 98 mg/dl. In group 2, the background level of apo-B was quite high (163 mg/dl), it remained at the same level in three hours, lowered moderately to 151 mg/dl in six hours, was restored in nine hours – 161 mg/dl, and in 24 h was 159 mg/dl [64, 183, 250, 330, 373, 519].

The ratios apo-B/apo-A-1 in healthy persons of groups 1 and 2 differed accordingly, If in group 1 this ratio was less than unity, the background parameter was 0.72, in three hours it was 0.67, in six hours it was 0.55, in nine hours it was 0.63, and in 24 h it was 0.71, then in group 2 the background parameter was increased considerably to 1.17, in three hours it was 1.22, in six hours it was 1.05, in nine hours it was 1.07, and in 24 h it was 1.20.

An increased level of apo-B (especially an increase in the ratio apo-B/apo-A-1) is a criterion of atherogenesis both in healthy persons and in patients with IHD.

The ratio characterizes the intensity of transportation of cholesterol into the tissues or out of them.

A growth in the level of apo-B and especially of the ratio apo-B/apo-A-1 indicates the increased transportation of cholesterol into tissues and the membranes of cells. The liver triglyceridelipase is the enzyme affecting the apo-B-LP transformation.

In recent years, both clinicians and biochemists have been attaching major significance in the regulation of lipid metabolism, especially under conditions of fasting or taking fat food, to NEFA and consider them to be regulators of the content of both triglycerides and apo-B in the blood [444, 493, 500, 561].

When comparing the dynamics of the level of the antiatherogenic factors of lipid metabolism (HDLP cholesterol, apo-A-1), we noted the clear distinction in healthy persons of groups 1 and 2 under a lipid load (see Fig. 2). It follows from the results that the background level of HDLP cholesterol in group 1 is higher (52 mg/dl) than in group 2 (45 mg/dl). In response to the lipid load in healthy persons of group l, the background value of the HDLP cholesterol remained virtually constant; in three hours it was 51 mg/dl, in six it was 53 mg/dl, in nine it was 53 mg/dl, and in 24 hours it was 46 mg/dl. In healthy persons of group 2 the relevant figures were 43 mg/dl in three hours, 41 mg/dl in six hours, 45 mg/dl in nine hours, and 46 mg/dl in 24 hours. It is very important to note the growth in the apo-A-1 level in healthy persons of group 1: the background was 139 mg/dl, the value in three, six, nine, and 24 hours were 152, 159, 153, and 139 mg/dl, respectively. Such a positive dynamics was not noted in healthy persons of group 2: the background was 142 mg/dl, and the values in three, six, nine, and 24 hours were 137, 148, 156, and 133 mg/dl, respectively.

Thus, if the second group of healthy persons with dyslipidemia was distinguished by higher background parameters of the total cholesterol triglycerides, apo-B, a higher ratio apo-B/apo-A-1 and a higher fraction of $HDLP_{3B}$, then in three hours after a fat load the level of apo-A-1 increased authentically on an average by 14±6%, while in individual persons the deviations reached 47%, i.e. were within the limits of the physiological norm. In six hours after a lipid load, the level of HDLP cholesterol lowered only in persons of group 2 with dyslipidemia ($p < 0.05$). As regards the subfractions $HDLP_{2B}$ and $HDLP_{3B}$, their background value differs in both groups. If in group l the fraction of $HDLP_{2B}$ is 33.4%, in group 2 it dropped to 25%. The level of $HDLP_{3B}$, conversely, in group 1 is 21%, and in group 2 is 28% (the difference is authentic). Under the influence of a lipid load, these parameters did not practically change: in group 1 the fraction of $HDLP_{2B}$ in 3 h was 34%, and in group 2 it was 23%. The value of $HDLP_3$ in group 1 persons in three hours was 22%, and in group 2 was 31%. There is strict correlation here between the fraction of

$HDLP_{2B}$ and the triglyceride level in the plasma. when the value of $HDLP_{2B}$ is lower, the TG level in the plasma is higher and, accordingly, the activity of lipoproteinlipolysis is lower. In healthy persons of both groups, there is no negative correlation between the background level of the triglycerides and the fraction of the $HDLP_{2B}$.

It was shown in biochemical investigations that in the lipoproteinlipolysis of particles enriched in triglycerides, part of the cholesterol of their surface layer detaches and passes over into the composition of the HDLP. After a lipid load, the coefficients of negative correlation of the parameters of the triglycerides and $HDLP_{2B}$ cholesterol in group 2 reached significant values. This indicates the preservation and increase in the activity of the lipoproteinlipolysis in the intensification of the flowing off of cholesterol from the chylomicrons onto HDLP particles with the formation of large particles of $HDLP_{2B}$ that transport the cholesterol into the liver. It was noted that an increase in the triglyceride content in the nucleus of the $HDLP_{2B}$ particles in persons with hypertriglyceridemia makes such particles a convenient substrate for the liver triglyceridelipase also having phospholipase activity. After the hydrolysis of the phospholipase and triglycerides in the $HDLP_{2B}$ fraction, the latter diminish in size without losing cholesterol, however, and return into the blood flow in the form of $HDLP_{3B}$. All this contributes to lowering of the functional activity and disturbance of the antiatherogenic properties of HDLP [272, 342, 376, 604, 621].

Part of the apo-A-1 has an intestinal origin: 70% of the apo-A-1 formed after the fat load enters the blood flow with the chylomicrons and immediately passes from the latter to the HDLP. If a growth in the apo-A-1 level in healthy persons in three hours after a lipid load plays, as it were, a "protective" antiatherogenic role in the utilization of fats after food lipemia, the absence of such a reaction presumes damaging of this protective mechanism. Lowering of the apo-A-1 level in healthy persons with dyslipidemia indicates disorder in the system of removal of cholesterol from the peripheral tissues as a result of damage to the cascade of reactions of cholesterol capture by the high density lipoproteins.

What are the differences in the metabolism of lipoproteins in persons of groups 1 and 2 associated with?

Doubtlessly, an important role is played by genetic factors of the predomination of the "functional unit of switching over" nervous or humoral regulation. Some clinicians attempt to find the initial links of the pathogenesis of atherosclerosis in just this "switching over" of the neurogenic influences to humoral ones.

An important factor in the humoral metabolism in cells is the free endogenic heparin of the blood. It is considered biochemically as an activator of the enzyme LPL. Under conditions of stress with pronounced tension of the neurogenic structures, a considerable discharge of endogenic heparin is noted in healthy persons. It prevents thrombogenesis and the development of atheromatous changes in the vessels. An absolutely different picture is observed in atherosclerosis and DM. Under the influence of factors of the environment and internal medium, and also of genetic risk factors, in patients with atherosclerosis disintegration of the hormonal systems appears. It is manifest in polymorphism and disconnection of the functions of individual units, namely, a change in the reactions of an organism to a load (emotional, physical, lipid), disturbance of the sensitivity of the receptor apparatus, and distortion of the metabolic reactions of an organism [173, 214, 239, 273, 288].

The above facts suggest that the adaptational reorganization of the endocrine and metabolic status of a person in the direction of disadaptation is an integrative process and includes first of all an increase in the sympathico-adrenal and glucocorticoid activity and also the corresponding changes in the activity of carbohydrate and lipid metabolism (Fig. 3).

Hyperinsulinemia, hyperlipidemia, and hyperglycemia are important links in disadaptation. These metabolic disorders cause diseases such as IHD, hypertension, DM, and obesity.

Of major significance in the progress of metabolic disintegration of an organism (the development of dyslipidemia) is the level of physical activity of a person and his or her professional activities, especially in middle or old age because a physical load increases the expenditure of energy resources. Here the weight of the body lowers, the reserves in the fat depots diminish, the nature of distribution of the fat changes, and the inclusion of lipids into the striated muscles grows. Under a physical load, the metabolism of the adipose tissue alters: the lipolytic activity grows, the sensitivity to hormones increases, and the liability of reesterification of the NEFA diminishes [74, 101, 148, 155, 212].

Clinical studies showed that the NEFA can be mobilized in physical activity in two ways, namely, spontaneously (lipolysis not blocked by the beta blocking agents) and with the aid of adrenergic lipolysis. Under a physical load, special importance is attached to adrenergic lipolysis. At rest, the physiological influence of the sympathetic nervous system on the regulation of fat metabolism is insignificant. Under conditions of physical activity, sympathetic stimulation of the receptors facilitates the additional discharge of the FFA from lipocytes (adipocytes). Physical load is just what increases the mobilization, oxidation, and recirculation of the FFA. If in a healthy person at rest the free fatty acids are mainly (80%) spent on the processes of synthesis of triglycerides, phospholipids, and cholesterol, with physical activity the proportion changes – not 20%, but 80% (depending on the intensity of the physical load) is spent on the oxidizing processes. To date, the pathogenesis of this switching over of the processes of oxidizing FA under conditions of physical activity was established only partly. In most clinical and experimental studies, the main significance is attached to the increased oxygenation of the muscles and accordingly to the increased oxidation of the FFA.

What hormonal systems react to a lipid load and what are their dynamics?

Lipolysis in adipose tissue is an integrative process combining both the energy requirements of an organism and the metabolism of lipids. The hydrolysis of triglycerides yields glycerol and NEFA. Special significance is attached to the level of the NEFA concentration in the blood that regulates cell metabolism, in particular in the tissues of the cardiovascular system. The synthesis of ATP, the absorption of oxygen and the utilization of glucose in the smooth muscle cells of the vessels and heart depend on the content of NEFA and triglycerides in the blood, and the contracting properties of the indicated tissues depend on these features. Catecholamines, ACTH, STH, and TSH are the main hormonal stimulants of lipolysis. Insulin has an antilipolytic effect. It increases the transportation of glucose into a cell, and since glucose is an important component for the synthesis of triglycerides, the antilipolytic effect of insulin is partly associated with a growth in the reesterification of the NEFA. Under the influence of insulin, the level of cAMP in the lipocytes lowers, while the content of cGMP in them grows [310, 489, 551, 556, 622, 628].

Investigations (Fig. 3) showed that the background level of insulin in both groups of healthy persons is different – it was 5.2 μU/ml in group 1 and 10.9 μU/ml in group 2, i.e. moderate hyperinsulinemia was observed with dyslipidemia.

What is the influence of insulin on lipid metabolism?

The liver and adipose tissue are the main targets of the action of insulin for regulating fat metabolism in an organism. Biochemical investigations showed that an increase in the insulin level in the blood plasma stimulates the synthesis of fatty acids in the liver. This effect is ensured by the following mechanisms:

(1) insulin can have a direct stimulating effect on acetyl-CoA-carboxylase – the key enzyme of liposynthesis;

(2) diminishing of the supply of FFA (under the influence of insulin) eliminates their inhibiting effect on acetyl-CoA-carboxylase;

(3) since insulin stimulates the utilization of glucose along the pentose path, the amount of reduced NADP grows; it is the required hydrogen donor for many stages of fatty acid synthesis;

(4) the growth in the flow of glycolysis substrates in the Krebs cycle occurring under the influence of insulin increases the level of citrate that also stimulates acetyl-CoA-

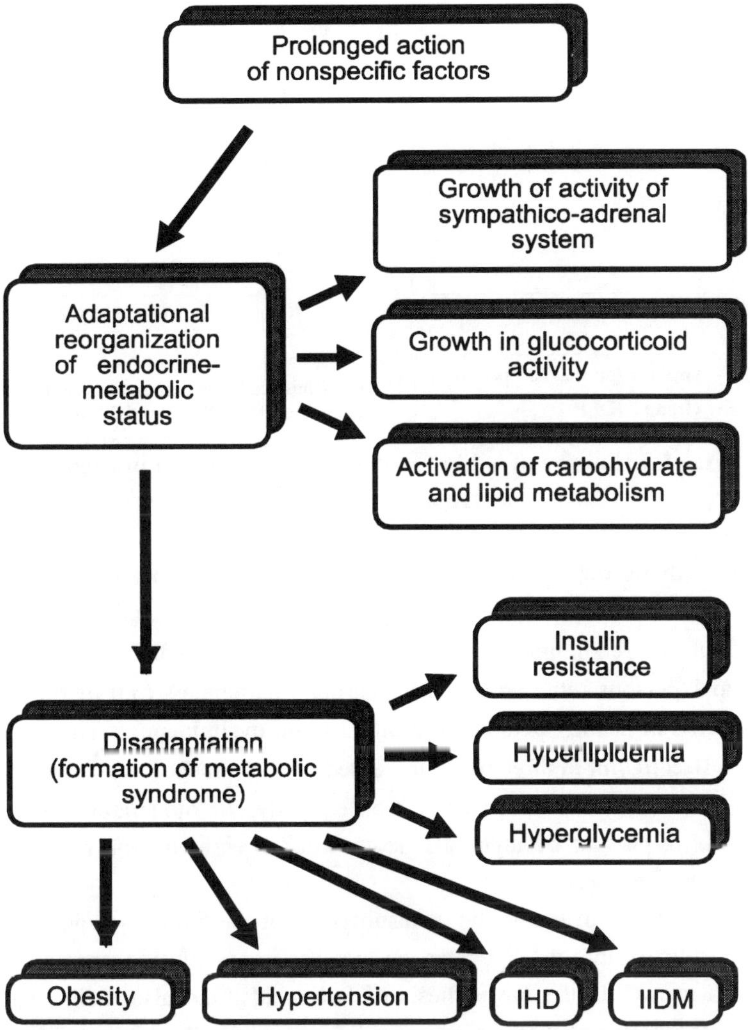

Fig. 3. Adaptational Reorganization of the Endocrine-Metabolic Status of a Person.

carboxylase. The fatty acids synthesized in the liver are then transported (in the composition of VLDLP) into adipose tissue where they accumulate in the form of triglycerides.

Insulin stimulates LPL – an enzyme synthesized in the lipocytes and acting on the surface of the endothelial cells of capillaries penetrating into adipose tissue. This accelerates the adsorption by the fat cells of the triglycerides circulating in the blood that form from endogenic or exogenic sources [120, 210, 292, 333, 365, 421].

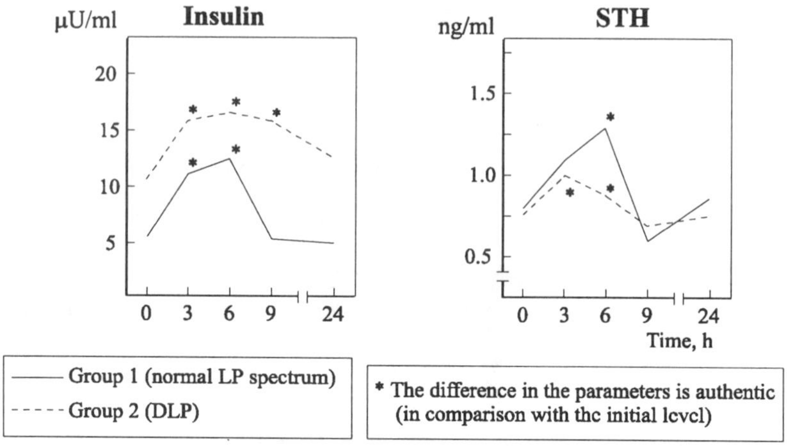

Fig. 4. Influence of Lipid Load on Hormone Level in Blood of Healthy Persons.

When studying the dynamics of the insulin content in the blood plasma under a lipid load, we noted that in persons of group 1 the insulin level increased moderately – it was 10.8 μU/ml in three hours and 11.6 μU/ml in six hours. After nine hours, the insulin level in group 1 persons returned to the background parameters (5.0 μU/ml). A different picture was noted in healthy persons of group 2 (with dyslipidemia). The background insulin value of 10.9 μU/ml in three hours increased to 15.6, in six it was 16.6, in nine it was 15.9, and in 24 hours it was 13.2 μU/ml. Consequently, the moderately increased insulin content in group 2 persons became still greater after three hours and remained at a high level during 24 h (see Fig. 4).

The main effects of the action of insulin on adipose tissue include a growth in the penetration of glucose into the lipocytes, an increase in glycolysis (especially of the pentose cycle), a growth in the biosynthesis of fatty acids, and also their esterification to triglycerides. Regardless of these effects, insulin is a powerful inhibitor of hormonosensitive lipase (HSL) that catalyzes the hydrolysis of the accumulated triglycerides and the release of fatty acids (i.e. lipolysis). The anticatalytic influence of insulin can be detected at very low concentrations of the hormone (about 0.1 μU/ml). Such an insulin level is much lower than the value needed for stimulating the penetration of glucose into a lipocyte. Moreover, the greater part of the glucose absorbed by the fat cells under the influence of insulin is used in the formation of alpha-glycerophosphate needed for the esterification of the fatty acids and the synthesis of triglycerides [382, 435, 469, 490, 563, 586].

As a whole, the adipose tissue is characterized by a considerable sensitivity to the action of insulin, and it is 10 times more sensitive to this hormone than muscle tissue.

In obesity, a typical disorder is a growth in the insulin concentration in the blood serum, i.e. the appearance of hyperinsulinemia. To date, the existence of basic and glucose-stimulated hyperinsulinemia in obesity is doubtless and is confirmed by numerous experimental and clinical studies.

What is the pathophysiology of hyperinsulinemia and dyslipidemia?

Under conditions of a surplus of insulin, the following occurs in an organism: (i) activation of LPL, (ii) a growth in the absorption of free fatty acids from VLDLP (contained in the blood), (iii) stimulation of lipogenesis, (iv) an increase in the transportation of glucose, (v) acceleration of the transformation of glucoso-6-phosphate along all the metabolic paths (oxidation to CO_2, synthesis of fatty acids, and oxidation by the pentosophosphate path), and (vi) the formation of a large amount of alpha-glycerophosphate needed for the synthesis of triglycerides. At the same time, the hydrolysis of triglycerides with the aid of hormonosensitive lipase is greatly inhibited by insulin.

Fat cells as their size grows become resistant to insulin. A drop in the number of receptors to insulin per cell is found in them (and also in hepatocytes and myocytes). When the body weight lowers because of limited food consumption, the fat cells again become sensitive to insulin, and the number of its receptors grows in all tissues [301, 358, 365, 435, 493, 584, 620].

Consequently, insulin facilitates the formation of a "store" of fat in a lipocyte, but when the stores are greater, its further accumulation proceeds more slowly. The appearance of insulin resistance in this situation is thus a physiologically protective reaction that prevents further obesity.

The accumulation of triglycerides in the fat cells is attended by a tendency of a growth in the level of the FFA in the blood, which ensures the supply of a large amount of substrates to the liver for the synthesis of VLDLP.

Consequently, a lipid load clearly demonstrates the adequate hormonal provision of lipoproteinlipolysis and catabolism of lipoproteins. The increase in the insulin concentration during the maximum lipemia (in three and six hours after a lipid load) lowered the decomposition of endogenic triglycerides in the fat depots and increased catabolism of the newly formed VLDLP, IDLP, and $HDLP_2$.

Do other hormones participate in these processes of lipid metabolism?

Of special interest is studying of the dynamics of the content of the somatotropic hormone (STH). The background level of it in both groups of healthy persons was practically the same (0.72 ng/ml in group 1 and 0.69 ng/ml in group 2). However, under the influence of a lipid load, the dynamics of the level of this hormone in the blood differed in the two groups. In group 1, the STH level was 1.06 ng/ml in three hours, 1.32 in six, 0.58 in nine, and 0.82 ng/ml in 24 hours. The dynamics was less pronounced in group 2, namely, 0.98, 0.90, 0.67, and 0.74 ng/ml in three, six, nine, and 24 hours, respectively.

The STH has a pronounced anabolic effect on protein metabolism (it increases the synthesis of protein, DNA, RNA), and also an insulin-like effect on fat metabolism (the content of NEFA in the blood grows). The functional activity of the STH is regulated by the hypothalamus.

Consequently, investigation of the metabolism of lipoproteins and the hormonal profile in healthy persons of both groups points to a broad range of lipid and hormonal parameters. As already indicated, in the formation of the groups (with normal parameters of the lipids and atherogenic disorders), we found that in persons of group 2 with atherogenic DLP (cholesterol over 200 mg/dl, triglycerides over 200 mg/dl, HDLP cholesterol under 35 mg/dl), disintegrative processes appear, evidently as a result of genetic or environmental factors, or both. They lead to disturbances of both the metabolism of lipoproteins and the hormonal status. The presence in group 2 of correlations between the lipid and hormonal parameters (similar to those for the group of IHD patients) makes this group (with DLP) a sort of intermediate one between healthy persons and IHD patients.

Studying of the influence of the lipid load on the lipid and hormonal status of persons of both groups showed that the dynamics of parameters such as the content of triglycerides, HDLP cholesterol, apo-A-1 and apo-B/apo-A-1 is more pronounced in persons of group 2 (a growth in the atherogenic shifts of the lipoprotein system). Here a delay in the return of the content of triglycerides and IRI to the initial level was noted (in 24 h after the taking of a lipid load).

Consequently, the application of a lipid load makes it possible to detect latent shifts in the lipoprotein system in healthy persons of group 2 (with DLP) in comparison with group 1 (normal lipid and hormonal spectra), namely, the absence of an increase in the content of apo-A-1 and a growth in the level of $HDLP_{3B}$ in three hours and a drop in the level of HDLP cholesterol by the sixth hour after a fat load.

1.2. Syndrome X (Hyperinsulinemia, Insulin Resistance)

Active studying of the pathogenesis of cardiovascular diseases allowed investigators to single out the special role of metabolic disorders. Individual aspects of the problem are continuing to receive attention by both cardiologists and endocrinologists.

The exceedingly pronounced interaction of all the risk factors of cardiovascular diseases, hyperinsulinemia or insulin resistance, or both, made it possible to interpret this state more likely as a syndrome than as a set of individual factors. In other words, a definite metabolic set of symptoms underlying the pathogenesis of many metabolic diseases was singled out. Reaven proposed to call them syndrome X [406, 417, 475, 477].

Hypertension, obesity, disturbance of tolerance to glucose, and dyslipidemia are the main risk factors of cardiovascular diseases. Currently, another factor was added to them – chronic hyperinsulinemia. A special, leading role is assigned to this factor in the formation of a metabolic syndrome leading in most cases to the development of cardiovascular diseases.

Syndrome X is thus a pathogenetic combined state of metabolism in which hyperinsulinemia may precede (in time) IHD, DM, and, consequently, coronary sclerosis and disorder of carbohydrate metabolism [110, 166, 225].

1.2.1. *P*athogenetic Mechanisms of Syndrome X Development

Syndrome X described by Reaven [474, 475] is based on the following disorders: resistance to insulin in the introduction of glucose, disturbance of the tolerance to glucose, a growth in the content of VLDL***P***, hyperinsulinemia, lowering of the HDL***P*** level, and hypertension. The author succeeded in combining the known risk factors of cardiovascular diseases and single out especially the significance of insulin resistance, although the relation between the indicated factors and their role in the appearance of cardiovascular diseases was discussed previously.

The influence of the environment, prolonged action of nonspecific factors, and a change in the function of the internal organs activate the adaptive mechanisms in an organism (see Fig. 3) [222, 289, 301, 327, 476].

Reorganization of the metabolic processes is one of the main conditions for maintaining homeostasis under the effect of nonspecific factors. Here the mechanisms of regulation of the carbohydrate and lipid metabolism are switched on especially actively. Metabolic reorganization reflects the state of the hormonal and mediator units in the process of adaptation to the new level of functioning. The adaptation reorganization of the endocrine and metabolic status is manifest in activation of carbohydrate and lipid metabolism on the background of an increase in the glucocorticoid and sympathico-adrenal activity.

Many investigators are interested in matters of the control of the metabolism and functional activity of cells, and determination of the molecular mechanisms of the intracellular reaction to hormones (their concentration in the blood). In recent years, most scientists acknowledge that hormones are the regulators and integrators of metabolism in organs and tissues. To understand the essence of metabolism, it is very important to study

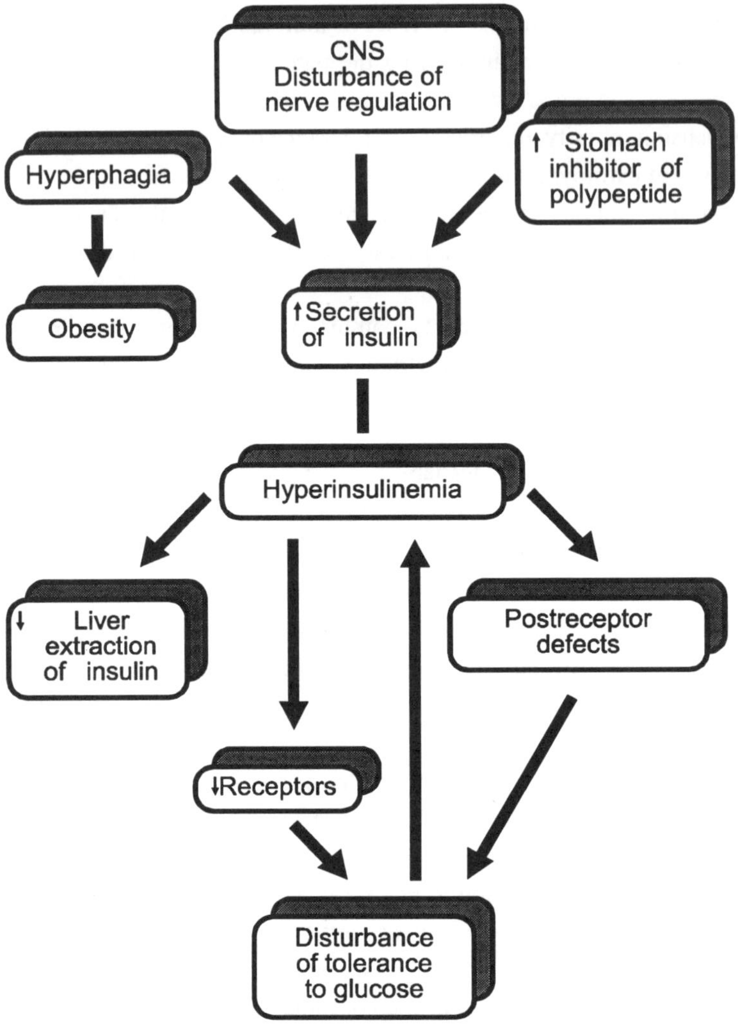

Fig. 5. Central Nervous System and Carbohydrate Metabolism.

the mechanisms of the reception and transformation of hormonal signals fed to the receptor apparatus of a cell [280, 330, 368, 472, 620].

A major role in metabolic processes in an organism is played by hormonal regulation, which is realized depending on the supply of a hormone to the tissues and the sensitivity of the latter to its effect. The efficiency of the action of a hormone on an organ depends on the sensitivity of the cell receptors, which for some hormones are on a cell membrane, for others – in the cytoplasm, and for still others – in a cell nucleus. These receptors are responsible for the hormonal action on the metabolism in a cell, the change in the activity of enzymes, and the synthesis of protein structures. A disturbance in the adaptive-trophic and modulating effect of hormones in diseases such as DM and IHD causes disintegration of the regulator and adaptive mechanisms in the activity of the cardiovascular system and carbohydrate metabolism in an organism. In a number of experiments, a correlation was noted between the amount of circulating hormone in the blood: and the concentration of receptors in the tissues. In experimentally induced DM, a drop in the level of the circulating insulin is attended by a growth in the number of receptors specifically binding insulin in the target tissues [301, 345, 421, 439].

Under the action of nonspecific factors (see Fig. 3), a healthy organism reacts by an increase in the sympathico-adrenal and glucocorticoid activity, and also by pronounced activation of carbohydrate and lipid metabolism. However, the prolonged action of various nonspecific factors (emotional stress, hypoxia, a physical load, etc.) produces disadaptation of the regulator mechanisms and the appearance of a metabolic syndrome whose main fragments are hyperinsulinemia, hyperlipoproteinemia, hyperglycemia, obesity, and hypertension. All these metabolic disorders lead to the development of diseases such as IHD, insulin-independent diabetes mellitus (IIDM), hypertension, and syndrome X [110, 405, 439].

In the pathogenesis of metabolic diseases, a special role belongs to the harmful consequences of the prolonged increased content of circulating insulin. The hypothesis was advanced of central regulation of the endocrine part of the pancreas. For instance, in an experiment a hypothalamic factor was separated of a polypeptide nature that affects the secretion of insulin. When obesity was studied in an experiment with animals, hyperinsulinemia was found to be preceded by hyperphagia (Fig. 5).

The insulin resistance syndrome includes IIDM, obesity, hyperinsulinemia, dyslipidemia, and hypertension (Fig. 6). The syndrome of insulin resistance is a marker of the syndrome X. Numerous clinical studies showed that the resistance to the insulin-stimulated capture of glucose may be one of the etiological links of IIDM, arterial hypertension, and IHD. It was shown that obesity is always associated with hyperinsulinemia and insulin resistance. ***P***atients generally note a lowered physical activity, they smoke and overeat. Hyperinsulinemia in combination with insulin resistance leads to disturbance of lipid metabolism (Fig. 7). Factors of dyslipidemia such as an increased content of VLDL***P*** and LDL***P*** and lowering of the level of HDL***P*** facilitate the development of IHD. Another aspect of the manifestation of hyperinsulinemia consists in the increased reabsorption of sodium ions, the pronounced activity of the sympathetic nervous system (SNS), an in-

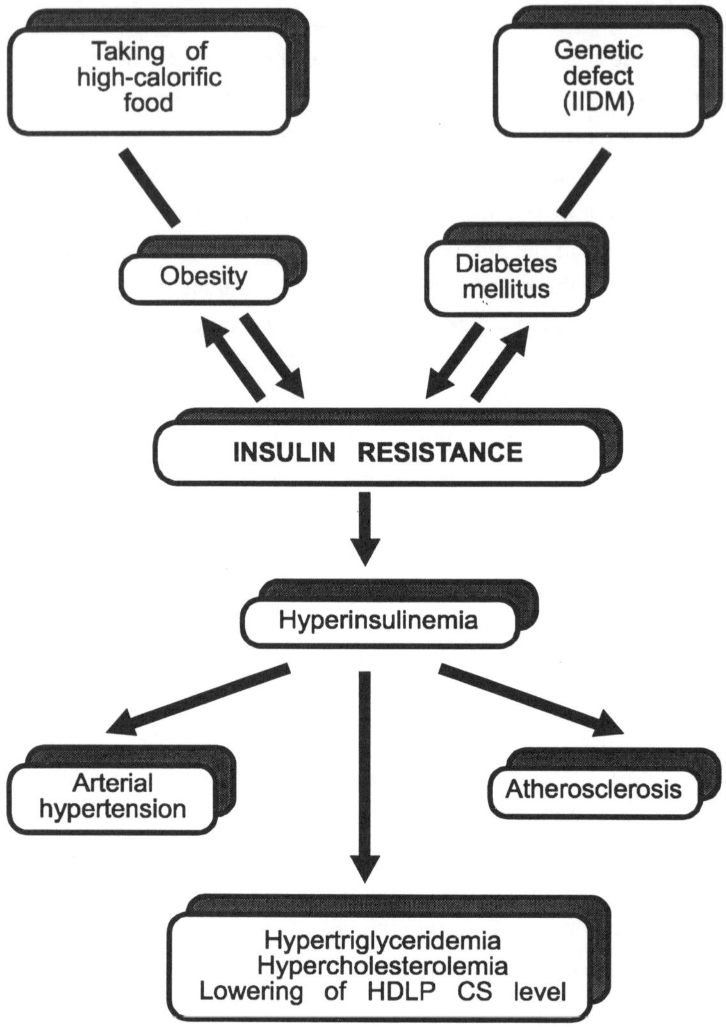

Fig. 6. Disturbance of Insulin Metabolism.

creased level of angiotensin II, and a natural growth of the arterial pressure [32, 111, 273, 345].

In obesity, an increased content of endogenic opioides both in the CNS and in the pancreas is significant in the appearance of hyperinsulinemia. Their stimulating effect on the secretion of insulin was also established.

Studying of the interaction of insulin, the CNS, and various neuropeptides revealed new aspects of the activity of insulin and the ways of its action on the CNS, insulin recep-

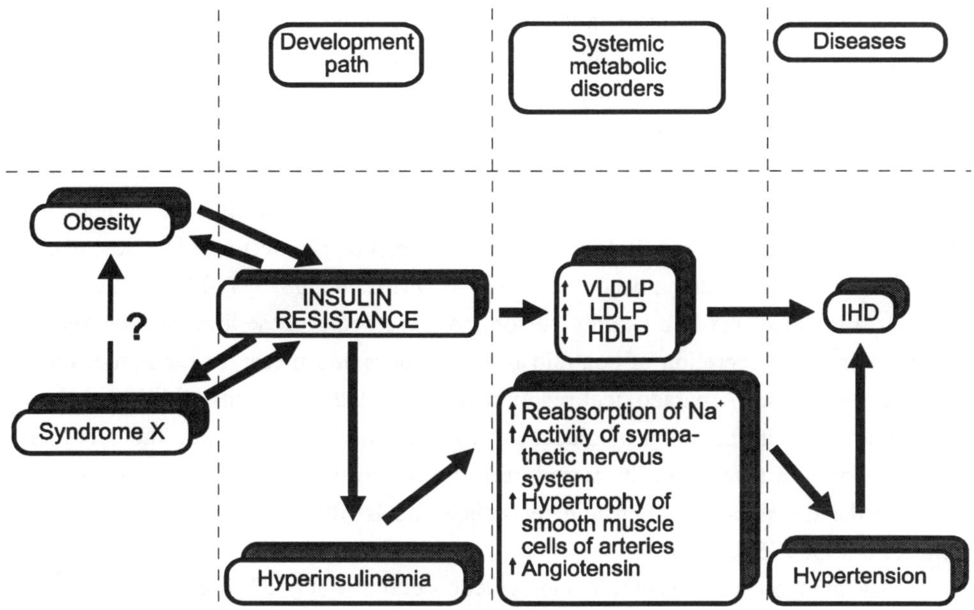

Fig. 7. Syndrome of Insulin Resistance.

tors were discovered on the membrane of nerve cells, and the suppression by them of the delivery of noradrenaline to the neurons was established. Insulin participates in the metabolism of the neuromediators because the content of the latter in DM patients lowers, while in the compensation of DM their level is restored. Therefore, insulin has a modulating effect on the state of the neuromediator systems [368, 472, 516, 555, 616].

Insulin is an anabolic hormone facilitating the utilization of glucose and the biosynthesis of glycogen, lipids, and proteins.

The metabolic transformations of glucose follow several paths:

(1) The transformation of glucose into glucoso-6-phosphate with subsequent resynthesis into glucose;

(2) transformation of glucose into glycogen;

(3) transformation of glucose into pyroracemic acid, lactic, then into oxalic-acetic acid and further into acetyl-Co-A;

(4) participation of glucose in a pentose cycle, transformation of glucose into AT***P*** (the cycle of tricarboxylic acids, or the Krebs cycle);

(5) transformation of glucose (glycolysis or the Embden path) via the key compound – acetyl-Co-A, and then the synthesis of lipids (cholesterol, fatty acids, and ketone bodies).

Substantial changes occur in the metabolism of lipids with a lack of insulin in DM patients. First of all, lipolysis in adipose tissue increases and accordingly the amount of FFA in the blood grows. The supply of the FFA into the liver increases, where the synthesis of triglycerides, the formation of cholesterol and ketone bodies grow [96, 285, 294, 567, 573]. The reason is that the increased amount of FFA does not have time to be oxidized in the Krebs cycle, and in DM patients with a lack of insulin, the content in the blood of FFA, triglycerides, cholesterol, beta-hydroxybutyric, acetoacetic acids, and acetone increases. This results in decompensation of DM [280, 358, 397, 432, 551, 601].

It was noted that in a healthy person, an increased glucose level (its excess in the blood) inhibits the secretion of contrainsular hormones and preserves the homeostasis of insulin in the blood. A paradoxical reaction was revealed in patients with hyperinsulinemia, namely, the introduction of glucose did not inhibit, but stimulated the secretion of contrainsular hormones. The increased secretion of insulin antagonists is an important factor in the development of insulin resistance and is often observed in many diseases [120, 131, 153, 176, 271, 318].

The antagonists of insulin include glucocorticoids (GC), STH, and hormones of the thyroid gland that cause hyperinsulinemia in various ways. Special importance in this process is attached to glucocorticosteroids that increase glyconeogenesis, promote the development of hyperglycemia leading to hyperinsulinemia, and also lower the affinity of the receptors for insulin. On the one hand, STH increases the secretion of insulin by stimulating the insular apparatus of the pancreas, while on the other, insulin in physiological doses stimulates the secretion of prolactin by the hypophysis. Experiments showed that when human insulin acts for five days on a culture of hypophysis tumor cells, the secretion of prolactin increased by 30% both in the control culture and in the tumor cells.

Consequently, hyperinsulinemia may be a result of dysfunctioning of the hypothalamus-hypophyseal system, and pronounced chronic hyperinsulinemia enhances it [313, 321, 328, 399, 472, 541].

Chronic endogenic hyperinsulinemia, apart from disturbances of the function of the hypothalamus-hypophyseal system, is maintained in an organism by a number of other factors.

The biochemical adaptation of an organism to the effect of extremal factors is thus closely related to the reactivity of the endocrine system, and also to the sensitivity to insulin. The performance of muscular work by an organism depends on the change in the content of glucose and insulin. Experiments with animals showed the hypoglycemic effect when insulin is introduced to be more pronounced in physically well trained animals than in untrained ones. Therefore, the sensitivity of the cAM***P*** receptors to insulin in trained animals is much higher.

The experimental studying of the biochemical transformations of insulin in an organism of trained and untrained animals revealed that in the former the sensitivity of the tissues to insulin does indeed grow, but simultaneously the rate of inactivation of this hormone in the blood grows, i.e. it reacts very rapidly with cAM***P*** and is then decomposed in the organism [83, 109, 199, 314, 358, 597].

Every biological homeostatic system is characterized by a multitude of contours that ensure various forms of hierarchic regulation with a differing fraction of the autonomous nature of functioning at different levels of the complicated dynamic system.

The homeostasis of glucose is maintained with a constant reaction of insulin and glucagon. The direct effect of these hormones is realized via cAM***P***.

An important principle of glucose homeostasis is the stable activity of the insular apparatus. On the one hand, fluctuations of the glucose level in the blood both excite and inhibit the insular apparatus of the pancreas, while on the other hand they stimulate the functioning of the adrenal glands and hypophysis. It is exactly hypoglycemia that is a powerful factor in activating the function of the adrenal glands with the discharge of catecholamines and contrainsular hormones.

It was shown in a number of experiments that even without regulation by the hypophysis and adrenal glands, the liver is an important regulating factor (ensuring a constant glucose level in the blood) that maintains the homeostasis of glucose in an organism [316, 384, 559, 574].

Clinical observations showed that chronic hepatitis and hepatic cirrhosis are characterized by pronounced dystrophic, necrobiological and necrotic processes in its parenchyma that lead to the replacement of the main part of its functional elements with connecting tissue. Such a liver extracts less insulin from the blood of the vena porta as a result of lowering of the specific binding. The latter is the only factor causing the degradation of insulin. This is why the development of hyperinsulinemia with lesion of the liver can be explained primarily by the lowering of the degradation of this hormone. The relation between lesion of the liver and hyperinsulinemia, and also insulin resistance may be due to the following factors [77, 137, 148, 271, 301, 365]:

(1) a drop in the number of functionally active cell elements in the liver;

(2) lowering of the glycogen-forming function of the liver (accordingly a high level of glycemia after taking food and a drop in the tolerance to glucose);

(3) diminishing of the volume of the liver parenchyma (this results in hyperinsulinemia);

(4) a growth in the secretion of insulin in response to pronounced glycemia after the taking of food.

1.2.2. Syndrome X and Its Clinical Manifestations

Syndrome X is based on metabolic disorders that can be revealed only in laboratory studies. Clinical symptomatology forms in time, i.e. well known metabolic diseases, namely, hypertension, atherosclerosis, and obesity.

DM, arterial hypertension, and hypercholesterolemia are undoubted risk factors of cardiovascular diseases, while according to Modan *et al.*, hyperinsulinemia is the main element combining obesity, DM, and arterial hypertension.

Major prospective investigations conducted recently by Haffner *et al.* [225] and devoted to the observation for many years of persons with metabolic disorders revealed that hyperinsulinemia is what can predetermine other metabolic disturbances that are risk factors in the appearance of cardiovascular diseases. They proposed using the term "insulin resistance syndrome" instead of metabolic syndrome (syndrome X) [32, 58, 70, 431, 460, 558].

We revised somewhat and supplemented the scheme proposed by DeFronzo [111]. Insulin resistance was shown to precede atherosclerosis, dyslipidemia, and hypertension (see Fig. 6). It is apparently expedient to note the influence of insulin resistance on the development of obesity and DM. According to Reaven, only the active function of beta cells normalizes carbohydrate metabolism in insulin resistance, i.e. the latter is a state of relative insulin insufficiency when hyperglycemia (or disturbance of the tolerance to glucose) is combined with a high insulin level in the blood (hyperinsulinemia). At the same time, when such patients are stimulated with glucose, an increased sensitivity of the insular apparatus of the pancreas is noted. Lowering of the sensitivity of tissues to insulin is the pathogenetic basis of this relative insulin insufficiency. Insulin resistance of the tissue is what leads to compensatory hyperinsulinemia. The variant of the formation of relative insulin insufficiency into absolute insufficiency is quite possible, when the compensating possibilities of the beta cells of the pancreas are exhausted [302, 303, 476, 542, 558, 595].

The reason for the development of insulin resistance still remains to be established completely. One can think of the existence of many mechanisms leading to the development of relative insulin insufficiency. They include (i) a change in the chemical structure of the insulin, (ii) disturbance of the transformation of proinsulin into insulin, (iii) the increased secretion of insulin antagonists into the blood (as noted previously), and (iv) the formation of antibodies to insulin or disorder of the function of the insulin receptors.

There are currently little proofs of the disturbance of the chemical structure of insulin, and also of the change in the transformation of proinsulin into insulin.

A major role in the appearance of insulin resistance is played by the state of the receptors and the mechanisms of the postreceptor effect of insulin. Moreover, as already noted, stimulation by glucose increases the sensitivity of the beta cells of the pancreas [314, 358, 515, 574, 584].

Insulin resistance is often accompanied not only by hyperinsulinemia, but also by disturbance of the tolerance to glucose and by basal glycemia. Overstressing of the function of the pancreas beta cells results in their exhaustion, and absolute insulin insufficiency appears.

Relative insulin insufficiency is characterized by improper carbohydrate metabolism, and it also leads to a change in lipid metabolism.

Insulin has a powerful antilipolytic effect. It stimulates lipogenesis processes and acts like an antagonist of hormones ensuring lipolysis. Insulin increases the permeability of the plasmatic membranes of the lipocytes and ensures the delivery of glucose into adipose tissue.

Insulin plays a quite important role in the genesis of dyslipidemia. It is one of the main factors regulating fat metabolism and facilitating lipogenesis. The inverse relation between the content of HDL***P*** in the blood and insulinemia was indicated previously. In persons with increased HDL***P*** content in the plasma, hyperinsulinemia is a lower risk factor of cardiovascular diseases. A growth in the insulin content in the plasma increases the synthesis of VLDL***P*** leading to the development of hypertriglyceridemia, i.e. hyperinsulinemia or insulin resistance or both combine with the atherogenic nature of the blood plasma [56, 91, 238, 267, 302, 313, 345].

Triglycerides play an important role in the regulation of insulin sensitivity. We found that in healthy persons the concentration of insulin and level of the free fatty acids correlate negatively with the metabolism of triglycerides, i.e. there is a relation between the decomposition of the latter and the resistance of an organism to insulin. A growth in the concentration of the FFA in the plasma in obesity facilitates the development of insulin resistance.

Recent experimental studies showed that brief hyperlipidemia stimulates, and prolonged hyperlipidemia inhibits the glucose-stimulated secretion of insulin. The inhibition is specific only for glucose and evidently associated with the oxidation of the FFA and also with the reduced oxidation of glucose. All these changes occur both under normal conditions and in hyperglycemia [253, 264, 281, 330].

Thus, when the triglyceride content in the blood plasma grows, redistribution of the assimilation in an organism of the main energy substrates – glucose and lipids – occurs, and the content of insulin in the plasma therefore depends on the state of the energy balance and initial level of these substrates in the plasma.

Insulin resistance was shown to have a metabolic nature with an accompanying growth in the FFA level in the plasma, i.e. substrate competition plays a role in the mechanism of its development. When fat is introduced, insulin does not alter the glucose oxidation process. Redistribution of the assimilation of the glucose in a cell occurs. According to Thorburn *et al.* [573], both hyperinsulinemia and hyperglycemia stimulate the assimilation of glucose and increase the intracellular metabolism of glucose at the expense of the increase in its accessibility as an energy substrate. However, insulin also regulates the oxidation of glucose by suppression of the formation of FFA and the metabolism of fats.

Nevertheless, with the maximum suppression of the formation of FFA and the oxidation of fats, the paths of intracellular metabolism of glucose are determined to a greater extent by the rate of absorption of glucose than by the level of insulin [120, 280, 341, 358, 431, 576].

Lowering of the rate of intracellular oxidation of glucose in IIDM (without obesity) occurs independently of lowering of the rate of glucose utilization by the tissues and of the intracellular absorption of fat.

The activity of the glycolysis enzymes also depends on the level of the free fatty acids in the blood. Randle's glucose-fatty acid cycle coordinates biochemically the reaction of glucose and fatty acids. For example, in fasting, when the content of glucose in the blood diminishes, fatty acids are mobilized from the adipose tissue because of a growth in lipolysis. As a result, the concentration of the FFA in the blood grows together with their oxidation in the muscles.

It was shown in experiments that the FFA block glycolysis in the stages of its regulation. There is an inversely proportional relation between the level of FFA in the blood and the activity of hexokinase. Moreover, the FFA inhibit the oxidation of pyruvate in the Krebs cycle and simultaneously increase the production of glucose in the liver, activate the pentosophosphate path of glucose metabolism.

There are thus a number of factors regulating individual units of carbohydrate metabolism. However, the chief role in this process undoubtedly belongs to insulin [137, 310, 449, 472, 498, 601].

Consequently, we can note (see Figs. 5-7) that the formation of an insulin resistance syndrome is an integrative process including both genetically determined defects of metabolism (carbohydrate and lipid metabolism) and acquired pathological syndromes (obesity and hyperinsulinemia). An important result of these metabolic disorders is the growth in the content of VLDLP and LDLP in the blood, and also the drop in the level of HDLP. These disorders of lipid metabolism cause the development of IHD, atherosclerotic changes in the vessels of the brain and lower extremities. Hyperinsulinemia leads to disturbance of the resorption of sodium ions, an increase in the activity of the sympathetic nervous system, and to development of hypertension. Hypertrophy of the smooth muscle cells of the arteries is a morphological manifestation of these metabolic disorders [49, 75, 87, 133, 271, 284, 314].

Considering the great significance of insulin resistance in the formation of syndrome X, we studied the sensitivity to insulin of 10 healthy persons and 15 youths with obesity of a hypothalamic and hypophyseal genesis. To appraise the state of the insulin resistance, we ran a test with insulin that was introduced in a dose of 0.15 units per kg of body weight. The insulin test reveals most vividly the sensitivity of receptors to insulin and disorders of carbohydrate metabolism. It was proved biochemically that glucose is a very important regulating factor of physiological homeostasis. In different organs and tissues, glucose is transported by various biochemical structures. For example, in adipose tissue, the transportation of glucose is associated with two glycoproteins. In the brain, apart from transmembrane cell transportation, there is a system of stereospecific vectors through the capillaries of the hematoencephalic barrier that carry glucose from the blood into the tissues. In human erythrocytes, the vector of glucose is a protein with a molecular weight of 55,000 and a high lipid and carbohydrate content. Numerous biochemical studies revealed that the rate of glucose transportation correlates without fail with its intracellular metabolism. The level of the latter is determined by the amount of glucose oxidized to CO_2 and by the biosynthesis of lipids in isolated lipocytes [96, 111, 131, 153, 213, 258].

After the introduction of insulin in healthy persons, a physiological drop in the glucose level in the blood plasma was noted from 4.88 mmol/l (background) to 2.04 mmol/l in 30 min. The glucose level gradually increased to 3.6 mmol/l (in 60 min), and after 120 min returned to its initial level (4.6 mmol/l). A similar reaction proceeds in erythrocytes: the background amount of glucose was 3.4 mmol/l; in 30 min after the introduction of insulin, the glucose level dropped to 1.55 mmol/l; in 60 min it increased to 2.4, and in 120 min returned to the background figures of 3.5 mmol/l (Fig. 8).

An absolutely different picture was observed in the obesity patients, namely, a background value of 5.1 mmol/l of glucose in the blood plasma and 3.6 in the erythrocytes. After the introduction of insulin, the glucose level did not practically lower. In 30 min, the glucose content in the plasma diminished to 4.5 mmol/l, and in the erythrocytes to 3.1. In 60 min, the glucose level in the plasma increased to 4.8, and in erythrocytes to 3.4;

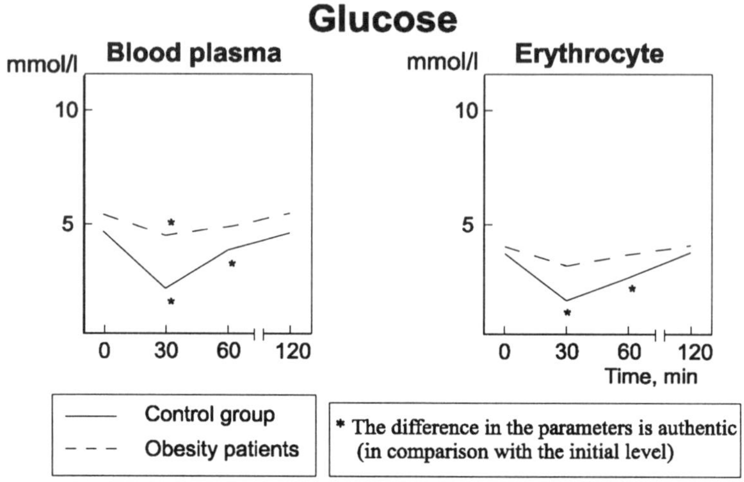

Fig. 8. Change in Glucose Content in the Blood When Conducting an Insulin Test in Patients with Obesity.

in 120 min it returned to the initial values – 5.26 mmol/l in the plasma and 3.7 in the erythrocytes.

The results indicate disturbance of carbohydrate metabolism in the obesity patients and the development of the insulin resistance syndrome in patients with a dysfunction of lipid metabolism. Hyperinsulinemia undoubtedly plays an important role in these pathological shifts.

Currently, numerous experimental and clinical data confirmed the influence of hyperinsulinemia on the development of atherosclerosis of the coronary vessels.

Most authors consider hyperinsulinemia to be another independent risk factor like hypercholesterolemia and arterial hypertension When studying the pathophysiological fundamentals of hyperinsulinemia, our goal was to see whether hyperinsulinemia is associated with ischemia of the tissues in atherosclerosis or with disorder in the permeability of the pancreas arteries. Clinical and experimental investigations showed that there is no relation between lesion of the pancreas arteries and hyperinsulinemia. The latter is a precursor of the development of IHD, and not a secondary factor. This reasoning is based on numerous studies in populations with a varying frequency of atherosclerosis development. During many years of observations of patients with a high insulin level in the blood, it was shown that IHD develops in them much more frequently than in persons not having such a level, and that an increased basal level of insulin makes it possible to predict the development of IHD in the future. The course of atherosclerosis and the degree of manifestation of

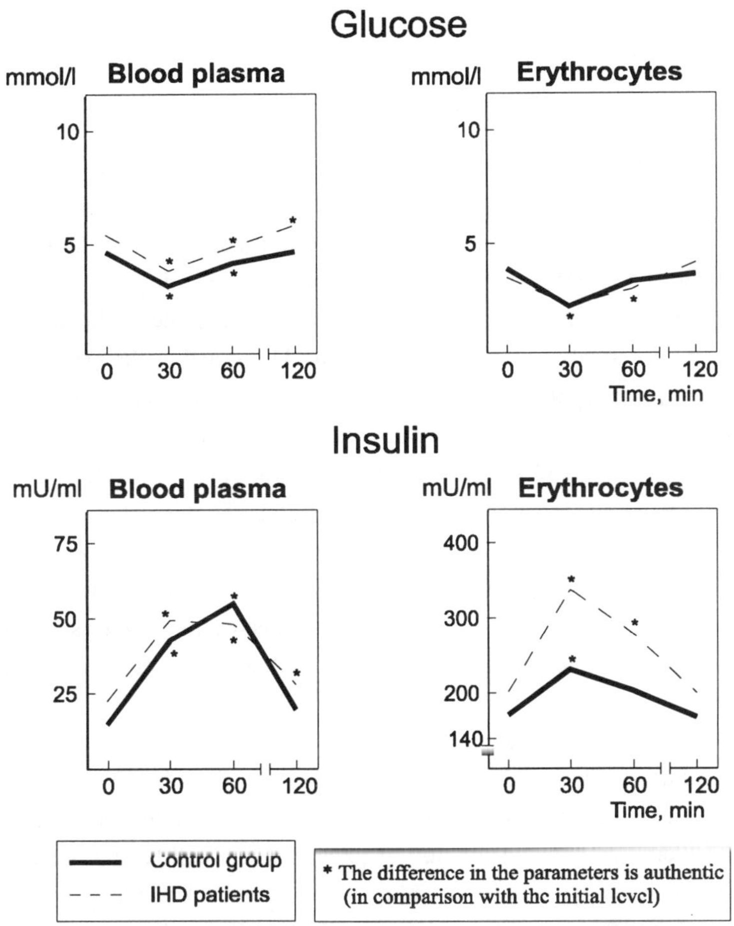

Fig. 9. Change in Glucose and Insulin Content in the Blood when Conducting an Insulin Test in Patients with IHD.

IHD symptoms depend to a considerable extent on the increased content of insulin in the blood plasma [289, 305, 400, 437, 469, 495, 547].

Hyperinsulinemia plays a very important role in the development of syndrome X. This is confirmed by a number of investigations in which it was shown that the biochemical adaptation of an organism to the effect of external extremal factors is closely related to the reactivity of the endocrine system of an organism, including its sensitivity to insulin.

When comparing these experimental results with clinical data on the diagnosis of disorders of insulin metabolism in IHD patients, we determined the sensitivity to insulin of the glucose in the blood plasma in erythrocytes in healthy persons and IHD patients. The insulin test, which determines the sensitivity of glucose to the action of insulin, is the most successful model. We ran it with 52 healthy persons and 31 IHD patients. First we determined the initial level of glucose and insulin in the blood plasma and in the erythrocytes, and then measured these parameters in 30, 60, and 120 min after the introduction of insulin (Fig. 9). In healthy persons, the introduction of insulin caused a clear physiological change in the glucose level. The background level of glucose in the blood plasma was 4.82 mmol/l, in the erythrocytes was 3.78, the initial insulin concentration in the blood plasma was 14.5, and in the erythrocytes was 146.8 mmol/l. In 30 min after the introduction of insulin, hypoglycemia appeared in the plasma (3.05 mmol/l) and in the erythrocytes (2.09 mmol/l). The insulin level in plasma grew to 41.0 μU/ml, and in the erythrocytes to 226.0 μU/ml. In 60 min, the hypoglycemia diminished – the glucose content in the blood plasma increased to 3.9 mmol/l, and in the erythrocytes to 3.1. The insulin level in the plasma increased somewhat (up to 53.3 μU/ml), and lowered in the erythrocytes (to 203.4 μU/ml). In 120 min. all the parameters approached the background values.

In the IHD patients who were given the insulin test, the background values of the glucose and insulin increased. The glucose level in the blood plasma was 5.13 mmol/l, and in the erythrocytes was 3.56 mmol/l. The hyperinsulinism syndrome was very pronounced: the level of insulin in the blood plasma increased to 22.6 μU/ml (in healthy persons it was 14.5 μU/ml), and in the erythrocytes to 182.9 (in healthy persons it was 146.8 μU/ml).

What are these changes associated with?

Many clinicians consider that in IHD a relative insulin insufficiency is often noted with an increase in the glucose level in the blood and with hyperinsulinemia.

Insulin increases the consumption of oxygen in the cells of the internal membrane of vessels, and its local hypoxia is possible with a subsequent increase in the lipid content in it. Of major significance is the presence in IHD patients of insulin resistance, the increased discharge of glucocorticoids being one of its causes.

Clinicians proved the stimulating influence of glucose on the secretion of insulin.

What is the pathophysiological mechanism of this stimulation?

Two theories appeared to answer this question, namely, the glucoreceptor and the metabolic ones. The first theory postulates that glucose binds with the specific membrane receptors of the beta cells which send information to the mechanisms releasing insulin. The adherents of the second theory consider that insulin is stimulated not by glucose itself, but by its direct and more remote metabolites [87, 113, 205, 280, 301, 358, 596].

A major role in the mechanism of the stimulating effect of glucose on insulin secretion is given to cAMP. Experiments showed that substances stimulating or inhibiting the decomposition of cAMP correspondingly inhibit or stimulate insulin secretion. The cAMP activates proteinkinase – an enzyme catalyzing the phosphorylation of definite function proteins. The results of endocrinological studies indicate that STH stimulates the secretion of insulin and has a certain influence on biosynthesis [145, 157, 223, 259, 269, 324].

There was an opinion that hormones enter a cell by diffusion. Recent studies showed the intracellular penetration of hormones to be a very involved biological process, and an equality sign cannot be placed between the concentration of the hormone in the blood and the efficiency of its biological effect.

Analysis of the change in the glucose and insulin content in the blood of IHD patients after the introduction of insulin showed substantial differences between these parameters in healthy and ailing persons and also the presence of insulin resistance in patients with syndrome X and IHD. If in healthy persons in 120 min the background level of glucose is restored, in IHD patients in 60 min it increased, while in 120 min the glucose content authentically exceeded the background level, being 5.56 mmol/l. The hyperinsulinemia syndrome stood out very clearly in IHD patients: in the plasma in 30 min the insulin level increased from 22.6 to 48.3 μU/ml, and in the erythrocytes from 182.9 to 330.0 μU/ml, In 60 and 120 min. these quantities gradually became smaller, but a trend to hyperinsulinemia and insulin resistance (inhibited hypoglycemia) was preserved.

The close relation between hyperinsulinemia and the development of atherosclerosis was proved morphologically. It was established in experiments on animals that the artery wall is a tissue very sensitive to insulin. It was noted that insulin ensures proliferation of the cells in the smooth muscles of arteries, and also increases the synthesis of lipids. Experiments with isolated arteries showed insulin to stimulate the growth of a culture of fibroblasts [447, 469, 491, 505].

1.2.3. Obesity and the Syndrome X

The disturbance of fat metabolism – obesity – is another important metabolic disorder that stimulates the development of syndrome X. Its incidence is based on involved mechanisms leading to a change in lipid metabolism.

Clinicians noted long ago that DM is encountered most frequently in persons with obesity, and the metabolism of carbohydrates and fats is disturbed in them. Biochemists showed adipose tissue to be the main object in which insulin transforms carbohydrates into fats [51, 75, 87, 115, 128, 171].

How does this occur?

Major importance is attached to hereditary factors in the development of obesity (in particular, an increased amount of fat cells and a changed nature of the hypothalamic function is transmitted from generation to generation), and also to disturbance of processes of tissue metabolism. Clinical studies of obesity patients showed the presence of hypertrophy of fat cells, their inadequate metabolic adaptation with lowered oxidation in the Krebs cycle, a lower activity of the lipolytic enzymes, and an increased content of the atherogenic fractions of the blood. Biochemical investigation of the blood of obesity patients indicated an increased content of lactate, pyruvate, and betaglycerophosphatedehydrogenase. The clinical relation between obesity and the development of DM is of special importance. Most scientists consider that 60% of DM patients also suffer from obesity. Generally in about 50% of obesity patients, a disturbed tolerance to carbohydrates is determined. We must stress that the presence of hyperinsulinism and also disturbance of the secretion and deposition of insulin are noted in obesity patients.

The pathophysiological interpretation of the increased insulin level in obesity consists in that most clinicians associate it with the resistance to the effect of endogenic insulin, an increase in the volume of the fat cells, and disturbance of metabolism (e.g. lipid metabolism), and also with a growth in the content of uric acid [38, 472, 561, 585, 596, 604]. We must note improper fat metabolism in obesity patients, in particular a growth in the level of NEFA. A biochemical relation was established between a growth in the level of NEFA and the insulin content. It is considered that an increased amount of NEFA inhibits the stimulating influence of insulin on the assimilation of glucose by tissues, and this results in the development of a state of hyperinsulinism. Simultaneously, an increased glucocorticoid activity of the adrenal cortex is noted in obesity patients.

Hyperinsulinemia with a view to its influence on lipid metabolism may enhance the course of cardiovascular diseases.

To prove the relation between obesity and hyperinsulinemia, we studied the tolerance to glucose of 10 healthy persons and 35 youths with obesity (Table 3). The average values of the anthropometric parameters in the group of obesity patients was height – 178.4 cm, body weight – 99.03 kg, and body surface area 2.17 m^2.

Clinically, the obesity patients exhibited first of all an increased appetite, emotional instability, irritation, disturbance of sleep, fatigue, pains in the joints and muscles, thirst. Auscultation of the heart showed tonelessness. Moreover, an increased arterial pressure was established.

The obesity patients were divided into two groups depending on the degree of obesity. Group 1 included 15 patients (BWI=25.0–30.0) and group 2 included 20 (BWI=30.1–35.0). There was a clear difference in the groups in the insulin content after the introduction of glucose. Pronounced hyperinsulinemia was noted in the obesity patients. If in healthy persons the background value of the insulin was 5.3 μU/ml, in obesity group 1 it

Table 3. Results of Test for Tolerance to Glucose in Obesity Patients.

Group	Glucose, mmol/l			IRI, μU/ml		
	background	in 1 h	in 2 h	background	in 1 h	in 2 h
Control	3.7±0.4	6.8±0.6[a]	3.9±0.3[b]	5.3±0.7	14.2±3.8[a]	4.8±0.5[b]
1st	3.9±0.3	6.6±0.4[a]	4.4±0.6[b]	15.5±3.6[c]	53.5±8.4[a, c]	36.0±8.6[a, c]
2nd	4.0±0.3	6.3±0.5[a]	4.1±0.3[b]	16.9±1.5[c]	91.8±11.4[a, c, d]	59.7±16.3[c]

The difference of the parameters is authentic:
[a] at the given time and before the load;
[b] at the given time and in one hour after the load;
[c] in the given and control groups;
[d] in the given group and in group 1.

was 15.5 and in group 2 – 16.9 μU/ml. After a glucose load, the insulin content increased, namely, in healthy persons to 14.2 μU/ml, in group 1 patients to 53.7 and in group 2 to 91.8 μU/ml. A slow drop in the insulin level was noted in two hours after the introduction of glucose. If in the healthy persons it returned to the normal value (4.8 μU/ml), in group 1 patients it was 36.0 μU/ml (the background was 15.5) and in group 2 it was 69.7 (the background was 16.9 μU/ml).

These comparisons show that hyperinsulinemia and obesity are mutually related metabolic disorders in obesity patients.

Experimental results obtained by Stout indicate that chronic hyperinsulinemia leads to the development of systemic atherosclerosis because of the proliferation of the smooth muscle and connecting tissue membranes of an artery wall stimulated by insulin.

It is shown that if glucose is perfused as a substrate in combination with insulin, the greatest amount of triglycerides are synthesized in an artery wall. We already noted that insulin increases the consumption of oxygen in the intima cells of human vessels. This; in turn, leads to development of the rigidity of the intima, narrowing of the lumen of an artery and arterioles over a considerable length, a growth in the total peripheral resistance, and to development of hypertension with subsequent hypertrophy of the myocardium. Cardiovascular pathology forms in this way, and one of its pathogenetic units is hyperinsulinemia or insulin resistance, or both, enhanced by DLP. Hyperinsulinemia and disturbance of the tolerance to glucose may be attended by the development of DM that affects adversely the course of cardiovascular diseases. Their prophylaxis requires early diagnosis of metabolic disorders [97, 111, 145, 166, 222, 243].

Consequently, metabolic disorders closely related to hyperinsulinemia underlie syndrome X whose inheritance is doubtless, while its pathogenetic mechanisms lead to the development of clinical symptomatology of many metabolic diseases.

The metabolism of adipose tissue and the secretion of insulin are closely related to the vegetative nervous system. The hypothalamic structures and adipose tissue are the points of their contact. Catecholamines are a very important regulator unit in the metabolism of adipose tissue. The FFA content in the blood is regulated by adrenaline and noradrenaline, which via the beta-l-adrenergic receptors stimulate lipolysis in the adipose tissue. The influence of catecholamines is very important. Through the system of cAMP they activate the hormonosensitive lipase that catalyzes triglycerides to free fatty acids and glycerol [237, 248, 281, 328, 396, 524].

The lipolytic effect of the catecholamines is achieved because of their simultaneous inhibition of insulin secretion via the alpha-adrenoreceptors.

Adipose tissue is one of the largest target tissues for catecholamines in an organism whose innervation is represented exclusively by adrenergic endings. The central vegetative centers and adrenal glands regulating the content of neuromediators and hormones play a substantial role in maintaining the FFA level in the blood.

The interaction of insulin and the sympathetic nervous system at the level of the hypothalamic structures is based on their possibilities of affecting diet-dependent thermogenesis (DT). Fasting suppresses, while the taking of carbohydrate or fat food stimulates the activity of the central structures of the SNS owing to the metabolism of glucose in the intermediate region of the hypothalamus (the ventromedial nucleus) promoted by insulin. The taking of carbohydrates and the growth in the content of glucose and insulin boost the metabolism of glucose in the nuclei of the hypothalamus. This results in deactivation of the inhibitor path connecting the ventromedial nucleus of the intermediate region of the hypothalamus to the central sympathetic structures increasing sympathetic activity [19, 237, 534, 567, 616].

Under conditions of the normal sensitivity of tissues to insulin, the growth in the activity of the SNS increases DT and thus helps retaining the energy balance withstanding the storage of a surplus amount of energy in the form of fat and helping its liberation in the form of heat.

We can assume that obesity develops when consuming a surplus amount of food combined with disturbance of DT. We already indicated that with an excessive body weight, insulin resistance appears as a sort of protective mechanism preventing its further increase. Under conditions of insulin resistance, the SNS as a means of restoring energy equilibrium cannot be stimulated normally by insulin. The physiological price of such a compensating mechanism is the increasing hyperinsulinemia and stimulation of the SNS, which subsequently facilitate the development of arterial hypertension [328, 400, 436, 447, 469, 488].

To verify our assumptions on the change in vegetative regulation in obese youths, we compared the excretion of adrenaline, noradrenaline (NA), dihydroxyphenylalanine (DOPA), and dopamine during various times of the day (9 a.m.–12 p.m., 12–3 p.m., 3–7 p.m., and 7 p.m.–9 a.m.) and the change in the level of catecholamines after the introduction of alpha-DOPA (Fig. 10). We studied two groups of youths 16-18 years old suffering from various degrees of obesity (the first group with a BWI of 25.1–30.0 and the second with a BWI of 30.1–35.0). The ratio of the waste circumference to the volume of the hip in group 1 was 0.94±0.02, and in group 2 was 0.93±0.02, thus exceeding the generally adopted critical value of 0.9 for men. The results were compared with those for healthy youths of the same age. We compared the levels of IRI and cortisol in these groups. We clearly established the presence of hyperinsulinemia in the obese youths. In the healthy ones, the IRI level was 5.1 μU/ml, in group 1 (with obesity) it was 13.0, and in group 2 it was 15.7 μU/ml. The cortisol content also differed: in healthy youths it was 659.9, and in obese patients it was 724.4 nmol/ml.

The results of determining the daily rhythm of excretion of catecholamines and their precursors with the urine in the studied groups indicated an authentic drop in the excretion of adrenaline, dopamine, DOPA in the evening (3 – 7 p.m.) and at night (7 p.m. – 9 a.m.) with moderate obesity (BWI=25.1–30.0) and a drop in the excretion of adrenaline, NA, dopamine, and DOPA in the evening and at night with pronounced obesity (BWI=30.1–35.0). In the healthy youths, a similar daily rhythm was observed only for adrenaline and DOPA.

When comparing the parameters of catecholamine excretion with the urine in youths of the control group and in patients with moderate obesity, we established an authentic drop in the excretion of dopamine at night (182.3±4.02 and 128.8±17.7 ng/min) and an authentic increase in the excretion of NA in the morning (6.1±0.9 and 12.9±1.7 ng/min). With a pronounced degree of obesity, we also observed an authentic drop in the night excretion of dopamine in comparison with that in healthy persons (128.3±4.02 and 122.2±9.3 ng/min), while the excretion of NA authentically exceeded the value in the control group both in the morning (6.1±0.9 and 12.8±1.8 ng/min) and at night (5.1±0.4 and 7.7±1.1 ng/min).

The established trend of an increase in the excretion of adrenaline and DOPA with the urine in obesity in comparison with excretion in healthy persons is not authentic owing to the high variability of the individual values.

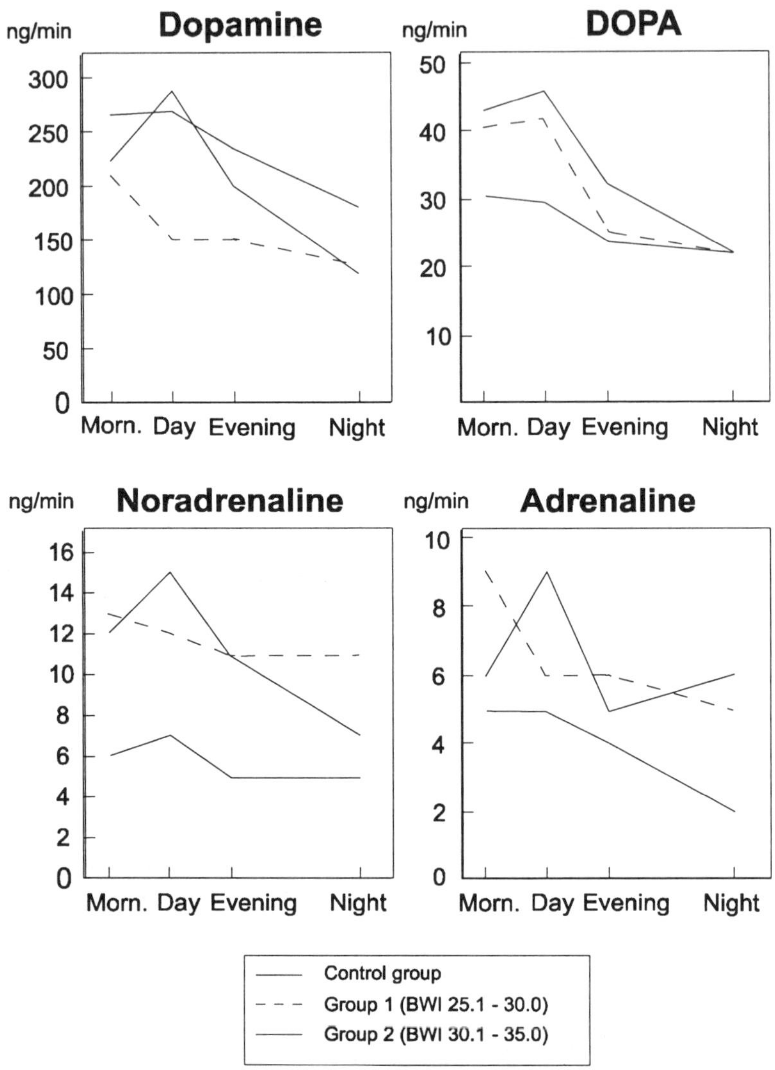

Fig. 10. Daily Excretion of Catecholamines with the Urine in Persons with a Various Degree of Obesity.

To estimate the synthesizing and reserve possibilities of the sympathico-adrenal system, we studied the excretion of catecholamines and their precursors with the urine with the use of the functional test with alpha-DOPA proposed by Vasiliev and Chugunov. The excretion of NA during the first three hours after taking 0.1 g of alpha-DOPA creased authentically by 964.8% in group l and by 387.5% in group 2 in comparison with 2294%

Table 4. Excretion of Noradrenalin with Urine after Taking Alpha-DOPA.

Noradrenaline, ng/min	Morning, 9 am–12 pm	Noon, 12 pm–15 pm	Evening, 15 pm–19 pm	Night, 19 pm–9 am
Control group				
Background	6.13±0.94	7.78±1.51	4.99±0.98	5.12±0.42
After taking α-DOPA	146.75±21.36[a] (+2294%)	27.22±10.34 (+250%)	11.90±3.44 (+138%)	4.97±0.66[e] (-2.9%)
1st group				
Background	12.96±1.67	12.27±3.17	9.09±2.17	9.82±3.39[e]
After taking α-DOPA	138.0±30.92[a] (+964%)	48.8±10.78[a,b] (+298%)	16.43±3.17[c] (+80.7%)	6.79±1.5[e] (-30.9%)
2nd group				
Background	12.79±1.82	15.29±1.47	9.75±1.07[c]	7.71±1.08[e]
After taking α-DOPA	49.56±9.97[a] (+387%)	47.73±7.27[a] (+212%)	16.01±3.58[c] (+64.2%)	7.86±1.43[d,e] (+1.9%)

The difference of the parameters is authentic:
[a] before and after taking α-DOPA;
[b] in the daytime and in the morning;
[c] in the evening and in the daytime;
[d] at night and in the evening;
[e] at night and in the morning.

in healthy youths (Tables 4 and 5). Similar dynamics were registered with respect to adrenaline, namely, an authentic increase by 86.7% in group 1 in comparison with 112.8% in the control group. With pronounced obesity, an authentic growth in the excretion of adrenaline after the introduction of alpha-DOPA was not established authentically.

Inspection of Tables 4 and 5 reveals that there is a clear relation between the degree of obesity and the metabolic disorders.

The adaptation of the organism of animals and humans to oxygen fasting (hypoxia) was studied by many authors. Experiments with animals showed that adaptation to hypoxia occurs with a sharp intensification of anaerobic glycolysis with a growth in the concentration of lactic acid in the blood and an increase in the lactate/pyruvate ratio. The lactic acid level is a very important criterion of adaptation of an organism under conditions of hypoxia, and its increased level under there conditions is inversely proportional to the physical workability of a person. Persons with a high level of anaerobic glycolysis have a high ratio of the content of lactic acid to the level of pyruvic acid [37, 64, 109, 125, 591, 554, 601, 608].

The relation between the glucose level in the blood and the absorption of glucose by the myocardium was studied by many authors, mainly experimentally. It was found under clinical conditions that glucose is used in glycolysis for supporting anaerobic metabolism. It was proved experimentally that the rate of absorption of glucose by the cardiac myocytes from the extracellular space is limited by the rate of tissue transportation of glucose through a cell membrane. There is a clear relation between the rate of transportation and that of absorption of glucose under the influence of insulin, and also under conditions of hypoxia. Much attention in the experiments was devoted to studying glycolysis, i.e. the decomposition of glucose and glycogen in the myocardium, when lactate forms under anaerobic conditions. In normal oxidative metabolism, glycolysis results in the formation of pyruvate, which under aerobic conditions decomposes in the Krebs cycle. Under anaerobic conditions, glycolysis is the only source of ATP. In the absence of oxygen, the rate of glycolysis grows several times owing to the Pasteur effect. In deciphering the biochemical essence of the Pasteur effect, we can say that it consists in an increase in the rate of lactate formation under anaerobic conditions. When an animal inhales a hypoxic mixture, the rate of glucose absorption grows, and instead of the absorption of lactate by the myocardium (which occurs under normal conditions), the emergence of lactate from the cardiac myocytes is observed.

In confirming the significance of glycolysis in the energy metabolism of the myocardium, we must note that the main role in regulating the rate of glucose absorption by the myocardium cells is played by the rate of glucose transportation. The latter may change (grow) under the effect of insulin, and also under conditions of hypoxia. Here the flow of substances grows because of the higher activity of phosphofructokinase that limits the total rate of glycolysis. Under aerobic conditions with a normal oxygen concentration, pyruvate entering the Krebs cycle is the final product of glycolysis. Under anaerobic conditions, lactate is the final product of glycolysis. In moderate ischemia of the myocardium, the rate of glycolysis grows. The myocardium utilizes lactate in a conventional quiet situation, absorbing 10% of oxygen from the entire oxygen of the air. If much work is being done, the oxygen consumption grows to 60%. The lactate and FFA compete for the oxygen absorbed by the myocardium in connection with the fact that each of these compounds may be the main nutritive substance depending on its level in the blood. The main metabolic paths of pyruvate transformation are aerobic oxidation with the aid of pyruvatedehydrogenase and in the citric acid cycle, and anaerobic transformation into lactate [83, 114, 206, 277, 346, 486, 619].

What is the increased level of pyruvic acid in the blood associated with?

Table 5. Excretion of Adrenalin with Urine after Taking Alpha-DOPA.

Adrenalin, ng/min	Morning, 9 am–12 pm	Noon, 12 pm–15 pm	Evening, 15 pm–19 pm	Night, 19 pm–9 am
Control group				
Background	5.31±1.01	5.68±1.98	4.53±1.19	2.67±0.43[e]
After taking α-DOPA	11.30±1.53 (+112.8%)	8.44±2.75 (+48.6%)	3.21±0.62 (-29.2%)	1.82±0.37[a,e] (-31.8%)
1st group				
Background	9.24±2.24	6.58±1.08	6.79±2.89	6.06±1.78[d]
After taking α-DOPA	17.25±6.49[a] (+86.7%)	10.72±1.65[a] (+62.9%)	5.46±0.85[c] (+19.6%)	2.63±0.51[a,d,e] (+56.6%)
2nd group				
Background	6.34±1.24	9.35±1.48	5.04±0.72[c]	6.52±1.96
After taking α-DOPA	8.33±1.23 (+31.4%)	10.20±1.29[b] (+9.1%)	4.66±1.01[c] (-7.5%)	4.98±1.29 (-23.6%)

The difference of the parameters is authentic:
[a] before and after taking α-DOPA;
[b] in the daytime and in the morning;
[c] in the evening and in the daytime;
[d] at night and in the evening;
[e] at night and in the morning.
The authenticity of the difference of background levels in two groups is not determined.

In the oxidation of pyruvate by oxidative phosphorylation, ATP forms, and the pyruvate level drops accordingly. If a high level of pyruvic acid is preserved, this indicates that the formation of ATP in the Krebs cycle is inhibited.

Under anaerobic conditions, glycolysis proceeds with the participation of a large number of enzymes. The glucose decomposes into pyruvic acid. The following are the basic biochemical stages of glycolysis:

1. Glucose $\xrightarrow{\text{Hexokinase}}$ Glucoso-6-phosphate + ADP-ATP

2. Fructoso-6-phosphate $\xrightarrow{\text{Phosphofructokinase}}$ Fructoso-1.6-biophosphate + ADP-ATP

3. Phosphoenolpyruvate + ADP $\xrightarrow{\text{Pyruvatekinase}}$ Pyruvate + ATP

It was noted in experiments that insulin affects the activity of glycolysis under anaerobic conditions: under the influence of insulin glycolysis grows, while with a lack of it (especially in DM patients) it weakens drastically. The biochemical effect of insulin is

Table 6. Results of an Insulin-Tolerance Test in Practically Healthy Persons.

Glucose, mmol/l		STH, ng/ml	
background	in 30 min	background	in 30 min
4.71	4.44	2.80	62.00
4.40	3.20	1.30	5.10
2.39	1.68	0.40	7.40
3.85	2.77	0.80	2.10
5.55	4.44	0.87	16.20
5.05	3.80	1.40	32.00
4.99	3.60	1.45	25.00
5.55	5.05	0.33	32.00
5.05	4.05	0.75	46.00
4.71	4.44	2.40	52.56

directed at the enzymes hexo(gluco)kinase, phosphofructokinase, and pyruvatekinase [86, 144, 189, 271, 333, 364].

We consider that in this disorder of metabolism an important role is played by insulin resistance, with the simultaneous existence of hyperinsulinemia and hyperglycemia. Consequently, the presence of insulin resistance is confirmed clearly.

How can we prove the presence of insulin resistance in obesity patients?

To solve this problem, we compared the results of an insulin-tolerance test in healthy persons and obesity patients. We selected 10 healthy persons (without dyslipidemia) and 15 patients with obesity caused by adolescent pituitarism. In all the persons, we determined the glucose level before the introduction of insulin and in 30 min after its introduction. We obtained the following results (Tables 6 and 7): in healthy persons, the glucose level dropped quite clearly after the introduction of insulin. Of 10 healthy persons, considerable hypoglycemia was noted in 7. This points to the good reaction of the receptor apparatus to the introduction of insulin. We noted an absolutely different picture in the 15 obesity patients. In three of them, moderate hypoglycemia could be noted. In 10 patients, the glucose level either increased or did not change. The dynamics of the STH level varied.

What is the pathophysiological basis of these disorders in the sensitivity of obesity patients to insulin?

We must stress first of all the fact that we noted hyperinsulinemia in the obesity patients. Numerous clinical and experimental studies showed that very often the disturbance of the tolerance to glucose combined with hyperinsulinemia. It is the insulin resistance of the tissues that leads to the development of compensating hyperinsulinemia. Many

Table 7. Results of an Insulin-Tolerance Test in Obesity Patients.

Glucose, mmol/l		Relative change	STH, ng/ml		Relative change
back-ground	in 30 min		back-ground	in 30 min	
4.39	4.78	+1.09	0.22	6.86	+30.60
5.06	4.28	-1.18	0.20	1.12	+5.57
5.11	5.22	+1.02	0.28	1.60	+5.79
4.44	4.94	+1.11	0.151	0.149	-1.01
4.44	5.11	+1.15	0.25	1.06	+4.24
3.83	3.72	-1.03	0.32	0.22	-1.46
4.78	4.39	-1.09	0.61	1.10	+1.80
4.56	3.56	-1.28	0.52	0.55	+1.05
4.44	4.17	-1.06	1.12	0.42	-2.65
5.44	3.39	-1.61	0.30	2.31	+7.78
3.94	3.61	-1.09	0.67	12.67	+18.94
4.28	4.22	-1.01	0.34	0.32	-1.04
4.28	4.78	+1.12	0.50	1.16	+2.31
4.00	5.39	+1.35	0.18	0.15	-1.18
4.89	5.22	+1.07	1.13	3.06	+2.72

years of observations of patients with a high insulin level in blood showed that IHD developed much more often in them than in persons without a high insulin level in the blood.

The following factors can produce hyperinsulinemia [157, 174, 213, 280, 619]:

(1) a change in the chemical structure of the insulin itself;

(2) disturbance of the transformation of proinsulin into insulin;

(3) the presence in the blood of a large amount of insulin antagonists or the presence of antibodies to insulin;

(4) improper functioning of the insulin receptors.

It was proved experimentally that the diverse effects of various hormones can be represented as the schematic chain reaction "hormone – receptor – adenylatecyclase – cAMP – proteinkinase – phosphorylation of proteins". A change in the sensitivity (increase or decrease) of the link "receptor – adenylatecyclase" alters the activity of a hormonal pulse. The functional interaction of insulin with a cell occurs at the level of the specific receptors of the cytoplasmatic membranes. The insulin receptors are a complicated group of proteins which made it possible to conclude that the system of receptors in the adipose tissue and transportation system of glucose have different physicochemical characteristics, but both the glucose vectors and the insulin receptors are polypeptides differing in their chemical structure [49, 74, 120, 314, 358, 595, 597].

The transportation and metabolic effects of insulin are a result of binding of the hormone to the membrane receptors. In hyperinsulinemia of any etiology, the number of receptors in the adipose tissue and muscles lowers appreciably, while with normalization of the insulin level the number of receptors is restored. The clinical picture of obesity combines with hyperglycemia, hyperinsulinemia, a reduction of glycogen synthesis in the liver, and enhancement of the glycogenesis process. It is the reciprocal nature of the reaction between the growth in the activity of the enzymes of glycogenesis and the inhibition of the activity of the glycolysis enzymes that disturbs lipid metabolism. Already in the early stages of obesity, hyperlipidemia, hypercholesterolemia, hypertriglyceridemia, and a growth in the FFA level are noted in the patients.

When studying the problem of insulin resistance, some authors give attention to hypertriglyceridemia. It was noted that in persons with hypertriglyceridemia, the utilization of glucose diminishes, the secretion of insulin grows, and the level of IRI increases. It was proved experimentally that VLDLP lower the utilization of glucose and, accordingly, stimulate the secretion of insulin, thus causing hyperinsulinemia. As a result, the high level of the IRI stimulates the synthesis of the triglycerides contained in the VLDLP, and the vicious circle closes. Glyconeogenesis is the synthesis of glucose and glycogen from lactic acid. It is not only one of the main adaptive mechanisms of carbohydrate homeostasis, but also an important factor in providing glucose for the energy cycle of an organism.

We can assume that glyconeogenesis is an important regulating facility in the integration of syndrome X. It is the intensification of glyconeogenesis that underlies hyperglycemia and hyperinsulinemia. The excess glucose with the aid of the increased amount of insulin is utilized in the tissues. A result is activation of lipogenesis and obesity. We consider that the enhancement of glyconeogenesis is what triggers off the disturbance of metabolism and the development of syndrome X [65, 70, 203, 279, 405].

It is not so much carbohydrates as fats that are of a major energy significance in an organism. Half of the energy formed in oxidative processes proceeding in the cells (heart, liver, kidneys) is provided by the oxidation of fatty acids. The conversion of carbohydrates into fats occurs in the adipose tissue and is an insulin-dependent process. Fatty acids are oxidized in the cells (mitochondria) which they penetrate into in the form of esters of carnitine and transform there into the active form – acyl-Co-A, which is subjected to beta-oxidation [395, 404, 412, 415, 449, 458].

An important role in the metabolic transformations of lipids is played by LPL, on whose activity the supply of fatty acids to the cells depends.

Fats are also synthesized from glucose with the aid of circulating fatty acids, which reach the fat depots in the form of triacylglycerol in the composition of lipoproteins and chylomicrons or of lipoproteins synthesized in the liver. Consequently, a fat cell uses either endogeneously generated fatty acids or fatty acids delivered with food and synthesized in the liver for the synthesis of fats. However, in either of these variants, the simultaneous supply of glucose is needed to synthesize alpha-glycerophosphate needed, in turn, for the synthesis of fats. The supply of carbohydrates with food stimulates the synthesis of fats in the adipose tissue considerably [129, 270, 473, 487, 498, 500, 515].

Insulin facilitates the penetration of glucose into the lipocytes, increases its transformation into glycerophosphate and fatty acids. It not only enhances synthesis, but also diminishes the decomposition of fats and the release of fatty acids from the adipose tissue. Fatty acids are synthesized under the influence of insulin as follows: the penetration of glucose into a cell is facilitated, glycolysis increases, this leads to the formation of glycerophosphate, and the latter is used in fatty acid esterification.

Lipolysis is also inhibited under the influence of insulin. This occurs by regulation of the cAMP level. In DM, the activity of lipase is increased, and under the influence of insulin its activity drops with the simultaneous stimulation of phosphodiesterase that destroys cAMP. In DM, lipogenesis attenuates, while lipolysis in the adipose tissues increases with the corresponding growth in the amount of fatty acids. The latter, in turn, have an unfavorable effect on the metabolism of carbohydrates.

The distorted metabolism of carbohydrates in DM consists in suppression of carbohydrate oxidation in the liver and enhancement of fat oxidation. The increased synthesis of carnitine in the liver in DM facilitates the greater oxidation of the fatty acids [87, 101, 271, 302, 333, 345, 365, 460].

In recent years, the important role of glucagon in DM as a contrainsular hormone is being widely discussed. Clinicians noted that in decompensated DM when ketoacidosis is quite pronounced, the glucagon content in the blood is sharply increased. The STH suppresses the secretion of glucagon and insulin. It was shown experimentally that the introduction of somatostatine to animals drastically lowers the glucose content because of the drop in glucagon secretion. In insulin-dependent diabetes, the suppression of glucagon secretion with the aid of somatostatine lowers hyperglycemia.

The relation between the somatotropic hormone (STH) and insulin is exceedingly interesting. Biochemical studies revealed that the secretion of the STH increases with postinsulin hypoglycemia. It was also proved that the STH increases the secretion of insulin. Clinicians noted that in DM patients, the STH concentration in the blood is increased. In this situation, it yields a diabetogenic effect that, in particular, is conducted in three ways [205, 271, 301, 494, 548, 620]:

(1) an increased discharge of glucose because of glycogenolysis;

(2) lowering of the absorption of glucose by hepatocytes;

(3) an increased inactivation of the insulin.

The STH disturbs the sensitivity to insulin of the system transporting glucose and the phosphorylation of glucose in the muscles. In adipose tissue, the STH inhibits the transformation of glucose into fats and causes lipolysis.

A comparison of the dynamics of the content of glucose and the STH after the introduction of insulin (see Tables 6 and 7) reveals that in healthy persons the background value of glucose fluctuates from 5.5 to 2.39 mmol/l, while the background level of the STH ranges from 0 to 2.80 ng/ml. The introduction of insulin causes a hypoglycemic state to appear with a considerable drop in the glucose level in 30 min. Of the 10 healthy persons, in three we could note a tendency to insulin resistance. In the remaining seven healthy persons, a considerable drop in the glucose level was noted. The dynamics of the STH level varies.

What occurs in obesity patients?

A glance at Table 7 shows that insulin resistance is noted in 11 patients (of the 15). It is interesting to note the following paradoxical phenomenon: in seven patients after the introduction of insulin, the glucose level increased, e.g. from 4.0 (background) to 5.39 mmol/l. When comparing this dynamics of the glucose level (insulin resistance) with the quantitative changes in the STH, we also noted paradoxes. First of all, the background level of the STH in obesity patients was much lower: in most patients, the STH content was within 0.22–0.52 ng/ml, and only in one patient was this quantity higher than unity (1.13 ng/ml). The introduction of insulin did not cause the STH level to rise in all patients. It lowered in five of them, and increased very moderately in the others. Only in one patient did the figure exceed 12.67 ng/ml.

The results give us the grounds to assert that syndrome X forms with complete disintegration of the regulatory hormonal processes and violation of the physiological relation between insulin, STH, and catecholamines. A pronounced insulin resistance promotes to a considerable extent the metabolic disorders in the formation of IHD [71, 110, 165, 503, 521, 558].

1.2.4. Syndrome X and Heart Pathology

Obesity causes changes in virtually all the organs and systems of an organism, but the cardiovascular system is the first to be involved in a pathological process, The excess accumulation of fat is attended by definite hemodynamic disorders that together with the metabolic one lead to the development of functional and organic pathology of the cardiovascular system. The following variants of lesion of the given system in obesity are singled out clinically: adipose dystrophy of the myocardium, hypertension, IHD, and the Pickwickian syndrome. These diseases frequently combine with one another and, moreover, each of them may be accompanied by the development of cardiac insufficiency to some extent or other [171, 186, 222, 225, 405, 430].

Functional disorders of the cardiovascular system, changes (lowering) of the tolerance to physical loads, and disturbance of the vascular tonus form already in the initial stages of obesity. Special attention is given to the fact of the appearance of arterial hypertension with an excessive body weight. The interest is due to the fact that the growth in the arterial pressure (AP) is closely related to hyperinsulinemia and insulin resistance and thus together with dyslipoproteinemias participates in the development of a set of mutually related metabolic disorders that may subsequently cause the appearance of atherosclerosis in obesity patients. However, the question of the etiological factors and mechanisms of the appearance of arterial hypertension with a surplus body weight remains to be answered. The following are considered to be possible causes of the rise in the arterial pressure [400, 406, 430, 436, 447, 469]:

(1) disturbance of the central and intracardiac hemodynamics, a change in the total peripheral resistance (TPR);

(2) an increased tonus of the sympathetic nervous system leading to vasoconstriction, an increase in the cardiac ejection and the excretion of renin by the kidneys that facilitates, in turn, the production of aldosterone;

(3) lowering of the excretion of sodium by the kidneys because of hyperinsulinemia and insulin resistance;

(4) a change in the intracellular concentration of cations and the subsequent increase of the sensitivity of the cells to vasoconstricting agents as a result of hyperinsulinemia and insulin resistance disturbing the functioning of the membrane pump;

(5) the direct stimulation by insulin or insulin-like growth factors of the hyperplasia of the smooth muscle cells and cardiac myocytes leading to narrowing of the arterioles and hypertrophy of the heart.

To determine the pathological changes in the cardiovascular system in obesity, we compared the tolerance to a physical load in 30 healthy youths and 35 youths with obesity.

The physical load was dosed with the aid of a Siemens Elema veloergometer. In the obesity patients, the BWI was 30.7 with an average body weight of 98.3 kg and a height of 179.4 cm. The load was increased in steps, beginning with 50 W with a further increase by 50 W every three minutes up to the maximum tolerated load. The test was terminated when there was a complaint on physical fatigue.

Table 8. Veloergometer Parameters in Healthy Persons and Obesity.

Parameter	Control group	Obesity patients
Background heart rate, bpm	71.3±2.2	88.8±1.9*
Threshold heart rate, bpm	162.1±2.3	166.2±1.3
Chronotropic heart reserve, bpm	91.1±2.5	77.4±2.0*
Pulse tolerance coefficient, %	102.7±1.3	103.4±0.8
Arterial pressure, mmHg:		
background systolic	115.2±2.7	137.5±1.2*
threshold systolic	186.0±4.6	201.6±4.0*
average initial	91.5±1.6	100.9±1.3*
average threshold	120.7±2.9	124.1±4.2
background diastolic	78.9±1.7	83.3±2.3
threshold diastolic	88.0±3.5	87.2±4.4
Isotropic heart reserve, mmHg	70.1±4.3	64.1±3.6
M_{th}, W/min	197.3±6.7	168.2±4.6*
Energy tolerance coefficient, %	126.0±2.5	73.4±2.5*
Total power of applied load, kJ	94.7±6.4	67.1±3.8*
Maximum oxygen consumption, l/min	3.7±0.2	3.1±0.08*
Max. oxyg. con./norm. oxyg. cons., %	125.1±2.9	80.7±1.5*
Double product index, units		
initial	82.8±2.7	122.3±3.2*
threshold	302.7±9.2	333.1±7.4*
Energy expenditure index, units	3.5±0.3	5.2±0.2*

* The difference of the parameters is authentic (in comparison with the level in the control group).

The data in Table 8 show that the hemodynamic parameters in the two groups differ both at rest and, especially, under conditions of a physical load. First of all, the heart rate at rest in healthy persons was 71 bpm, and in the obesity patients was 88 bpm. The level of the background systolic arterial pressure in healthy persons was 115.2 mmHg, and in obesity patients was 137.5 mmHg; the diastolic pressure was 78.9 and 83.3 mmHg, respectively. Of special importancc is thc reaction of the arterial pressure under a physical load: if in healthy persons, the threshold systolic arterial pressure was 186.0 mmHg, in obesity patients it was 201.86 mmHg. The values of the diastolic threshold arterial pressure in the two groups coincided (88.0 and 87.2 mmHg).

What do these parameters indicate?

First of all, they confirm the information on the disorder of metabolism in obesity and, moreover, witness a change in the metabolic processes in the vegetative nervous system, especially in the metabolism of catecholamines.

Adaptation to a physical load is combined with an increased activity of the adenylatecyclase system and a growth in the number of adrenergic fibers per unit of myocardium

Table 9. Parameters of Hemodynamics in Healthy Persons and Obesity Patients (Data of Echocardiography).

Parameter	Control group	Obesity patients
Heart rate, bpm	57.5±3.1	71.2±1.9
Arterial pressure, mmHg:		
systolic	119.8±2.7	122.6±1.9
diastolic	76.1±2.1	77.6±1.4
average	90.6±3.1	92.6±1.3
Minute volume, l/min	5.7±0.3	7.1±0.2*
Cardiac index, $l/min \cdot m^2$	3.2±0.2	3.3±0.1
Stroke volume, ml	85.6±4.9	100.2±3.3*
Stroke index, ml/m^2	47.4±2.4	46.2±1.5
Peripheral resistance, $dyn \cdot s \cdot cm^{-5}$	1340.6±78.2	1085.7±39.7*
Unit power, $J/min \cdot m^{-2}$	39.2±2.0	40.2±1.0
Myocardium weight, g	115.7±7.2	152.4±6.0*
Myocardium weight index	64.5±3.8	70.1±2.4
End systolic volume, ml	35.2±3.4	40.9±2.3
End diastolic volume, ml	121.2±7.2	141.2±4.4*
End diast. vol./end. syst. vol.	1.1±0.07	0.9±0.03*
Ejection fraction, %	70.4±2.1	70.6±1.2
%ΔS	40.1±1.6	41.1±1.1
V_{cf}, s^{-1}	1.4±0.06	1.6±0.06*
Contraction rate, cm/s	5.1±0.06	5.3±0.2
Relaxation rate, cm/s	13.2±1.5	11.8±0.4

* The difference of the parameters is authentic (in comparison with the level in the control group).

weight. This increases the adrenoreactivity of the myocardium. A growth in the number of light H-chains is observed in the "heads" of the myosin molecules. They are carriers of the ATPase activity of the myosin. Simultaneously the power of the "calcium pump" of the sarcoplasmatic network increases, and as a result – the rate and depth of the diastolic relaxation of the myocardium. The growth in the power of the energy supply system lawfully results in a greater resistance of the heart to hypoxia during a physical load [75, 109, 129, 277, 344, 389, 544, 619].

From our viewpoint, of major importance are Meerson's experimental data [386, 387, 388, 389] showing that the energy potential of a cell regulates the activity of its genetic code, i.e. rigorous structural organization of the regulatory mechanisms of a human organism is carried out. An important role in the functioning of the regulatory mechanisms in a cell is played by cAMP. All external factors (a physical load, hypoxia) activate the adrenergic regulation in an organism and promote the formation of cAMP, which is a powerful inductor stimulating the genetic apparatus of a cell. The human myocardium has an increased adrenoreactivity, i.e. in response to the secretion of noradrenaline, the myocardium reacts by an increase in the inotropic effect (increasing of its contracting ability) . Biochemical studies revealed that under the influence of noradrenaline, the activity of adenylatecyclase grows, and it steps up the synthesis of cAMP. In adapted animals, the inotropic effect of catecholamine is increased by the power of the system "adrenoreceptor – adenylatecyclase". The membrane cell systems in an organism (in the myocardium) responsible for the processing of the control signal ensure its increased effectiveness. Adaptation of an organism that is too stressed may terminate either in exhaustion (functional) of the relevant system, or in disintegration of the regulatory adaptive (hormonal) mechanisms of various organs and systems [2, 71, 534, 543].

What is the drop in the parameters of physical activity in obesity associated with?

There is no doubt that the level of the central hemodynamics is an important factor of the ability to perform physical work. To study this matter, we compared the state of the central and intracardiac hemodynamics for 29 youths suffering from obesity of various degrees, and for 30 healthy youths with the aid of the echocardiographic method.

It follows from Table 9 that in youths with obesity the minute volume of the blood is increased to 7.1 l/min (in healthy persons it is 5.7 l/min), while the stroke volume of the blood is increased to 100.2 ml (in healthy persons it is 85.6 ml). A comparison of the parameters indicates that obesity patients have a hyperkinetic type of blood circulation in comparison with the eukinetic type in healthy persons. It should be noted, however, that in obesity patients the cardiac and stroke indexes are not increased and equal the values for healthy persons. Consequently, the hypercirculation of the blood in obesity patients is caused not by the high contracting ability of the left ventricle, but by other factors.

That in connection with the excess body weight the metabolic demands, the required amount of oxygen and nutrient substances grow may be one of the causes of development of the hyperkinetic syndrome. An increase in the volume of the circulating blood and, therefore, in the minute volume of blood regardless of the arterial pressure level is the factor for the compensating supply of oxygen to the tissues under the given conditions.

Simultaneously with the hyperkinetic type of blood circulation, in obesity patients we noted a drop in the total peripheral resistance to 1085.7 $dyn \cdot s \cdot cm^{-5}$ (in healthy persons it is 1340.6).

What is the pathogenesis of this syndrome?

As a result of an excessive amount of fat deposits, the vascular channel grows and expands, and this lowers the total peripheral resistance.

Inspection of Table 9 shows that in obesity patients the weight of the myocardium of the left ventricle is considerably increased – up to 152.4 g against 115.7 in healthy persons.

The compensating hypertrophy of the myocardium in obesity patients is a heart activity regulating factor for maintaining a normal minute volume. The pathophysiological hypertrophy of the myocardium is based on activation of the synthesis of nucleic acids and proteins and an increase in the heart volume. In obesity patients, myocardium hypertrophy is a result of mobilization of the energy and structural resources of the heart. The hyperfunction of the myocardium, manifest in a growth in the stress, entails a pronounced increase in the consumption of oxygen, the resynthesis of ATP, and a no less pronounced activation of the synthesis of nucleic acids and proteins. The continuous hyperfunction in obesity is attended by the consecutive mobilization of the energy and structural reserves of the cardiac myocytes with subsequent wear (early) of the myocardium and disturbance of its function. The morphofunctional basis of the compensating hypertrophy of the myocardium in obesity is formed by the increased function of the myocardium that activates the genetic apparatus of the cardiac myocytes. This increases the synthesis of nucleic acids and proteins. Hypertrophy of the myocardium is a consequence. In obesity, the myocardium experiences additional damaging factors, in particular a lack of oxygen, and separation of the oxidation and phosphorylation processes [28, 109, 356].

What is the mechanism of action of these damaging factors?

It is assumed that it may be closely related to the influence of catecholamines. It was shown in experiments that moderate ischemia of the myocardium (caused by the dosed narrowing of the coronary arteries or the introduction of large doses of noradrenaline) first lowers the concentration of creatine phosphate and ATP, and then activates the synthesis of nucleic acids and proteins. As a result, the weight of the myocardium grows by up to 30–80%.

What is the pathophysiological mechanism of hypertrophy of the myocardium in obesity?

The following aspects of the effect of lipid metabolism on the metabolism of the myocardium can be noted [45, 109, 190, 333, 338, 366, 486, 543]:

(1) a large amount of triglycerides concentrate in the myocardium in obesity; they hydrolyze with the liberation of fatty acids;

(2) the membrane phospholipids may be hydrolyzed by activated phospholipases of the cytosol, which causes the liberation of fatty acids and, possibly, of lysophosphatides;

(3) the intracellular content of fatty acids may grow as a result of their increased absorption by ischemized myofibrills; inhibition of fatty acid oxidation and condensation of dicarbon fragments also occur.

These biochemical processes result in an excessive accumulation of fatty acids and intermediates of their metabolism inside and around the myofibrills. The fatty acids disturb the functioning of the membranes of the myofibrills: they disrupt the transport of Ca^{2+} and the structure of the membrane.

Disruption of the Ca^{2+} transport affects the contraction and relaxation of the myocardium. Lowering of the rate of relaxation of the myocardium may be due not only to disorders in the absorption of Ca^{2+} from the microfibrills, but also to the lowering of its concentration acting in the myoplasma and activating the Ca-dependent ATPase of the sarcoplasmatic network.

The energy deficiency of the myocardium in obesity is also associated with the excessive activation of catecholamines. Their excessive accumulation in the myocardium, in addition to the hyperfunction of the heart, leads to the activation of a set of metabolic mechanisms that may inhibit biological oxidation or lower its efficiency.

The increased decomposition of lipids and phospholipids with improper utilization of their decomposition products is due to the excessive activation of the lipases and phospholipases together with the accumulation of Ca^{2+} in the myoplasma. These enzymes are activated by catecholamines via the adenylatecyclase – cAMP system. The increased decomposition of the neutral lipids under the influence of lipases is attended by the excessive accumulation of fatty acids that are not subjected to beta-oxidation, are not utilized in the Krebs cycle, but are accumulated in the cardiac myocytes. The accumulation in a cell of the fatty acids and lysophosphatides formed in the decomposition of the phospholipids points to the high significance of the lipid mechanism of lesion of the cardiac myocytes. The increased decomposition of the phospholipids and the sharp swelling of the mitochondria inhibit the respiratory train. Lipid damage to the mitochondria separates the processes of oxidation and phosphorylation and disturbs ion transport. The accumulation in the myocardium of the products of decomposition of phospholipids – lysophospholipids (lysophosphatidylcholine, etc.) disturbs the function of the ion channels of membranes and the loss of K^{+} by the cells. Electrical instability of the myocardium appears, arrhythmias develop, as well as contractures of the myofibrills with suppression of the contractive function of the myocardium [10, 333, 542, 619].

The pathophysiological basis of the numerous pathological processes consists in disturbance of the structure and function of the cell and intracellular membranes. This is why the restoration of the disturbed membrane relations is an important component of therapeutic measures. The activity of the membrane structures is of special significance in the development of the pathology of the myocardium. In particular, such basic functions of the myocardium as the excitability of the cells, the conjugation of excitation and contraction, neurohumoral and hormonal regulation of biochemical processes are all associated with the activity of the heart's membrane structures. When studying the pathology of the myocardium, four main problems have to be solved that are closely related to investigating the lesion of cell membranes [131, 256, 306, 332, 389, 537]:

(1) studying of the structure and function of the mitochondria and (accordingly) the mechanism of the incidence of ischemia of the myocardium cells, in particular, in IHD;

(2) with syndrome X, studying of the influence on the membranes of cholesterol, peroxide oxidation of lipids, and the relation between the pathology of the membranes and the development of IHD;

(3) studying of the excitability of the myocardium cells associated with the functioning of the ion channels and the sodium pump;

(4) studying of electrical phenomena in the myocardium, the appearance of disturbed relations of the calcium metabolism and, accordingly, the process of relaxation of the myocardium and contractility.

All these problems are solved at the junction of several specialities: biochemistry, biophysics, morphology, and internal diseases. A number of scientists showed that in syndrome X there appear tissue hypoxia and ischemia of the myocardium cells pointing to the great significance of the membrane phospholipases in the irreversible lesion of cells. The lipid bilayer of membranes acquire electrical instability and lose their barrier functions. In recent years, major significance in the pathogenesis of atherosclerosis is attached to the peroxide oxidation of lipids. The use of antioxidants inhibiting the peroxide oxidation of lipids was found to affect deeply the lipid composition and physical properties of all membrane structures of the cells, and this may be the theoretical basis for the use of antioxidants as an important therapeutic factor in atherosclerosis.

Experiments showed that the hyperfunction of the myocardium appearing under the influence of sympathetic neurohormones is not attended by a corresponding increase in the coronary blood circulation, and in this connection a discrepancy appears between the requirement of oxygen by the myocardium and its consumption. Suppression of the utilization of oxygen by the myocardium under the influence of catecholamines appears as a result of inhibition of the activity of the cytochromes, The accumulation of catecholamines in the myocardium increases the peroxide oxidation of lipids in the membranes of the cardiac myocytes, which leads to destabilization of the membranes, liberation of proteolytic

Table 10. Content of Glucose, **IRI**, Pyruvic and Lactic Acids in the Blood Plasma in Type II DM and IHD P a-tients.

Parameter	Control group	Type II DM patients	IHD patients
Glucose, mmol/l	4.82±0.13	10.33±0.42[a]	5.13±0.11[b]
IRI, μU/ml	13.92±1.37	25.12±4.4[a]	12.92±3.78[b]
Pyruvic acid, μmol/l	59.03±4.03	105.43±12.2[a]	70.8±3.64[b]
Lactic acid, μmol/l	0.8±0.041	0.556±0.12	1.025±0.055[b]

The difference of the parameters is authentic:
[a] in type II DM patients in comparison with the level in the control group;
[b] in IHD patients in comparison with the level in type II DM patients.

enzymes, and autolysis of the subcellular structures of the myocardium [227, 406, 409, 416, 609].

In a number of clinical studies, the changes in the ECG and the excess body weight were compared. We noted that an excess weight enhanced disturbance of the bioelectrical activity of the myocardium, the atrial conduction changed, the atrioventricular and intra-ventricular conduction was disturbed, and the amplitude of the T tooth in the chest leads lowered. Apart from lesion of the myocardium itself, obesity also causes disorder of the activity of the cardiovascular system in connection with the influence of numerous ex-tracardiac factors such as excessive development of adipose tissue that increases the ab-dominal pressure, a high diaphragm, and its restricted mobility. This diminishes lung elas-ticity, and increases the intrapleural pressure, which leads to hypoventilation, the retention of carbon dioxide and lowering of the sensitivity of the respiration center, the appearance of hypoxia. All this affects the state of the myocardium.

An important indication of a disorder of metabolism in obesity is a pronounced growth in the weight of the myocardium (its hypertrophy) in youths 16–18 years old. These changes suggest that obesity is attended by a disorder in metabolism (especially of hormonal, carbohydrate, and fat metabolism) that predisposes the development of hyper-trophy of the myocardium and the appearance of syndrome X [65, 110, 152, 392, 395, 462].

To study the metabolism disorders in IHD and type II DM, we analyzed the content in the blood plasma of glucose, insulin, pyruvic and lactic acids in 14 patients with IIDM, 52 healthy persons, and 31 IHD patients (Table 10).

In comparison with the healthy persons, the IIDM patients had an increase in the glucose level to 10.33 mmol/l (in healthy persons it was 4.82 mmol/l), in **IRI** to 25.12 μU/ml (in healthy persons 13.92), and in pyruvic acid to 105.4 mmol/l (in healthy persons 59.03). The lactic acid content lowered somewhat to 0.55 mmol/l (in healthy persons it was 0.8). In IHD patients, all these parameters differ insignificantly from their level in healthy persons.

What are these changes associated with?

We must recall that hypertrophy of the myocardium in obesity prevents the appearance of a deficiency of energy in the myocardium and the development of acute cardiac insufficiency, i.e. hypertrophy is a compensating state. Physiologists proved that the maximum strength and rate, and also the energy efficiency of a contraction in a hypertrophic myocardium lower, and its functional reserve diminishes with the development of pronounced hypertrophy. A state of an increased wear of the hypertrophic myocardium appears.

Morphofunctional studies showed that in hypertrophy of the myocardium, the density of the sympathetic innervation of the myocardium is disturbed, the consumption of noradrenaline grows and exceeds its resynthesis, and its concentration in a hypertrophic myocardium is from one-third to one-fifth of the initial value. This is just what causes a drop in the positive inotropic influence of the sympathetic section of the nervous system on a hypertrophic myocardium, i.e. the inotropic function of the heart is disturbed. Moreover, in hypertrophy of the myocardium, the unbalanced growth of the myofibrills is attended by a lag in the growth of the arterioles and capillaries (in comparison with the growth in the muscle cells). As a result, the number of coronary capillaries per unit weight of the myocardium diminishes.

In hypertrophy of the myocardium, on the background of lowering of the total number of coronary capillaries, diffusion of oxygen is disturbed and, accordingly, hypoxia of the myocardium appears. As a result, already when resting most of the reserve capillaries are open, and the coronary reserve drops appreciably. In the hypertrophic myocardium, the increased consumption of oxygen is not ensured by its timely greater transportation into the intercapillary spaces, and ischemia zones appear [261, 350, 388, 405, 479].

All this was confirmed under clinical conditions by echocardiography. The volume and weight of the myocardium in IHD patients (not associated with hypertension) grow by 20–30%, i. e. in an ischemic lesion of the myocardium, hypertrophy is manifest early and progresses as ischemia grows.

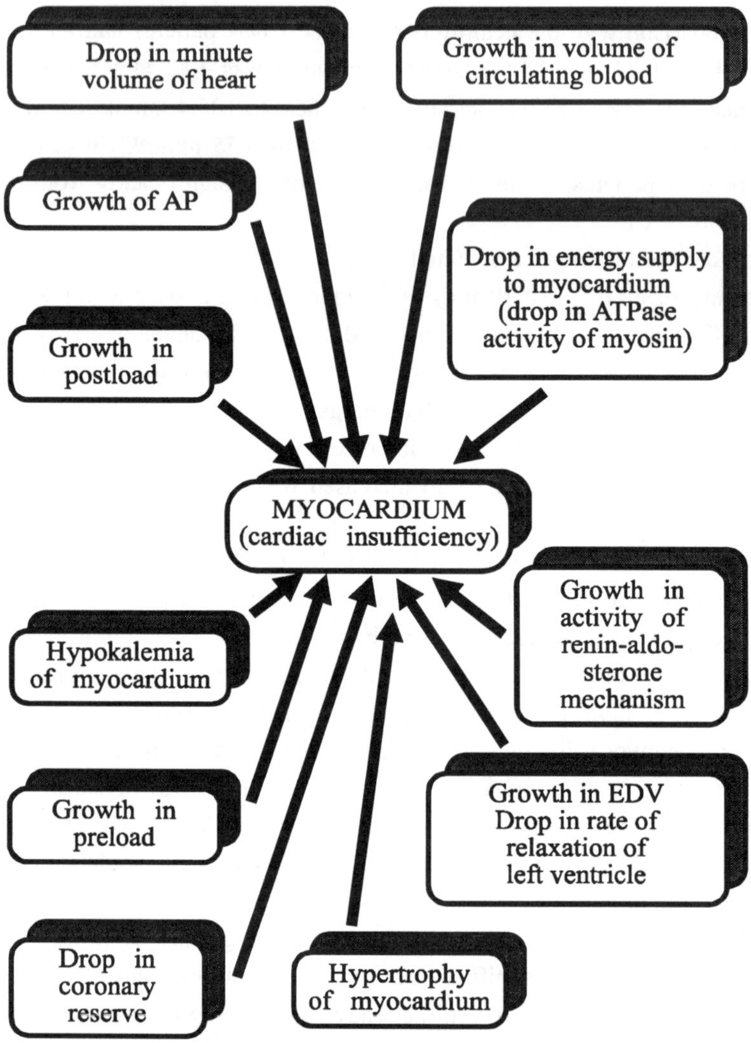

Fig. 11. Development of Cardiac Insufficiency in Syndrome X.

In obesity, hypertrophy of the adapted myocardium develops when the adaptation reaction transforms into a pathological one. In this disease, improper functioning of the mechanisms responsible for synchronization of excitation and contraction occurs.

The physiological systems of an organism are determined genetically for the timely change of the period of intense functioning during an interval of relative physiological rest [17, 28, 129, 338, 592].

The drop in the rate of relaxation results in the heart not having time to relax completely during the short intervals between contractions, and a phenomenon called the diastolic defect develops. With myocardium hypertrophy in obesity patients, a diastolic defect appears at a lower heart rate and is expressed very sharply in a quantitative respect. This syndrome of an uncompleted diastole is the initial link in the pathological chain of development of insufficiency in blood circulation in obesity patients. It is very important that the efficiency of utilizing oxygen in hypertrophy of the myocardium in obesity patients drops substantially. The reason is a drop in the transportation ability of the coronary arteries with a corresponding drop in the capacity for the aerobic resynthesis of ATP.

The development of insufficiency of a hypertrophied heart is based on the combined loss of capacity of the system of energy supply and of utilization of energy. This lowers the rate of transformation of the chemical energy of ATP into the mechanical energy of contraction, i.e. becomes the cause of depression of the contracting function of the heart (Fig. 11) [102, 152, 388, 392, 434].

After the discovery of insulin receptors, biochemists and clinicians received the possibility of studying the molecular mechanisms of the action of insulin, i.e. of investigating the molecular-biological processes underlying the changes in the activity of intracellular enzymes after insulin is introduced into an organism.

The amount of insulin receptors in persons with obesity and hyperinsulinemia was found to be diminished considerably. It was considered that insulin itself regulates the concentration of the insulin receptors with hyperinsulinemia. The introduction of considerable insulin doses to healthy persons lowered the amount of insulin receptors of blood cells. Correspondingly, fasting during from two to four days was accompanied by a growth in the concentration of the insulin receptors.

The insulin content is of major significance in the development of myocardium hypertrophy. Hyperinsulinemia in obesity patients results in insulin stimulating protein metabolism. It was shown experimentally that the introduction of insulin is attended by a greater inclusion of amino acids into the proteins of the tissues (especially muscles). In alloxan diabetes in animals, the inclusion of methionine into the muscles lowers, while when insulin is introduced it grows.

When comparing the hemodynamic parameters in healthy persons and obesity patients, we found considerable differences. If in the former the minute volume of the blood was 5.7 liters, in youths with obesity it was 7.1 liters. Simultaneously, the stroke volume of the blood in obesity patients increased to 100.2 ml (in healthy persons it was 85.6 ml).

The volume of the circulating blood is an important factor for the compensating supply of tissues with oxygen and various nutritive substances. An increase in the volume of the circulating blood in obesity is attended by constant hypoxia because the total body weight grows. The coefficient of the effectiveness of circulation is drastically decreased, and this is also associated with the high body weight. Growth of the vascularized adipose tissue considerably expands the vascular canal, and this lowers the total peripheral resistance. We noted a drop in the total peripheral resistance in obesity patients to 1085.7 $dyn \cdot s \cdot cm^{-5}$ (in healthy persons it was 1340.6).

Most clinicians consider that the development of blood circulation insufficiency in obesity is associated with the increased demand of an organism for oxygen and nutritive substances. This demand is compensated to a certain extent by a growth in the volume of the circulating blood, and also in the stroke and minute volume of the blood. However, the large volume of circulating blood produces a considerable preload on the left ventricle, and its capacity and stroke work grow. A growth in the index of useful power of the left ventricle is noted in patients. This factor constantly stimulates the myocardium and causes hypertrophy. As the disease progresses, the volume of circulating blood, the end-diastolic rigidity of the myocardium, and cardiodilatation grow, while the contractive ability of the heart lowers. All these factors cause cardiac insufficiency to develop.

When comparing the basic parameters of central hemodynamics in healthy persons and in obesity patients (see Table 9), we noted clear distinctions in such an important criterion as the end diastolic volume that in obesity patients is increased to 141.2 ml (against 121.2 in healthy persons).

What do these changes indicate?

First of all, progressing of cardiac insufficiency. The functional characteristic of the myocardium is expressed by the following relation: the contractive ability of the heart is inversely proportional to the ratio between the rate of muscle contraction and the load on the myocardium. The muscular work of the heart is due to components such as a contracting element that readily stretches in rest, can develop a definite force and contract when loaded, and a passive elastic element that retains the same•length in a state of rest as the contracting one. The interrelation and functional integration of these structures reflects the state of contractibility of the myocardium. The preload is determined by the length of the myocardium fibers in the diastole period. The quantitative characteristic of the preload depends on the initial length of a muscle fiber at rest, the venous inflow of blood to the heart, and the volume of circulating blood, which grows in obesity patients. Moreover, the venous inflow in them grows, and the preload grows quite logically, which is expressed in an increase in the end diastolic volume [110, 152, 221, 302].

When studying other parameters of the central hemodynamics, we noted that the ejection fraction in healthy persons and obesity patients is the same (70.4 and 70.6%), as is the degree of lowering of the front and rear size of the left ventricle (%ΔS) during systole (40.1 and 41.1%). However, these coincidences indicate that in obesity patients the myocardium compensates its contracting ability mainly at the expense of a growth in the mass of the myocardium and in the energy expenditures, especially under conditions of a physical load. The sharp drop in the power of the threshold load in veloergometry to 168.2 W/min (197.3 in healthy persons) points to disturbance of the contracting ability of the myocardium and the progressing development of cardiac insufficiency.

In concluding, we can assert that syndrome X is based on definite metabolic disorders that lead to pathological changes in the myocardium. They grow progressively and are manifest clinically in the form of IHD and cardiac insufficiency. In syndrome X, integrative symptoms form that are typical of various metabolic disorders (DM, hypertension, obesity, hyperinsulinemia, and insulin resistance). They all lead to the incidence of atherosclerosis. DM, hypertension, and hypercholesterolemia are undoubted risk factors of cardiovascular diseases, while hyperinsulinemia facilitates their development in combination with obesity and hyperglycemia [184, 227, 457, 484].

Clinical studies of recent years revealed that a growth in the insulin concentration in the blood plasma is an independent risk factor of coronary diseases.

1.3. Pathophysiological Relation between Obesity, Diabetes Mellitus, and Ischemic Heart Disease

An important role in the pathogenesis of atherosclerosis, hypertension, and DM is played by disorders in the metabolism of lipids, in particular by obesity.

Obesity is the excessive growth in body weight because of the surplus accumulation of fat in the depots. Fat is deposited in lipocytes in the form of triglycerides.

The growth in the volume of the lipocytes is attended by a growth in the amount of intra- and extracellular water and protein. The surplus body weight forms at the expense of one-third of water and two-thirds of fat.

The etiopathogenesis of this disease is integrative and consists of disorders in the hormonal balance of an organism, the increased taking of food, energy imbalance, etc. A clinical feature of obesity is the growth in the weight of the adipose tissue as a constituent part of an organism. Obesity generally develops gradually under the influence of a positive calorie balance, a reduced consumption of energy, and in the disintegration of the regulatory and adaptive mechanisms of basic metabolism.

In the economically developed countries, about 20–30% of the population suffer from obesity. It was found that 34 million Americans have a body weight index above

27.8, while up to 20% of the children and adolescents suffer from obesity [128, 270, 295, 539].

Adipose tissue is a metabolically very active substrate and performs vital functions as the main donor of energy-supplying substances. It is the presence of a sufficient amount of adipose tissue that allows an organism under different extremal conditions to adapt itself to an increased energy metabolism. The metabolic processes in lipocytes are regulated with the aid of hormonal and nerve mechanisms. The basic metabolic functions of lipid tissue (lipogenesis, lipolysis, biosynthesis of fatty acids, glycolysis) are closely related to both carbohydrate and protein metabolism. The amount of fat in an organism can be regulated by an increase in the size of the lipocytes (hypertrophy) or by increasing their amount (hyperplasia). We can accordingly distinguish hypertrophic and hyperplastic obesity. Actually, the lipocytes in most persons with pronounced obesity are simultaneously hypertrophied and hyperplastic [247, 272, 316, 373, 474].

Let us consider in greater detail what pathogenetic mechanisms underlie these processes.

The most important physiological regulators of energy metabolism are the hormones of the hypophyseal-adrenal system and the pancreas. A very significant factor in regulating energy metabolism is the synchronous interaction of hormones having both synergism and antagonism. The hormones with a synergic effect include glucocorticoids and catecholamines. Those with an antagonistic effect include glucocorticoids and insulin, Different physiological states of an organism (e.g. fasting) produce an increased activity of the glucocorticoids and catecholamines. Simultaneously, the insulin-like activity of the blood plasma: lowers. Inhibition of carbohydrate metabolism (blocking of the key enzymes of glycolysis and glycogenolysis) with the accompanying activation of the key enzymes of glyconeogenesis and lipid metabolism occurs at this ratio between the hormonal relations. Lowering of the insulin-like activity of the blood ensures the energy stability of an organism under fasting conditions. The physiological significance of hypercatecholemia under conditions of an energy stress in an organism ensures the rapid mobilization of the carbohydrate reserves in the liver and muscles, thus creating conditions for adaptation. The fat-mobilizing effect in the adipose tissue grows under the influence of catecholamines. It is these hormonal relations with an increased energy activity of an organism that make it possible to switch over the carbohydrate type of metabolism to the fat one. This function is achieved by mobilization of the key enzymes of carbohydrate-lipid metabolism. Hypercholesterolemia in fasting develops by an increasing production of glucocorticoids and a simultaneous drop in the insulin-like activity of the blood. At this ratio of the hormones (under conditions of an emotional load), the content of free fatty acids in the blood

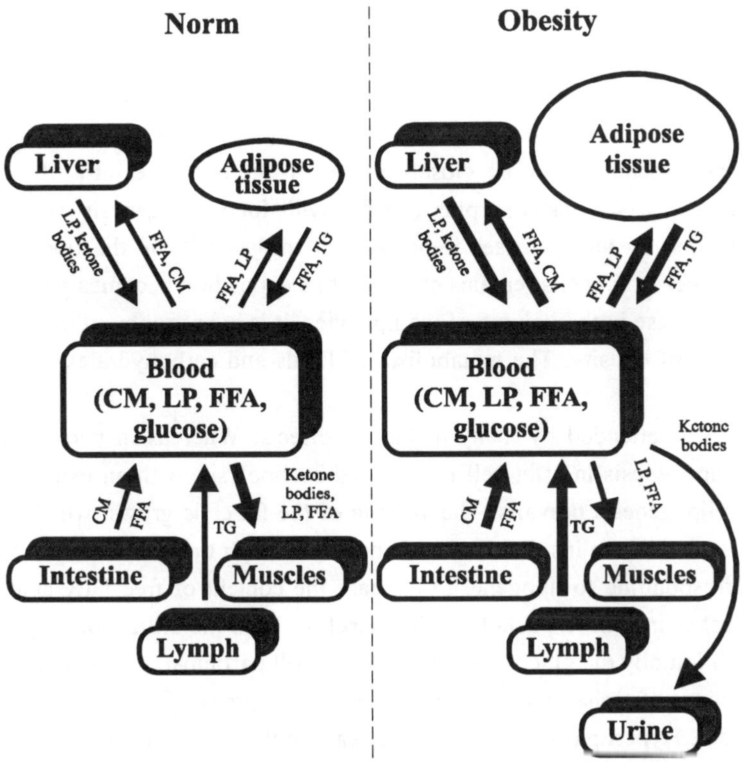

Fig. 12. Metabolism of Lipids and Carbohydrates Normally and in Obesity.

grows, and hypercholesterolemia, hyperlipoproteinemia, and hypertriglyceridemia appear. Glucocorticoids are very important for controlling fat and carbohydrate metabolism. They amplify the synthesis of VLDLP in the liver. The growth in the level of the VLDLP in the blood is compensated by the energy production of the adipose tissues [490, 515, 537, 549, 551].

A considerable amount of energy is reserved in an organism in the form of fat depots. The deposited triglycerides of the adipose tissue decompose under the effect of the hormone-sensitive triglyceridelipase and other lipases to glycerol and free fatty acids, and the latter enter the blood flow. Here the fatty acids combine with albumin and are carried in such complexes to the liver. In the liver cells, in the endoplasmatic network, the fatty acids are used to synthesize triglycerides The endogeneously synthesized triglycerides, and also those synthesized by the intestine cells from fatty acids and monoglycerols (absorbed from the food), are transported in the form of complexes enriched with triglycerides –

lipoproteins. Of the total energy needed by an organism 90% is provided by the free fatty acids that have a very high rate of metabolism [45, 103, 113, 148, 189, 283, 620].

Under normal conditions, the adipose tissue receives (under the influence of insulin) only 5% of the initial amount of glucose, and nevertheless the formation of triglycerides and their accumulation are closely related to the adequate intracellular supply of glucose. When a person takes surplus calories with food, lipogenesis exceeds lipolysis, which is attended by the increased deposition of triglycerides in the lipocytes. Since in adults the number of fat cells remains constant, this deposition becomes possible only as a result of am increase in the volume of the lipocytes. It is hypertrophy of the fat cells that is the main factor of obesity. The metabolism of lipids and carbohydrates is shown in Fig. 12.

Obesity is attended by many metabolic effects. Whereas in persons with normal metabolism lipogenesis in a fat cell is balanced by lipolysis, with an excessive supply of triglycerides lipogenesis prevails. The volume of the fat cells grows, which underlies the drop in the sensitivity to insulin. These changes (although they do not ensure a growth in lipolysis corresponding to lipogenesis) increase the content of free fatty acids, which inhibit the effect of insulin (especially in the muscles). Here the utilization of glucose is limited, although initially the glucose in a fat cell can still participate in metabolism, and there is a large amount of alpha-glycerophosphate as an acceptor of fatty acids for the synthesis of triglycerides. Hyperlipacidemia, i.e. an increase in the amount of free fatty acids, causes a diminished sensitivity to insulin, which is leveled out by an increased consumption of glucose in the adipose tissue. However, if a fat cell approaches its maximum volume, lipogenesis no longer grows. Lipolysis reaches the same extent as lipogenesis. The space occupied by triglycerides can no longer expand. With pronounced hyperlipacidemia, the peripheral effect of insulin is inhibited, which lowers the sensitivity to this hormone. Strong stimulation of the secretion of insulin occurs, but it is restricted by the genetically caused secreting ability of the islet apparatus of the pancreas.

Since the glucose pool cannot be utilized completely in either the adipose or the muscle tissue, a diabetic situation arises proceeding with glycosuria. With pronounced hyperlipacidemia, insufficient burning up of fats occurs at the periphery, and next their input to the liver grows. The synthesis of triglycerides in the liver increases, which leads to metabolically caused fat dysfunction of the liver. A part of the fats synthesized in an excessive amount is excreted from the liver, so that hyperlipacidemia is followed by hyperlipoproteinemia. The peripheral utilization of glucose lowers, and clearing of the blood serum is slowed down because of the lack of lipoproteinlipase. Hyperlipacidemia results in the appearance of ketonemia and ketonuria. Fig. 13 shows various models of the reactions

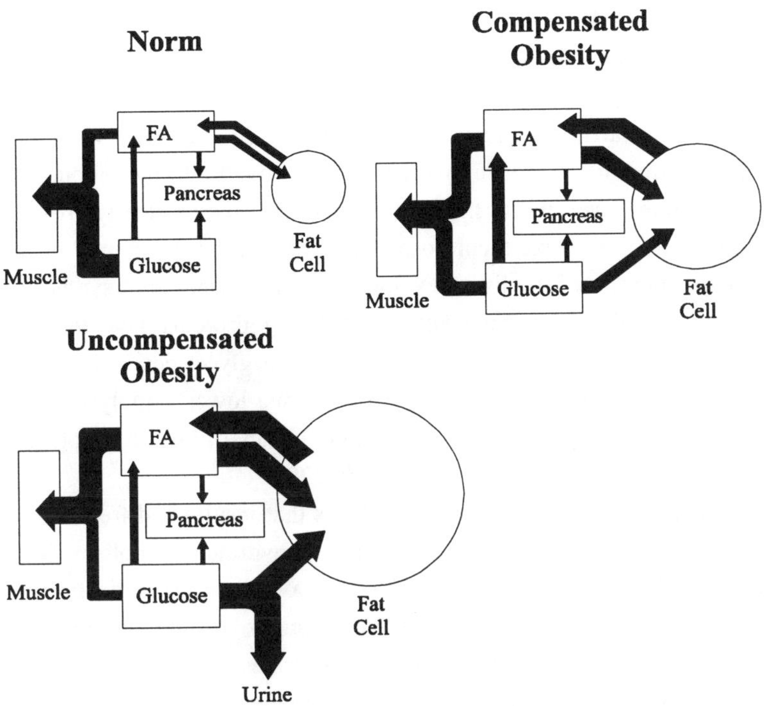

Fig.13. Relation of the Fat Cell Size to the Kinetics of Fatty Acids and Glucose.

of glucose and fatty acids, namely, in healthy persons and with compensated and decompensated obesity.

In 1963 P.Randle discovered the glucose-fatty acid cycle and showed that an increase in the glucose level is accompanied by the inhibition of lipolysis in the adipose tissue. If glucose is utilized actively in the muscle tissue, the level of free fatty acids in the blood grows. The dynamic equilibrium of these processes can be represented as follows: normally 50% of the glucose is oxidized to carbon dioxide and water, 40% accumulates in the form of triglycerides, and 5% in the form of glycogen. In DM, these proportions change, which causes disorder of fat metabolism [155, 159, 173, 213, 244].

The membranes of both muscle and adipose tissues have a specific carrier system for D-glucose that is affected by insulin in the transportation of glucose and free fatty acids into cells. An increased inflow of glucose increases the pool of alpha-glycerophosphate. As a result, more fatty acids are bound on these acceptors, and lipogenesis grows.

Further, insulin influences a cell membrane, the activity of adenylatecyclase is inhibited, and the cAMP level lowers. Consequently, insulin is a hormone of fat accumulation and synthesis. In DM, i.e. with a deficiency of insulin, a growth in lipolysis is noted. In connection with the lack of insulin, glucose cannot be supplied to the fat cells in a sufficient amount.

If glucose (the carrier mechanism) is not assimilated in a cell in DM, the major part of the fatty acids through the intercellular space is removed into the blood. The free fatty acids are carried to the liver, where the synthesis of lipoproteins clearly grows. If the ability of an organism to synthesize them does not correspond to the supply of the free fatty acids, pronounced obesity of the liver develops. We can assume that synthesis of apoproteins is disturbed, and this limits the synthesis of the lipoproteins. Diabetic hyperlipacidemia in several hours leads to a growth in the level of triglycerides or lipoproteins. When insulin is introduced, the degree of the hyperlipacidemia lowers, lipolysis diminishes, and lipogenesis increases. Therefore, in DM (with a deficiency of insulin) lipogenesis is inhibited and lipolysis is activated [247, 264, 272, 330, 379].

All clinical and experimental results witness that insulin and free fatty acids play a very important role not only in lipid, but also in carbohydrate metabolism. DM is attended by a growth in the level of the free fatty acids, and a drop in the utilization of glucose in the muscles because the muscle cells switch over to the use of free fatty acids. The glucose accumulates in fat cells and transforms into triglycerides if there is sufficient insulin. If hyperinsulinism develops, lipogenesis becomes excessive. The fraction of the triglycerides in the cells grows, which leads to hypertrophy of the adipose tissue and obesity [384, 418, 422, 450].

Intermediate metabolism of glucose, whose main substrate is lactate, occurs in a human organism in most tissues, where the produced lactate is oxidized in the Krebs cycle and does not enter the blood channel. The main sources of lactate are erythrocytes, the medullary substance of the kidneys, and, possibly, the brain. Under a physical load, the growth in the lactate content in the blood is due to the sharp increase in anaerobic glycolysis in the muscles. In the restoration period after a physical load, after hypoxia 10% of the lactate formed in a person's muscles gets into the blood channel, 15% is oxidized to $C0_2$, and 75% of the muscular lactate transforms into glycogen directly in the muscle tissues. The main pool of lactate undergoes metabolism in the Cori cycle. The lactate formed in the tissues is transferred to the kidneys and livers where it transforms into glucose. The latter again enters the tissues and is oxidized to lactate. In the long run, all the circulating lactate is utilized by the organism – 50% in the skeletal muscles and the heart, 30% in the liver, and 20% in the brain and cortical substance of the kidneys.

An important role in regulating the glycolysis enzymes in an organism is played by the synthetic fatty acids. Fatty acids and glucose react in the Randle glucose–fatty acid cycle whose direction is determined by the concentration and utilization of its substrates. Fatty acids inhibit not only glycolysis, but also the entering of pyruvate into the Krebs cycle by inhibiting the pyruvatedehydrogenase reaction combining cytoplasmatic glycolysis with mitochondrial oxidation processes. Glucose is not only an energy and plastic factor, but also a regulating one. Autoregulation of carbohydrate metabolism is manifest in the increase in the glucose synthesis of glycogen, and also in the inhibition of glycogenolysis and glyconeogenesis by modification of the activity of the relevant enzymes [458, 474, 490]. Lactate blocks the utilization of glucose by tissues and its oxidation in glycolysis and in the Krebs cycle.

Numerous biochemical studies showed that the tissues most sensitive to insulin include the skeletal muscles, myocardium, adipose tissue, liver, cortical substance of the kidneys, connecting tissue, mammary glands, skin, adrenal glands, uterine, and lungs. In the central and peripheral nervous system and in the erythrocytes, insulin does not affect glucose metabolism. The degree of sensitivity of the tissues to insulin is probably determined by the activity of hexokinase or the presence of glucokinase. The tissues absolutely dependent on insulin include the myocardium, skeletal muscles, adipose tissue, liver, and island apparatus of the pancreas. In the myocardium, skeletal muscles, and adipose tissue, up to 96% of the total enzyme activity falls to the share of hexokinase II. The latter is absent or is present in miserable amounts in tissues independent of insulin. At the same time for tissues not sensitive to insulin (the brain, erythrocytes, medullary substance of the kidneys), glucose is the only energy substrate and 95% of it is oxidized by the glycolytic path or lactate with the maximum rate of transportation.

When the content of insulin in the blood is normal, it does not affect directly the transmembrane transfer of glucose in the tissues, but suppresses lipolysis and oxidation of fatty acids, and this, in turn, inhibits glyconeogenesis in the liver and accelerates the utilization of glucose in the muscles.

In accordance with the above mechanism (the Randle cycle), in the regulation of glycemia in a healthy person, the homeostatic and anabolic effect of insulin is 80–90 and 10–20%, respectively. The figures reverse in obesity patients – the homeostatic effect is 10–20%, and the anabolic one is 80–90%. It should be noted that insulin in an organism increases the synthesis of glycogen and protein from glucose 5–6 times, and of lipids – 10 times. This selective sensitivity of lipid metabolism to insulin underlies obesity and other pathologic states associated with hyperinsulinemia. Also, insulin regulates and modifies the activity of many enzymes only in the presence of glucose. Here enzymes such as hexokinase, glucokinase, glycogen-synthetase and glycogenphosphorylase, and pyruvatedehydrogenase are especially significant. The primary reaction of insulin with enzymes occurs on a plasmatic membrane without direct contact with the intracellular structures [493, 495, 515, 556].

The main cause of relative insulin insufficiency is lowering of the sensitivity of tissues to insulin (insulin resistance), and this is just what leads to compensating hyperinsulinemia. Insulin resistance associated with compensating hyperinsulinemia may be due to [565, 573, 584, 624, 628]:

(1) a changed product of secretion of the beta cells: (a) a change in an insulin molecule, and (b) disturbance of the transformation of proinsulin into insulin;

(2) circulation of the insulin antagonists: (a) an increased content of the hormonal antagonists, (b) the presence of antibodies to insulin, and (c) the presence of antibodies to insulin receptors,

(3) a defect in the target tissues: (a) a defect in the insulin receptors, and (b) a postreceptor defect.

We find this concept to be the most suitable.

The development of insulin resistance of the peripheral tissues in atherosclerosis is very important and the most significant as a result of the action of insulin antagonists: an increase in the level of the hormone antagonists and a defect of the target tissues. Glucocorticoids, hormones of the thyroid gland, and the somatotropic hormone are physiological antagonists of insulin. Special importance here is attached to glucocorticoids that affect the sensitivity of the cell receptors to insulin, block receptors competitively, and diminish their number.

The clearly pronounced disorders in the hypophysis-adrenal cortex system is possibly related pathogenetically to atherosclerosis of the brain arteries. The development of insulin resistance is associated not only with the influence of glucocorticoids, but also with lowering of the thyroxin level in the blood in atherosclerosis patients.

According to modern concepts, insulin acts physiologically on biochemical processes by means of receptors, which are sections of a cell membrane with a genetically caused molecular structural affinity for insulin. Morphologically, they are integral components of the cell membranes. The insulin receptors in various organs and tissues have a universal structure. The striking similarity of the insulin receptors in cells of adipose tissue, muscles, and the liver is noted. It is very interesting that a physical load affects the number of insulin receptors of erythrocytes and other cells. Similar changes probably occur in the cells of the muscle tissue, and in this connection the oxidation of glucose is enhanced in veloergometry. It was noted in experiments with animals that in genetically caused obesity with the presence of hyperglycemia and hyperinsulinemia, the number of insulin receptors in the myocardium muscle, liver, and adipose tissue diminishes. Similar

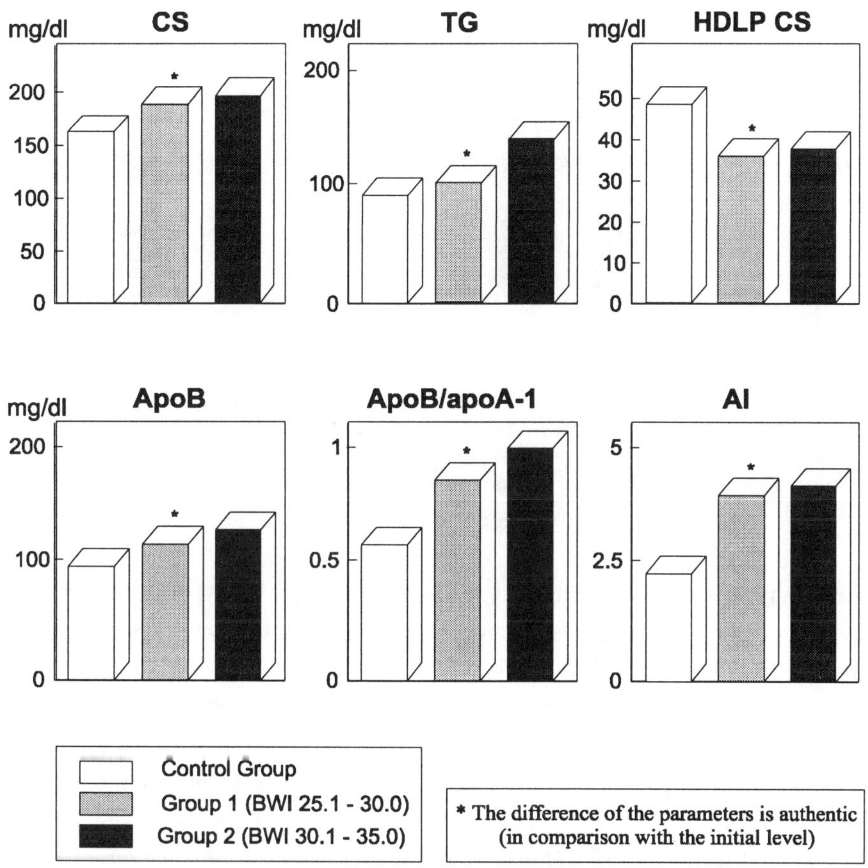

Fig. 14. Background Parameters of Lipoprotein Metabolism in Persons with Different Degree of Obesity.

changes are noted in patients with obesity and hyperinsulinemia. It was proved that the concentration of the receptors in the lipocytes and erythrocytes lowers in them [110, 133, 132, 141, 205, 280, 289].

We can thus conclude that the pathophysiological basis of obesity is the increase in the level of free fatty acids in the blood with their following use in the muscles. Energy forms at the expense of the acids, and not of glucose. Obesity is generally attended by hyperinsulinemia and a reduced tolerance to glucose. The appearing hypertriglyceridemia differs from that with a lack of insulin. Clinicians showed that the pathogenesis of hypertriglyceridemia in DM of types I and II differs. In type I (a lack of insulin), although the synthesis of triglycerides is suppressed, their catabolism lowers, and this results in the ap-

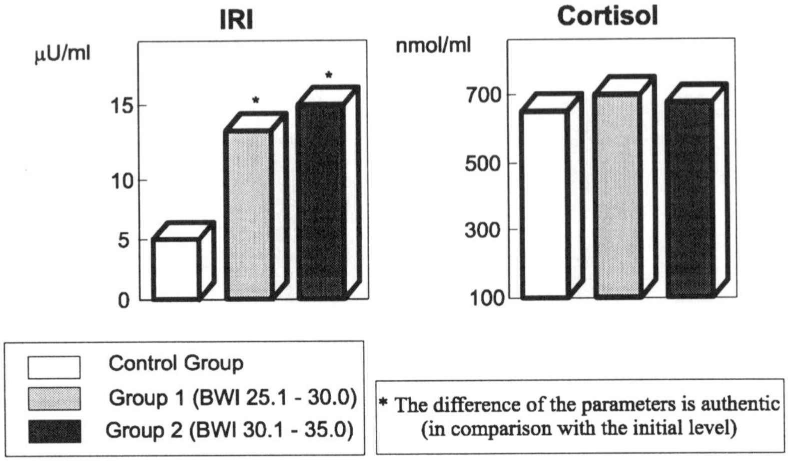

Fig. 15. Level of Hormones in Blood in Persons with Different Degree of Obesity.

pearance of hypertriglyceridemia. In type I, often attended by hyperinsulinemia and obesity, hypertriglyceridemia appears because of the growth in triglyceride synthesis in the fat cells and the liver [167, 192, 224, 238].

The metabolism of lipoproteins and the hormonal profile was studied in three groups of patients: a control group including 20 healthy persons from 18 to 20 years old, group 1 including 15 patients from 16 to 18 years old with hypothalamic obesity and a body weight index of 25.1–30.0, and group 2 including 20 persons with hypothalamic obesity and a BWI of 30.1–35.0 (Figs. 14 and 15).

When comparing the background parameters of cholesterol in these groups, we noted clear distinctions, namely, the average cholesterol value in the control group was 150.0 mg/dl, while in group 1 it was 172 and in group 2 – 181.1 mg/dl.

Although the cholesterol content in groups 1 and 2 was not beyond the limits of the normal values for persons of this age, in group 2 it was authentically higher than in the control group.

What are these differences associated with?

In persons with an increased level of total cholesterol in the blood serum (especially of LDLP) and a reduced content of HDLP, disorders in the lipid transportation system and a liability of atherosclerosis are noted. Numerous clinical studies showed that LDLP are the most dangerous. They carry cholesterol into the tissues, including an artery wall. They concentrate two-thirds of the cholesterol content in the blood serum.

When comparing the background values of the triglycerides in these groups, we also noted clear distinctions. In the control group of healthy persons, the triglyceride level in the blood was 72 mg/dl, in persons of group 1 it was 99, and of group 2 it was 136 mg/dl. And although (like the cholesterol content), the average values of triglyceride in the studied groups were not beyond the limits of the normal values for persons of this age group, in obesity patients of groups 1 and 2 the triglyceride level was increased in comparison with the value in the control group.

What is the pathophysiology of these changes in the triglyceride metabolism?

When studying the relation between obesity and disturbance of lipid metabolism, we noted that in obesity patients, the level of triglycerides, NEFA, cholesterol, and the total fraction of pre-beta-LP and beta-LP (VLDLP and LDLP) grows first of all. The growth in the TG level is most typical.

There is a clear correlation between the BWI and the TG level. This is associated with the increased synthesis of the triglycerides in the liver at the expense of the increased formation of FFA, the diminished metabolism of VLDLP, and hypertrophy of the fat cells. In obesity, the cortisol level is increased, and that of the STH is decreased. In the treatment of obesity, atrophy of the adipose tissue occurs (in connection with lipolysis), and 90% of the energy consumed by the organism is supplied by triglycerides [247, 272, 286, 342, 393].

Inspection of Fig. 14 reveals that the background value of the HDLP cholesterol in the control group was 48.2 mg/dl, in group l it was 34.9, and in group 2 it was 36.1 mg/dl. These results indicate that in the obesity patients of groups l and 2 the HDLP cholesterol level lowered. Numerous investigators showed there to be an inverse correlation between the content of HDLP and triglycerides: a growth in the HDLP level in the blood is attended by a drop in the triglyceride level. The suggestion that this is associated with a change in the activity of LPL because its increase leads to an increase in the content of HDLP cholesterol was confirmed completely in our investigations, namely, in patients of groups l and 2 the triglyceride level is increased substantially and the parameters of HDLP cholesterol in the blood are decreased in comparison with these parameters in the control group.

HDLP can transport cholesterol from the peripheral tissues to the liver. An increased concentration of apo-B-LP in combination with lesion of the endothelium facilitates the development of atherosclerotic changes in the vessels. The HDLP contain the main part of apo-A-1. An important role in the catabolism of apo-A-1 is played by the parenchymatous cells of the liver. Some investigators discovered a correlation between the level of LPL in the blood and the concentration of apo-A-1. Forming the main part of the protein component of HDLP, the apo-A-1 ensures the acceptance and transportation of cholesterol into the liver and, accordingly, its catabolism. No differences were found in the content of apo-A-1 in the blood plasma in persons of groups 1 and 2 in comparison with its level in the control group.

In going over to an analysis of the apo-B content in the persons being studied, we noted a considerable quantitative difference. In the control group, the apo-B level was 82.4 mg/dl, in group 1 it increased to 104.3, and in group 2 – to 115.3 mg/dl. The apo-B/apo-A-1 ratio changed accordingly in these groups. It was 0.62 in the control group, 0.88 in group 1, and sharply increased to 1.02 in group 2.

In view of the major significance of hormonal regulation of lipid metabolism in an organism, we compared both the background level of insulin and the cortisol content in the blood in these groups, and the values under conditions of a lipid load. We gave special attention to insulin. A glance at Fig. 15 reveals that in healthy persons the insulin level (background) was very moderate – 5.1 μU/ml, in group 1 it was authentically higher – 13.0, and in group 2 it was 15.7 μU/ml ($p<0.05$).

What physiological laws ensure the relation between insulin and lipid metabolism?

A growth in the insulin level in the blood is very important. Lipids have a contrainsular effect [397, 428, 439, 450, 474]. The background concentration of insulin in obesity is sharply increased, and when obesity is more pronounced, the insulin level is higher. This is especially important in the hypothalamic form of obesity, when a higher insulin content is noted in comparison with that in its exogenoconstitutional form. Some authors apart from hyperinsulinemia note a drop in the concentration of the thyroid-stimulating hormone, lower excretion of adrenaline and dopamine in the daily urine, and a growth in the excretion of 17-ketosteroids. The increase in the excretion of glucocorticoids enhances the processes of glyconeogenesis, promotes the development of hyperglycemia, which has a stimulating effect on the insular apparatus. Hyperinsulinemia, in turn, increases lipogenesis, which results in fat deposition. Consequently, obesity is attended by hyperinsulinemia with a drastically diminished biological efficiency of the hormone, i.e. insulin resistance.

It was noted that hyperinsulinemia mainly affects the triglyceride content. The level of total cholesterol correlates less with the insulin level. The content of HDLP cholesterol is inversely proportional to the background insulinemia [547, 555, 562, 573, 599].

The relation between insulin and the lipid metabolism processes was shown in experiments and under clinical conditions. It was noted that insulin, on the one hand, stimulates lipogenesis and thus facilitates the accumulation of cholesterol in the tissues of the vessels, and on the other, by inhibiting the processes of lipolysis, it lowers the mobilization of lipids from the wall of the vessels. It was noted that the degree of insulinemia regulates the content of total cholesterol, HDLP cholesterol, and triglycerides, i.e. the insulin level is not only a damaging factor for the development of atherosclerosis, but also affects the plastic processes in the cells of vessels. It was noted, in particular, that in hyperinsulinemia, the rate of proliferation of endothelium cells, muscle cells, and fibroblasts alters. Therefore, a relation appears between the degree of glycemia, the insulin content in the blood, and the lipoprotein level in the blood and in the endothelium of the vessels. There is a direct correlation between the level of insulin and triglycerides and an inverse one between the content of insulin and HDLP cholesterol [56, 75, 112, 153, 602, 614].

An important role in regulating the insulin level is played by the hypothalamus-hypophysis system. The possibility of the pathological influence of the hypothalamo-hypophyseal-adrenal gland system on glucose-insulin homeostasis via the growth hormone, which is an antagonist of insulin, was shown. The influence of cAMP and cGMP on the effect of insulin was studied insufficiently.

In the opinion of a number or authors, it is the receptor apparatus of the cells, the system of cAMP and cGMP (including in the liver cells) that determine the influence of glucocorticoids on lipid metabolism in an organism.

When studying the hormonal profile in obesity patients (see Fig. 15), we found that the level of cortisol in the control group was 659 nmol/ml. Its growth in groups 1 and 2 because of the large scatter of the results was not authentic.

Comparisons of the parameters of the metabolism of lipoproteins in healthy persons and obesity patients of groups 1 and 2 suggested that a prediction of the development of IHD can be confirmed by a summary analysis of the lipid and hormonal parameters. The atherogenesis index is such an integrating parameter. By comparing the index in different groups, we found it to be 2.2 in the control group, 4.1 in group 1, and 4.2 in group 2. Consequently, the development of IHD in patients of groups 1 and 2 is more likely.

After analysis of the background comparisons of the lipid and hormonal parameters in the control group and obesity patient groups, we studied lipid and hormonal metabolism under conditions of a lipid load in the persons of these groups to establish the influence of the taking of fat food on the dynamics of lipid disorders and the reserve possibilities of an organism for processing the lipids supplied with food, and also for determining the groups of risk of IHD development associated with DLP.

It is expedient first of all to consider the investigations devoted to studying not only the reactions of the lipid spectrum, but also the change in the hormonal parameters in response to a fat load. These studies mainly consider matters connected with the secretion of insulin, and the state of carbohydrate metabolism in response to the introduction of exogenic fat.

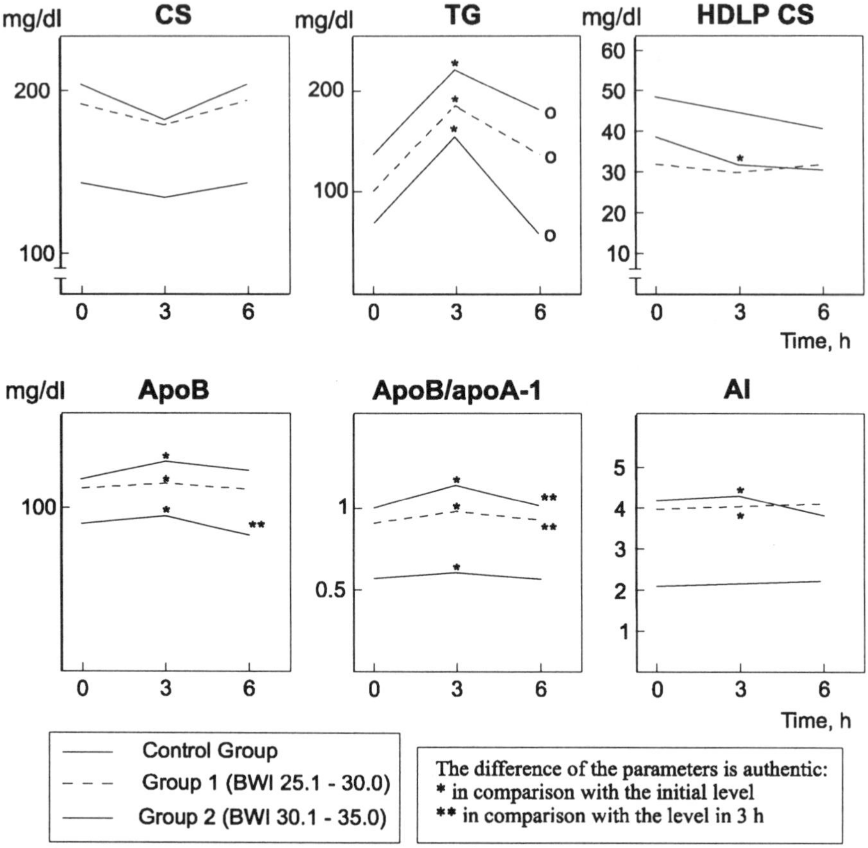

Fig. 16. Influence of Lipid Load on the Content of Lipids and Apoproteins in the Blood in Persons with Different Degree of Obesity.

A lipid load enables one to uncover the different disorders in the system of lipid transportation and study the processes of accumulation of cholesterol in an arterial wall directly after the taking of food.

Physiological studies showed the secretion of insulin to be closely related to the taking of food. Under the influence of the introduction of exogenic proteins, fats, and carbohydrates (that decompose to glucose, amino acids, and fatty acids), insulin is secreted depending on the amount of food taken. In the 1970s, the concept appeared that insulin acts on cell metabolism by means of the receptors of the cell membranes. Special significance was attached to the receptors of cells of the adipose tissue, liver, muscles, and blood cells.

In recent years, special attention is being given to cAMP as a mediator of biochemical processes. Under the influence of insulin, the suppression of adenylatecyclase occurs. The formation of cAMP lowers, but the content of cGMP grows. It was noted in numerous experiments that the introduction into animals of STH leads to a diabetogenic effect. In DM patients, decompensation (the appearance of ketoacidose) is attended by an increase in the STH level in the blood. It should be noted that the metabolism of some substances drastically alters the nature of the course of DM. For example, free fatty acids and ketone bodies are insulin antagonists, and their increased formation (occurring in DM) sharply changes the level of insulin in the blood [154, 164, 212, 238, 253, 269].

In going over to an analysis of the results of a lipid load (Fig. 16) in healthy persons of the control group, we note that the cholesterol level in three and nine hours did not practically change with respect to the background (150, 142, 152 mg/dl).

What is such a stability of the results associated with?

Apparently, with the good compensating possibilities of a young healthy organism.

Numerous studies revealed that the changes in the cholesterol content in the blood serum at the expense of food fats are chiefly due to changes in the concentration of LDLP cholesterol and to a smaller degree of VLDLP cholesterol. When saturated fats in the food are replaced with polyunsaturated ones, the rate of LDLP synthesis falls, while that of catabolism grows. A result is lowering of the intravascular level of LDLP. In healthy persons, the drop in the cholesterol content in the blood serum caused by polyunsaturated fats is explained by the increased excretion of it.

Moreover, a high content of HDLP is an important factor for stabilizing the cholesterol level in the blood with a lipid load.

With a lipid load, the triglyceride level changes considerably. The background value of triglycerides was 72 mg/dl, in three hours after the load it increased to 150, and in nine hours it dropped to 54 mg/dl.

What factors ensured these dynamics of the triglyceride level?

After the taking of fat food, the "food" triglycerides resynthesized in the cells of the small intestine pass out into the intestinal lymphatic vessels in the form of chylomicrons, and then through the pectoral lymphatic duct enter the blood flow. Normally over 90% of the triglycerides is absorbed.

The triglycerides taken with food in the lumen of the intestine decompose into free fatty acids and monoacylglycerols, and in this form are absorbed by the epithelial cells of the intestine, where they are included in the composition of the chylomicrons. The main function of the latter is the transportation of the food triglycerides via the intestinal lymphatic vessels and the pectoral lymphatic duct into the blood flow. Proteins are contained

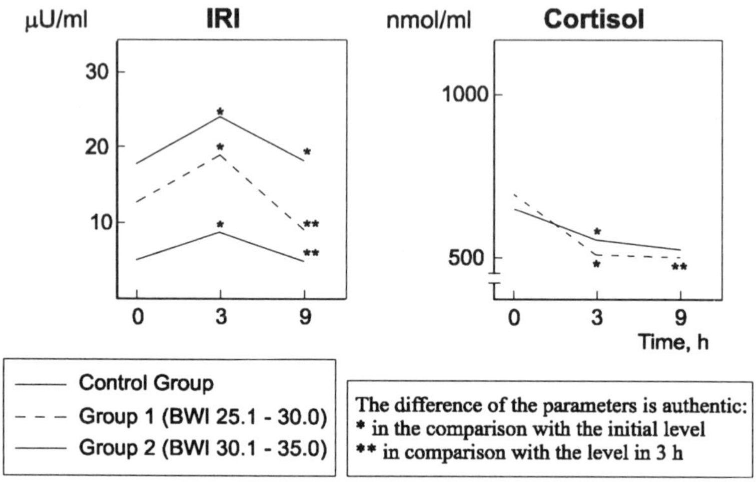

Fig. 17. Influence of Lipid Load on the Level of Hormones in the Blood in Persons with Different Degree of Obesity.

in the chylomicrons in a very small amount, but play the main functional role. The main protein component in the lymph chylomicrons is apo-B (up to 20%), the remaining part is apo-A and apo-C. The apoprotein B-48 is synthesized only in the cells of the intestine. If this protein is absent, the chylomicrons lose apo-A, but acquire apo-C and apo-E that are supplied to them by the HDLP. Triglycerides may be synthesized from endogenic fatty acids in the small intestine, but their main source is the liver, and also adipose tissue [316, 342, 412, 444, 452].

When analyzing the dynamics of the HDLP and apo-A-1 content, we noted the sufficient stability of these parameters and the absence of a reaction to the lipid load. The level of apo-B changed more actively under a fat load. Its background value was 82 mg/dl, in three hours the apo-B level increased to 93, and after nine hours it dropped to the initial value (83 mg/dl). It was shown in numerous clinical and experimental studies that the level of apo-B reflects very clearly the content of LDLP in the blood plasma. Under the conditions or a lipid load. the formation of lipoproteins occurs with the aid of chylomicrons that form in the small intestine.

Lipid metabolism both at rest and especially under conditions of a lipid load is due to hormonal regulation, mainly at the expense of insulin and glucocorticoids. Fig. 17 shows that in healthy persons of the control group, the IRI in the blood plasma increased in three hours after a lipid load from 5.1 to 9.04 μU/ml. In nine hours, the insulin level dropped to the initial values (4.4 μU/ml).

What is the relation between the activity of insulin and lipid metabolism?

This problem was discussed widely by both physiologists and clinicians because the prophylaxis and treatment of atherosclerosis, especially in DM patients, must be conducted with a view to these relations, Numerous investigations revealed that a high level of insulin in the blood plasma stimulates the synthesis of fatty acids in the liver. This effect is manifest in the following [490, 508, 523, 547, 555, 604]:

(1) the flow of substrates formed in glycolysis in the Krebs cycle increases; of special importance here is the growth in the activity of acetyl-CoA-carboxylase;

(2) insulin has a direct stimulating effect on acetyl-CoA-carboxylase;

(3) utilization of glucose by the pentose path in connection with the increasing insulin level leads to the appearance of an increased amount of reduced NADP, which is an important component of fatty acid synthesis. It was shown that fatty acids are chiefly synthesized in the liver, and not in the adipose tissue, and these acids are then transported in the composition of VLDLP into the adipose tissue, where they accumulate in the form of triglycerides. The adipose tissue absorbs the circulating triglycerides at an increased rate under the influence of insulin. Stimulation of LPL occurs;

(4) the inhibiting effect of insulin with respect to the hormonosensitive lipase in the fat cells was studied by many authors. The lipase catalyzes the hydrolysis of triglycerides and frees fatty acids. Such an antilipolytic effect of insulin increases the total amount of fat in an organism.

Considerable importance in matters of regulation of lipid metabolism is attached to the activity of the hypothalamo-hypophyseal complex, which can be estimated indirectly by the cortisol content in the blood. We can see from Fig. 17 that the level of cortisol in the blood of healthy persons of the control group dropped from 659.9 nmol/ml (background) in three hours after a lipid load to 559.2, and in nine hours to 534.6 nmol/ml. However, this drop was not authentic. Apparently, the physiology of this process is associated with the activity of the hormonosensitive lipase of the adipose tissue. Catecholamines increase the activity of the hormonosensitive lipase of the adipose tissue and enhance the fat-mobilizing effect. Under conditions of an emotional stress, synergism of catecholamines and glucocorticoids appears in the metabolism of proteins and fats. The major part of the FFA is absorbed by the liver and participates in the resynthesis of the triglycerides. It is interesting to note that in the regulation of fat metabolism, the glucocorticoids and catecholamines react like synergists, while in the regulation of carbohydrate metabolism, these hormones are antagonists. Under the influence of glucocorticoids, the most important enzymes of glycolysis – hexokinase and phosphorylase – are inhibited. Carbohydrate metabolism is switched over to fat metabolism by means of glucocorticoids. Under various conditions (emotional stress), an advanced growth in the content of glucocorticoids and catecholamines is observed [120, 193, 280, 600, 620, 622].

It is interesting to note that in healthy persons, the integrative reactions of the lipid and hormonal units of metabolism retain a stable atherogenic index under conditions of a lipid load (2.2, 2.2, 2.5) and a stable apo-B/apo-A-1 ratio (0.62, 0.67, 0.60). This points to a clear physiological relation between all the regulatory systems of an organism.

Thus, in analyzing the reaction of a healthy organism (control group) to a lipid load, we noted that an increase in the content of triglycerides and apo-B (atherogenic components of lipid metabolism) is compensated by a moderate increase in the level of apo-A-1 and stable values of the HDLP (antiatherogenic components of lipid metabolism). The regulating influence of insulin (a moderate increase in its concentration) and cortisol (a drop in its level) makes it possible to retain a stable atherogenic index within 2.2–2.9. Accordingly, it is very important to compare the reaction of healthy persons of the control group and obesity patients. Of special significance is the hormonal regulation of lipid metabolism under conditions of taking fat food.

When comparing the parameters for patients of group 1 (moderate obesity) and the control group, we noted a difference, primarily in the background figures – if the cholesterol level in the control group was 150 mg/dl, in group 1 it was 171.7. Accordingly the triglyceride content increased from 72 to 99 mg/dl. The drop in the level of HDLP was considerable: if in healthy persons it was 48 mg/dl, in group 1 it dropped to 35 mg/dl. Accordingly, the apo-A-1 level dropped to 117 (in the control group it was 132 mg/dl). These facts indicate that the antiatherogenic protection of the organism in obesity patients was reduced considerably. As should be expected, the content of the atherogenic components of the blood increased: if in healthy persons the apo-B level was 82 mg/dl, in group 1 patients it increased to 104.

In analyzing the reaction to a lipid load in patients of the control group, we made an assumption on the important role of insulin in regulating lipid metabolism, especially when taking fat food, A comparison of the background values of insulin in the control group and in group 1 confirmed this. If the background level of insulin in the control group was 5.1 μU/ml, in group 1 it was 12.97.

The growth in the insulin content in the blood in group 1 patients was attended by an increase in the cortisol level to 724.4 nmol/ml (instead of 659.9). All the disorders of lipid and hormonal metabolism were produced by a growth in the atherogenic index to 4.1 (in the control group it was 2.2).

In analyzing the reaction of group 1 patients to a lipid load, we noted that the cholesterol level did not experience any special changes (172, 169, and 184 mg/dl).

A clear dynamics of the triglyceride level was revealed: from the background value of 93 mg/dl it grew to 180.6 in three hours and failed to return to the normal value (128.5 mg/dl) in nine hours.

What is the pathophysiological mechanism of such a prolonged retaining of hypertriglyceridemia?

As already noted, the triglycerides of the chylomicrons are exogenic fats and contain food fats. The duration and level of the increased chylomicronemia depend, on the one hand, on the LPL concentration, and on the other, on the amount and composition of the introduced fats, and also on the amount of carbohydrates in the food. If carbohydrates are taken together with fats, the chylomicron level after a lipid load diminishes,

Many investigators note the relation between the insulin content in the blood and the triglyceride level. In some patients, this is explained by obesity that is attended by hypertriglyceridemia. In other patients obesity is absent, but there is resistance to insulin.

When analyzing in group 1 patients the reaction of the antiatherogenic fractions (HDLP and apo-A-1), we noted the absence of positive dynamics: the level of HDLP cholesterol was 35, 32, and 35 mg/dl, and that of apo-A-1 was 117, 122, and 121 mg/dl. This indicates that under conditions of taking fat food, the antiatherogenic function of the organism is lowered.

It was quite logical to assume that the atherogenic components of lipid metabolism are activated under conditions of a lipid load. When analyzing the dynamics of the apo-B content in group 1 patients, we noted that the background level of 104.3 mg/dl grows in three hours to 118.3 and in nine hours remains at the level of 114.7 mg/dl, unlike the dynamics in the control group, where the apo B content after nine hours dropped to the background value.

Apo-B is the main protein component of LDLP, VLDLP, and chylomicrons. Catabolism of apo-B-LP occurs in the peripheral tissues. By combining with the receptors on the plasmatic membrane of the cells, apo-LP is transported into the cells, where the hydrolysis reactions occur. A growth in the content of apo-B in the blood is one of the risk factors of IHD development. The pathogenetic significance of the growth in the level of apo-B is associated with disturbance of the reaction of lipoproteins with receptors because of the appearance of dyslipoproteinemia, and also because of the modification of apo-B or under the influence of insulin in the blood serum and esterified fatty acids. There are also numerous biophysical transformations in a vessel wall. First of all, electrostatic reaction of the apo-B particles with glycosaminoglycanes appears or hydrophobic reaction with elastin – the main component of the connecting tissue. These biophysical phenomena retain the LDLP in a vessel wall [347, 351, 365, 446, 484, 502].

When going over to an analysis of the dynamics of the insulin level in response to a lipid load in group 1 patients, we noted considerable changes: if the background level of insulin as 12.97 μU/ml, in three hours it grew to 20.4 and in nine hours dropped to 8.6 μU/ml.

The degree of lipacidemia and hyperlipidemia under the influence of insulin may lower rapidly, at the same time lipolysis lowers and lipogenesis grows. The deficiency of insulin in an organism retards lipogenesis and increases lipolysis. We can conclude from this that FFA, like insulin, are a decisive factor in regulating not only lipid, but also carbohydrate metabolism.

The deficiency of insulin increases the free fatty acid level and lowers the utilization of glucose in the muscles. This is why the muscle cells switch over to an energy donor – the free fatty acids. Accordingly, under conditions of a short supply of insulin, glucose accumulates in the adipose tissues and transforms into triglycerides.

We consider that in obesity, the number of receptors to insulin changes in all the organs and tissues, and this is what ensures the development of hyperinsulinemia under conditions of a lipid load.

Changes in the hypothalamic-hypophyseal system were noted in group 1 patients. The increased background cortisol value of 724.4 nmol/ml (659.9 in the control group) under the influence of a lipid load also dropped in three hours to 541.1 and in nine hours to 457.5 nmol/ml.

Pathophysiological studies revealed a direct relation between the glucocorticoid content and the level of hyperlipoproteinemia. A very important relation was established between hypertriglyceridemia and the excretion of free cortisol with the urine. We consider that the growth in the level of VLDLP cholesterol and LDLP triglycerides in IHD patients is associated with the high cortisol level, and pathogenetically, most likely, glucocorticoids disturb the metabolism of lipids by lowering the activity of heparin-dependent LPL. A growth in the cortisol level inhibits toe effect of the LPL. It was shown experimentally that the high level of glucocorticoids in the blood is the primary factor of lesion of the endothelium and intima of the vessels, and also of the growth in the adhesiveness of the thrombocytes, retards reparative processes in a vessel wall, and inhibits the growth of endothelial cells [50, 87, 139, 175, 542, 558].

In group 1 patients, the atherogenic index grows to 4.1. After a lipid load, it grows to 4.3 in three hours, and to 4.5 in nine. These changes appear in connection with a growth in the level of total cholesterol and a drop in the content of HDLP cholesterol. Consequently, we can conclude that the prognostic characteristic with respect to the development of IHD for group 1 patients is unfavorable.

When analyzing the parameters for group 2 of obesity patients (a BWI of 30.1–35.0), we noted further progress in disorders of lipid metabolism, First of all, the cholesterol level increased to 181.1 mg/dl (the background value for healthy persons was 149.7). The triglyceride level also changed considerably: it increased to 136 mg/dl against 72.2 in the control group. In group 2, as in group 1 patients, the background level of HDLP lowered to 36 mg/dl. Accordingly, the apo-A-1 level was only 114 mg/dl against 132 in the control group. The background increase in the apo-B level in group 2 patients to 115 mg/dl (against 82 in the control group) was exceedingly unfavorable, from our viewpoint. The hormonal regulation of lipid metabolism with the aid of insulin and cortisol was disturbed considerably: if in group 1 patients the background insulin level was 12.97 μU/ml, in group 2 it increased to 15.7. The high background level of cortisol (719.6 μU/ml) supplements the picture of the disturbance of the hormonal regulation of lipid metabolism in group 2 patients.

The relation between the activity of insulin and the disturbance of lipid metabolism was studied by many authors under both clinical and experimental conditions. Attempts to find a relation between the content of cholesterol and that of insulin were not successful. It was noted that in hyperinsulinemia, the level of the endogenic triglycerides grows, which facilitates the development of atherosclerosis. To confirm this assumption, we observed type 1 DM patients whom insulin was introduced to. The level of not only glucose, but also of triglycerides lowered [89, 103, 122, 136, 166].

In going over to an analysis of the response of group 2 patients to a lipid load, we noted further enhancement of the pathological changes. The cholesterol level did not virtually change (the background was 181 mg/dl, in three hours the level was 175, and in nine it was 183.9 mg/dl. Special attention was attracted by the changes in the triglyceride level. The background TG level was 136 mg/dl, in three hours it was 238, and in nine hours it remained at 196 mg/dl.

What is the pathogenesis of this phenomenon?

In the metabolism of lipoproteins, a very important role is played by the lipolytic enzymes LPL and liver triglyceridlipase. They affect not only the plasma concentration of the lipoproteins, but also the content of the products of their decomposition in the blood, i.e. of LDLP and remnants. The status of the receptor apparatus on the surface of the membranes is the biochemical basis of these physicochemical reactions [203, 222, 247, 264, 286].

The level of the antiatherogenic components of lipid metabolism (HDLP and apo-A-1) is lowered in the same proportion as in group 1, and does not change under the influence of a lipid load. The content of apo-B changes more significantly. If in group 1, the background value of 104 mg/dl under the influence of a lipid load increased to 118, in group 2 patients the background level of 115 increased to 138 and remained at the level of 131 mg/dl in nine hours after a lipid load. Clinical studies showed that the functional reaction of apoproteins with tissue elements ensures a system of lipid transportation in an organism, and accordingly the formation of atherosclerotic changes in the vessels. The directed transitions of apoproteins between lipoproteins of various classes in fasting or when taking food ensure in every person an individual nature of the physiological relations of lipid metabolism. The apo-B/apo-A-1 ratio determines the relation between the intensity of transportation of cholesterol into the tissues and the rate of its removal from the cells. An increase in the content of apo-B-LP is generally attended by the increased transportation of cholesterol into the tissues and the cell membranes [272, 288, 297, 342, 384, 422].

When analyzing the ratio in healthy persons (0.62) and in patients of group 1 (0.88) and group 2 (1.02), we noted that in disturbances of lipid metabolism in group 1 and 2 patients the transportation of cholesterol into the tissues and membranes of the cells grows.

When going over to an analysis of the hormonal regulation of lipid metabolism in group 2 patients, we noted the progress of hyperinsulinemia in them: the background IRI level was 15.7 μU/ml, in three hours it was 26.5, and in nine hours was 19.31 μU/ml.

These laws of the progress of hyperinsulinemia and dyslipidemia suggest a clear integrative relation between the content of insulin, triglycerides, and apo-B.

Insulin acts on cells in several stages [439, 450, 463, 482, 490]:

(1) binding of the hormones to the membrane receptors (it is very important that this relation depends on the insulin concentration in the blood, the number of receptors, and the affinity of the receptors for insulin;

(2) transformation of this hormone-receptor reaction into a transmembrane signal;

(3) the formation of an intracellular pulse;

(4) the subsequent chemical modification of various enzymes and transportation systems leading to the final biological effects of insulin.

Disorders leading to the development of insulin resistance may appear in each stage of this physicochemical train.

Insulin resistance is an integral concept characterizing all cases of a reduced biological effect of insulin with its normal concentration and activity. The mechanism of action of insulin may be disturbed even in hyperinsulinemia.

Insulin resistance develops very often in obesity patients, and the level of NEFA and the content of lactic acid rise. The normal tolerance to glucose in obesity requires an increased insulin level in the blood. The sensitivity of the receptors to insulin in obesity is reduced.

In summarizing, we can conclude that in the control group of healthy persons (and moreover in groups 1 and 2), the background parameters of the lipid and hormonal spectrum have a high variability. The lipid load enhances still more the mosaic nature of the picture and generally discloses in persons with a high degree of obesity more pronounced changes in the parameters being studied. Whether or not the changes in a number of parameters found in three hours after a load remain can be solved only in repeated observations. Nevertheless, even background disturbances of a lipid and hormonal nature are present in persons with a high degree of obesity, which suggests the probable further deepening and enhancing of the defects already existing in the metabolic units of the lipoproteins and hormonal status.

1.4. Disorders of Lipid Metabolism in Ischemic Heart Disease Patients

Most authors consider that the clinical picture of IHD is a result of manifestation of atherosclerosis of the coronary arteries of the heart. However, pathologo-anatomic studies did not confirm a direct correlation between the degree of stenosis of the coronary arteries and the extent of the clinical symptoms. The explanation is that many factors participate in the pathogenesis of IHD, and the integrative nature of the effect of hormonal, lipid, and coagulating factors suggests a mosaic interaction.

The participation of lipoproteins in the development of atherosclerotic changes in an arterial wall can be represented pathogenetically as consisting of several units [32, 36, 201, 234, 291, 310, 344]:

(1) the deposition of lipids in a vessel wall, and also processes of proliferation and necrosis, whose most important component are smooth muscle cells;

(2) synthesis of the main proteins of the connecting tissue of a vessel wall (collagen and elastin) occurring in smooth muscle cells; when atherosclerosis develops, structural changes occur in the smooth muscle cells that transform them into fibrocytes and disturb the production of collagen and elastin. As a result, the arteries become sensitive to mechanical effects;

(3) penetration of LDLP and VLDLP from a vessel lumen into an intima with the subsequent penetration into an arterial wall;

(4) penetration of the lipoproteins into an arterial wall (and their accumulation) that occurs more actively in disturbance of the intactness of the endothelium; damage to the endothelium nay be associated with both hemodynamic factors and with increased agglutination and adhesion of thrombocytes to an arterial intima;

(5) lesion of the endothelium leading to progressing proliferation of the smooth muscle cells and development of atherosclerosis.

The content of free fatty acids in the serum is very insignificant. After a fat food load, a considerable part of the FFA forms from the food fats along the path of their transformation into chylomicrons. Under the influence of the adipose tissue lipase, the triglycerides of the fat depots decompose into glycerol and fatty acids that either go to synthesize triglycerides or enter the blood.

The FFA are given up by the adipose tissue under the influence of hormonal and nerve regulation. Of especially great significance in this process is the level of glucose in the blood.

Cholesterol esters play the main role in the transportation of unsaturated fatty acids, especially if patients have a genetically determined deficiency of the enzyme LCAT (that forms in the liver), Here the level of unesterified cholesterol grows in the blood plasma and membranes of the erythrocytes of patients as a witness of the disturbed evacuation of cholesterol [58, 77, 90, 115, 128].

Lipolysis proceeds in an adipose tissue with the aid of a hormone-sensitive lipase whose activity is inhibited by insulin. When the blood contains insufficient insulin, the lipolysis of the adipose tissue grows, and the level of the circulating fatty acids increases. With an excess of insulin, the activity of the hormone-sensitive lipase lowers, and the level of the FFA in the blood drops accordingly. The supply of the FFA as a substrate is also regulated by insulin.

An important role in lipid metabolism is also played by LPL, which hydrolyzes lipoproteins rich in triglycerides in the capillaries of the adipose tissue and the skeletal muscles. where it is bound to the glycosaminoglycanes localized on the surface of a capillary endothelium facing the lumen of a vessel. The enzyme is activated by apo-C-11, and is inhibited by sodium chloride and protamine sulfate. If LPL hydrolyzes lipoproteins rich in triglycerides chiefly within the capillaries of a vessel wall, the liver lipase hydrolyzes them in the liver.

LPL is synthesized under the influence of insulin, and when there is a lack of the latter, there is also a deficiency of LPL. In the extreme forms of insulin insufficiency, the LPL level is very low, which leads to the accumulation of lipoproteins rich in triglycerides in the blood (diabetic lipemia). The process is compensated by the introduction of exogenic insulin [136, 148, 155, 159, 167, 173].

A number of authors consider that the atherogenic nature of lipoproteins enriched with triglycerides is due to the fact that their lipoproteinlipolysis occurs not only on the endothelium of the capillaries, but also on the surface of the larger arteries where a high local concentration of LDLP is created. Owing to the very small size of these intermediates of proteinlipolysis, they can penetrate Into an arterial wall through interendothelium gaps. Morphological studies of the vessels of atherosclerosis patients revealed that an increased LDLP concentration in the blood plasma in attended by an Increased amount of LDLP and apo-B in the intima of arteries and by increased hyperplasia of smooth muscle cells. It was established in biochemical studies that with definite types of hyperlipidemias, the rate of synthesis of the apo-B contained in VLDLP and LDLP is increased.

Clinico-biochemical investigations showed that the level of apo-B-100 in LDLP particles reflects the tendency of atherogenesis more exactly than the level of LDLP cholesterol. The inverse correlation between the lowering of the HDLP content and the frequency of developing IHD In also authentic. Accordingly, the parameter apo-A-1 shown the level of the atherogenic nature better than the HDLP cholesterol. As a result of the electrostatic reaction of apo-B with glycosaminoglycanes or hydrophobic reaction with elastin (components of the connecting tissue), particles of LDLP are retained in a vessel wall [224, 247, 260, 272, 282, 288,].

It was proved that the key role in atherogenesis is played by the reaction of lipoproteins with smooth muscle cells of the middle membrane of an arterial wall. An important role here is played by the cell membranes. The latter react with lipoproteins because a number of parameters of the cell membranes coincide with those of the surface layer of lipoprotein particles. Molecules of unesterified cholesterol are an essential part of cell membranes. It is cholesterol that affects the macromolecular structure, permeability, and electrical potential of a membrane. With an increased cholesterol content in a membrane, the latter becomes dense, its permeability both for ions and for nonelectrolytes diminishes, and the activity of the membrane enzymes drops, and which is especially important, including that of adenylatecyclase. The accumulation of cholesterol in a membrane disturbs its vitally essential transportation and metabolic functions. When animals are fed cholesterol, the proliferation of the smooth muscle cells of the medium membrane of the aorta was registered. Apo-B is an essential component of the atherosclerotically damaged arteries.

That the functioning of carbohydrate and lipid metabolism is related was assumed long ago, although the number of publications on this topic is insignificant. Biochemical studies showed that in atherosclerotic lesion of the vessels, patients showed an increased activity of the alpha-glycerophosphate shunt, an increase in the amount of metabolites and co-enzymes (malate, lactate, pyruvate, NADP, oxaloacetate).

All these changes are based on suppression of the activity of glyceraldehydophosphatedehydrogenase, malatedehydrogenase, and lactatedehydrogenase.

It is in connection with this relation between carbohydrate and lipid units that all clinicians consider DM as an important factor facilitating the development of IHD [68, 189, 210, 316, 326, 363].

The risk factors of IHD (hypercholesterolemia, increased arterial pressure, smoking, DM, excess body weight, IHD in close relatives, hypoalpha-cholesterolemia) play an active role in the formation of pathological changes in the coronary blood flow by disturbing the sufficiency of blood supply to the myocardium, creating a deficiency in the demand of the myocardium for oxygen, etc.

Lipid metabolism in IHD patients is closely related to the functional state of the coronary vessels. In the period of a coronary spasm, metabolism is sharply disturbed in the zone of ischemia, the catecholamine level grows, and the peroxide oxidation of lipids increases [116, 386, 387, 388, 389, 625].

To study the relation between the morphological lesion of the coronary vessels and disorder in lipid metabolism, we conducted clinical investigation of 39 patients with various forms of IHD from 27 to 60 years old, all men. The latter circumstance was due to two reasons: (i) in males the influence of the hormonal background on the level of lipid metabolism in less pronounced, and (ii) the effect of alcohol on the functioning of organs and systems in males is manifest more clearly than in females.

The influence of alcohol on the development of hypertriglyceridemia was described by many clinicians. The mechanism is probably as follows: ethanol is oxidized mainly in the liver, which leads to the additional formation of FFA participating in the synthesis of triglycerides. In alcoholism patients, the HDLP level grows (at the expense of $HDLP_2$ and $HDLP_3$), probably as a result of the increase in the activity of LPL with the regular taking of alcohol.

All the patients we observed had no accompanying diseases leading to secondary dyslipidemias (diseases of the liver, kidneys, arterial hypertension, DM, hyper- and hypothyreosis, obesity). The diagnosis of IHD was made on the basis of physical examination, electrocardiography, veloergometry, and selective coronarography. The coronarograms were appraised by the classification of Yu.Petrosian and L.Zingerman. Of the 39 patients, 33 had myocardial infarction in their case history; nine persons had it twice.

For control purposes, we studied 20 healthy persons from 20 to 25 years old without diseases, with normal parameters of lipid metabolism.

By the results of the coronarographic investigations, all the patients with IHD were divided into two groups, the first (17 patients) with moderate lesion of the coronary arteries, and the second (22 patients) with a pronounced lesion of them.

When comparing the results of studying lipid metabolism in healthy persons and IHD patients, we noted clear distinctions of the background parameters – the background level of cholesterol in healthy persons was 167 mg/dl, and in IHD patients it was 281 mg/dl.

Hypercholesterolemia in animals is one of the widespread models of development of atherosclerosis and determining the role of cholesterol in atherogenesis. However, why "atherogenic" factors appear in hypercholesterolemia still remains a puzzle.

It was proved that atherosclerosis appears when the coordinated regulation of the metabolism of lipoproteins in the blood plasma and cholesterol in the cells vanishes. In this case, the function of the key regulators in this coordination is played by LDLP receptors. This is why most clinicians consider that "genetic (family)" atherosclerosis appears on the basis of molecular defects of the lipoprotein receptors. Various biochemical processes appear when the content of cholesterol in the blood grows and when the speed of its forward and reverse transportation into the cells of the cardiovascular system organs alters. A major significance here is attached to activation of the peroxide oxidation of lipids in the cells and tissues. To prevent the damaging effect of hypercholesterolemia on various enzymatic systems (including the membrane ATPases), diverse antioxidants are very popular.

The peroxide oxidation of lipids is an important physiological process in an organism and constantly proceeds in the cell membranes. The process, together with renewal of the membranes, plays the role of a very important link in the biosynthesis of prostaglandins, leukotrienes, and also in the processes of phagocytosis and a change in the structure of the intracellular membranes [15, 185, 320, 373, 429].

The lipid triad consisting of three biochemical processes is an important biochemical link in the peroxide oxidation of lipids. The first of them includes the stage of a primary deficiency of ATP, inhibition of the cycle of tricarboxylic acids, glycolysis, ATP-dependent cation pumps, and the accumulation of fatty acids. This triggers off the second process with inhibition of the cycle of tricarboxylic acids and the appearance of a deficiency of energy-rich phosphorus compounds. The third process consists in activation and the subsequent inhibition of glycolysis, the accumulation of fatty acids that is attended by inhibition of the slow channel of the electrogenic entry of Ca^{2+} into the cardiac myocytes and depression of the contracting function of the myocardium. This lipid triad produces very important changes in the lipid bilayer of the membranes of the cardiac myocytes. Damage to the membranes, potentiation of the excess of catecholamines, activation of lipases and phospholipases appear. The surplus of fatty acids and lysophospholipids on the

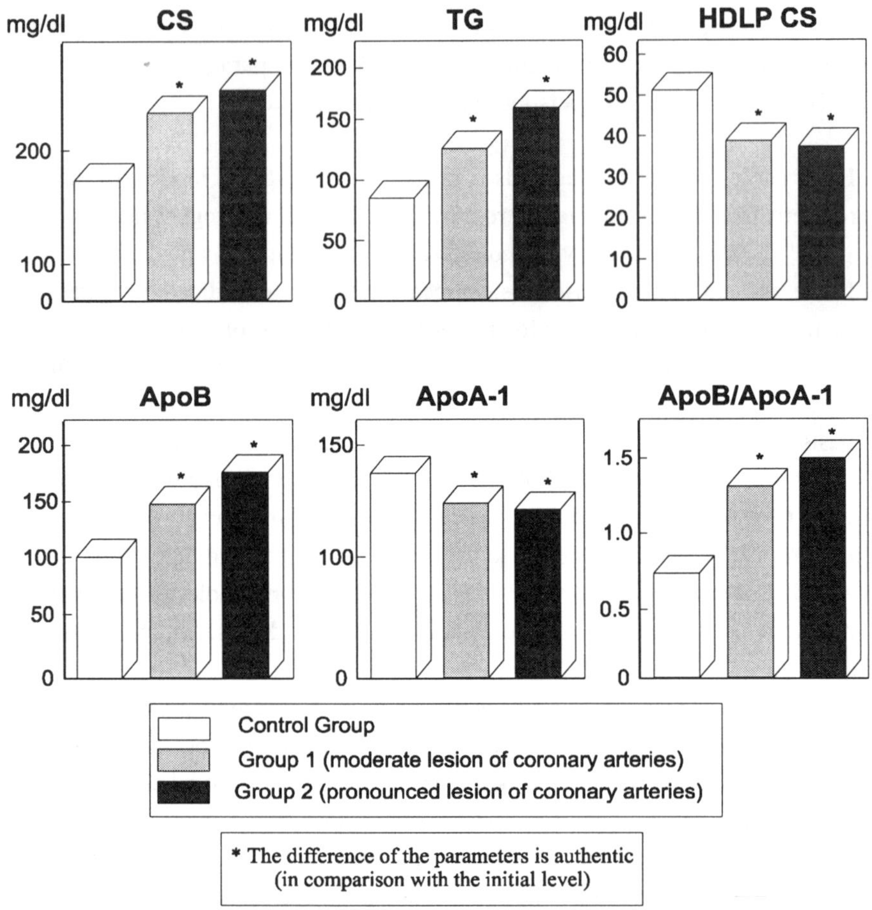

Fig. 18. Content of Lipids and Apoproteins in Blood in IHD Patients.

membranes is attended by activation of the peroxide oxidation of the lipids and damage to the membranes [458, 494, 498, 509, 555].

When comparing the triglyceride parameters in healthy persons and in patients with IHD in groups 1 and 2 (Fig. 18), we noted authentic differences. If in healthy persons the triglyceride level was 70 mg/dl (on an average), in group l IHD patients It was 124, and in group 2 it was 157 mg/dl.

The growth in the triglyceride level in the blood is an integrative parameter associated both with the nature of a patient's diet, the level of hormonal activity, and with physical activities. With a carbohydrates diet during a long time, the level of insulin in the blood increases, and this boosts the synthesis of endogenic triglycerides. The pathogenesis

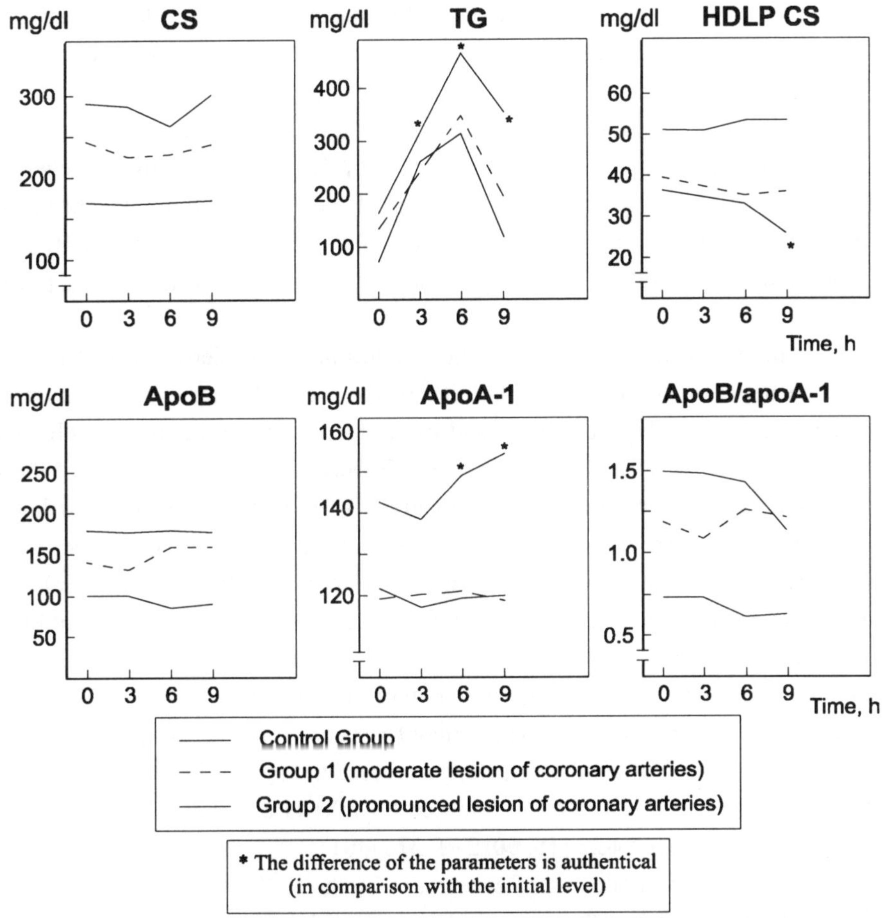

Fig. 19. Influence of Lipid Load on the Content of Lipids and Apoproteins in Blood in IHD Patients.

of hypertriglyceridemia is due, on the one hand, to disturbance of the excretion of triglycerides from an organism because of the drop in the activity of extrahepatic LPL, while on the other, resistance to insulin appears, and an a consequence lipolysis of VLDLP is disturbed.

With an increased physical load, the triglyceride level in the blood plasma lowers in connection with the fact that the activity of LPL in the muscles and adipose tissue grows, and accordingly the rate of removal of triglycerides from an organism increases. Moreover, physical activities increase the level of HDLP (especially $HDLP_2$), which is very important for DM patients [42, 193, 571, 572, 614, 628].

It was noted in many investigations [189] that there is a correlation between the triglyceride level and the frequency of development of IHD, although these correlations are weaker than in hypercholesterolemia. It was noted previously that in 82% of persons who had had myocardium infarction, the background level of triglycerides increased.

A major importance in the pathogenesis of IHD is attached to the level of the HDLP cholesterol. The antiatherogenic function of the HDLP inhibits the formation of atherosclerotic plaques. A glance at Fig. 19 shows that in healthy persons the HDLP cholesterol level was 52 mg/dl, in group 1 IHD patients it was 39, and in group 2 patients it was 37.

A significant feature of the metabolic changes in atherosclerosis is the maintaining of a balance between the inflow of cholesterol into the cells in the composition of the LDLP (especially of cholesterol esters) and its removal from the peripheral cells into the liver with the aid of the HDLP. An important role here is played by the rate of the intracellular synthesis of cholesterol and the sensitivity of the vessel wall cells to hormonal regulation. It was proved clinically and experimentally that the HDLP in atherosclerosis patients have a diminished ability or accepting cholesterol from the smooth muscle cells of the aorta. In these persons on a background of hypercholesterolemia, the amount of glucocorticoid receptors of the blood lymphocytes grows. An obvious law was noted, namely, in atherosclerosis patients with an increased content of LDLP and VLDLP, the number of glucocorticoid receptors and the glucocorticoid sensitivity of the peripheral tissue cells lower. The combination of all these factors increases the synthesis of cholesterol, and it accumulates in the cells. The rate of cholesterol synthesis correlates with the number of cellular glucocorticoid receptors [15, 50, 210, 270, 301].

It was noted that the degree of acceptance by the HDLP of the cellular cholesterol correlates with the activity of the intracellular prostacyclin that catalyzes the process. The drop in the ability of the HDLP to accept cholesterol from the cells of the arteries and other tissues is connected with a decrease in the synthesis of prostacyclin.

Since the disorders in the process of formation of triglycerides and HDLP cholesterol may be associated with the quantitative aspect of lipoproteinlipolysis, we obtained the separate fractions of the HDLP. In healthy persons, the fraction of $HDLP_{2B}$ was 33% and of $HDLP_{3B}$ was 21%. In IHD patients, accordingly, these parameters were 10.2% and 7.8% in group 2, and we noted that persons with a low level of $HDLP_{2B}$ (IHD patients) had an increased triglyceride level in the blood plasma.

The main part of the apo-A-1 is in the HDLP. A major role in the catabolism of apo-A-1 is played by the parenchymatous cells of the liver. Some authors discovered a correlation between the LPL level in the blood and the apo-A-1 concentration. Forming the main part of the protein component of the HDLP, apo-A-1 ensures the acceptance and transportation of the cholesterol into the liver and, accordingly, its catabolism.

Inspection of Fig. 18 shows that in healthy persons the apo-A-1 level was 141 mg/dl, in group 1 IHD patients it was 118, and in group 2 it was 119 mg/dl.

As already indicated, the apo-B/apo-A-1 ratio is an important criterion of atherogenesis in healthy persons and IHD patients. It determines the relation between the intensity of transportation of cholesterol into the tissues and the rate of its removal from the cells. A growth in the level of apo-B-LP is generally attended by the greater transportation of cholesterol into the tissues and cell membranes.

After comparing the apo-B/apo-A-1 values in healthy persons equal to 0.72 and in IHD patients (1.24 in group 1 and 1.50 in group 2), we concluded that the higher figures may indicate a greater intensity of cholesterol transportation into the tissues in patients with a pronounced lesion of the coronary arteries.

The average background level of insulin in healthy persons was 6.1 μU/ml, in group 1 IHD patients it was 7.0, and in group 2 it was 14.6. The transportation and metabolic effects of insulin are a result of binding of the hormone to the membrane receptors. In hyperinsulinemia of any etiology, the number of receptors in the adipose tissue, liver, and muscles diminished appreciably [286, 316, 326, 342, 487].

To establish the latent forms of disturbance of the metabolism of lipoproteins and their hormonal regulation, we used a single food fat load. The results obtained when using a functional load test will reflect more accurately the dynamic state of the mechanisms regulating the metabolism of the lipoproteins. As indicated, the group with atherogenic dyslipidemias included persons for whom on a hungry stomach the content of cholesterol was above 200 mg/dl, that of triglycerides was above 200, and that of HDLP cholesterol was under 35 mg/dl.

It follows from Fig. 19 that the cholesterol level under the influence of a lipid load remained almost unchanged both in healthy persons (background value 167 mg/dl, in three hours it was 161, in six hours it was 164, and in nine hours it was 171 mg/dl) and in IHD patients. In group 1, the dynamics of the cholesterol content was 231, 218, 225, and 228 mg/dl, and in group 2 it was 281, 281, 267, and 301 mg/dl.

Triglycerides showed a more outstanding dynamics of the metabolic disorders under a lipid load. The physiological reaction in healthy persons was clearly expressed. The background was 102 mg/dl, in three hours the figure was 244, in six hours – 308, and in nine hours it was 133 mg/dl. An absolutely different picture was noted in IHD patients. The reactions in group 1 were 124, 234, 342, and 285 mg/dl, while in group 2 the changes were sharper: 157, 280, 483, and 368 mg/dl.

We see that the main defect of the metabolic disorders in IHD is that the physiological norm of the triglyceride level in nine hours after a lipid load is not restored.

What are these disorders associated with?

After the taking of food, chylomicrons are detected in the intestine. They form in the cells of the mucous membrane of the small intestine and pass into the lymphatic tracts. After getting into the blood channel, the chylomicrons experience numerous changes, in particular phospholipids are detached or are exchanged for other lipid fractions. The triglycerides of the chylomicrons are exogenic fats and contain the fats from food. The duration and level of increased chylomicronemia depend on the amount and composition of the introduced fats, and also on the amount of carbohydrates in the food. If the latter are taken together with fats, the chylomicron level after a lipid load diminishes.

The level of endogenic triglycerides in the blood depends on the caloricity (energy value) of the food, the fraction of carbohydrates, the kind and amount of mono- and disaccharides in the carbohydrate part of the food, and on the consumption of alcohol [385, 399, 419, 474].

The increase in the triglyceride level is especially noticeable in patients with hyperlipoproteinemia induced by carbohydrates. The qualitative composition of them is very important here – the TG level increases especially clearly when fructose or alcohol are consumed.

Biochemical investigations showed that between the level of the lipids (triglycerides) in the blood and that of heparin a relation exists. A clarification factor was discovered that acts on VLDLP and hydrolyzes triglycerides with the formation of FFA and heparin. It is LPL. A role in its activation is played by insulin, catecholamines, and STH. The above data show that in IHD patients, the activity of lipoproteinlipolysis is lowered.

We attach great significance to the dynamics of HDLP cholesterol under conditions of a lipid load. In healthy persons, the content of this cholesterol remains at the background level of 52 mg/dl (in three hours it was 51, in six it was 53, and in nine hours it was 53 mg/dl). The picture was absolutely different in IHD patients – for group 1 the figures were 39, 37, 36, and 37 mg/dl, and for group 2 – 37, 36, 34, and 28 mg/dl.

Numerous clinical and epidemiological investigations showed that the risk of IHD development is closely related to the low level of HDLP and does not usually depend on the total cholesterol content in the blood plasma. Experiments with animals revealed that only HDLP has the unique ability of accepting cholesterol from the cell membranes. This is connected with the fact that the biochemical composition of HDLP differs from that of

HDLP and VLDLP. The high-density lipoproteins have a large fraction of phospholipids and apo-A-1, which is absent in LDLP (where apo-B is present) [508, 515, 547, 561, 572].

The antiatherogenic function of HDLP is due to the fact that it accepts unesterified cholesterol from the membranes of the smooth muscle and endothelial cells of vessels, and also from VLDLP and chylomicrons when their triglycerides are hydrolyzed in the blood plasma. Biochemical studies showed that the transformation of free cholesterol into the esterified compound occurs with the participation of the enzyme LCAT. Its substrates are free cholesterol and lecithin contained in HDLP. The esterified cholesterol forms on the surface of the HDLP particles containing apo-A-1 combined with apo-D and LCAT. Numerous studies showed that the esterified cholesterol is transported into the liver with the aid of $HDLP_2$, where it is transformed into bile acids and excreted from the organism.

The antiatherogenic function of HDLP consists of an integral system including several units:

(1) a protein-lipid structure of the surface monolayer that ensures incorporation of the nonesterified cholesterol into it;

(2) an enzyme system that with the aid of LCAT reacts with its activator – apo-A-1 (inhibitor apo-A-2);

(3) a system of nonenzymatic directed transportation of the cholesterol esters including lipid-transporting proteins.

Most scientists consider that this integral system under physiological conditions ensures the antiatherogenic function of the HDLP. It was noted that the activity of the LCAT and the intensity of removal of the esterified cholesterol from the surface of the HDLP particles depend on the inflow to the small spherical HDLP particles of nonesterified cholesterol both from lipoproteins enriched in triglycerides (chylomicrons and VLDLP) and from cell membranes [48, 64, 210, 594, 611, 614].

The activity of development of the atherosclerotic process depends on the relation of the intensities of the following nonenzymatic processes:

(1) transportation of cholesterol esters in the composition of HDLP into the liver for excreting cholesterol from the organism;

(2) transportation of cholesterol ethers in the composition of VLDLP and LDLP into the smooth muscle and endothelial cells of a vessel wall.

Disturbance of the normal ratio of the activities of these processes leads to the progressive development of IHD. On the surface of VLDLP, the amount of low-molecular hydrophilic proteins and high-molecular hydrophobic apo-B affects the physicochemical properties of the surface layer of these lipoproteins. Of special importance is the state (physicochemical) of the VLDLP for the intensity of penetration of the cholesterol esters through the surface layer into the nucleus of particles. The degree of activity of LCAT is closely related to this process. When the activity diminishes, the cholesterol esters are transported to a greater extent to the VLDLP. The rate of the reaction catalyzed by LCAT depends not only on the amount and activity of the enzyme, but also on the concentration and composition of the substrates, activators and inhibitors of lecithin, free cholesterol, apo-A-1, and cholesterol esters.

In IHD patients, clinicians noted a drop in the activity of LCAT whose combination with the low level of HDLP cholesterol is especially unfavorable. A drop in the fraction of polyunsaturated fatty acids together with a growth in the fraction of saturated ones was noted. It was assumed that the drop in the complexing of lecithin with cholesterol is probably due to the lower activity of the LCAT and results in diminishing of the amount of cholesterol transported by the HDLP, i.e. a drop in the level of HDLP cholesterol is noted.

When analyzing the dynamics of the apo-A-1 content under the influence of a lipid load (see Fig. 19), we noted a clear distinction between the groups of healthy persons and IHD patients. In the former, the background level of apo-A-1 was 141 mg/dl, in three hours it was 138, in six hours it was 148, and in nine hours it was 156 mg/dl, i.e. a moderate increase was noted. The situation was different for the IHD patients. The background in group 1 and 2 patients was 118 and 119 mg/dl, respectively. In group 1 the apo-A-1 level in three hours was 119, in six hours it was 121, and in nine hours it was 117 mg/dl. The relevant values for group 2 were 116, 118, and 118 mg/dl.

The level of atherogenesis of lipoproteins in the blood flow is closely related to the following biochemical processes: (i) the rate of synthesis of cholesterol and lipoproteins, (ii) the rate of separation of cholesterol from the blood flow, (iii) the rate of excretion of cholesterol from the organism. An important role in maintaining the cholesterol level in an organism is played by the liver and small intestine. Especially importance in these processes is attached to the liver, because it synthesizes and excretes over 50% of the atherogenic lipids in an organism. In the peripheral tissues, atherogenic lipoproteins are mainly utilized, while cholesterol is excreted chiefly with the aid of the HDLP. In other words, the peripheral tissues are "programmed" for the utilization of cholesterol from the HDLP circulating in the blood flow. The metabolism of cholesterol in an organism is carried out with the aid of numerous enzymes, especially LCAT [33, 116, 270, 278, 291].

The biochemical process of cholesterol evacuation with the aid of HDLP is carried out in two ways: (i) via an aqueous phase according to the gradient of concentration, and (ii) by direct contact with the plasmatic membrane of a cell, including with the participation of HDLP receptors.

The main mass of the cholesterol is transferred from the cells into the blood through the aqueous phase, the HDLP functioning like a pump in the cycle of cholesterol pumping. The reaction of the HDLP particles with the cells is accomplished with the aid of a receptor protein and is attended by the appearance of a whole cascade of reactions in a given cell, namely, (i) pumping of cholesterol not only from the plasmatic, but also from the internal membrane, (ii) secondary pulse-mediators appear in a cell and activate the synthesis of cholesterol and block its esterification, and (iii) the capture of lipoproteins and fragments of cells modified by lipoproteins by macrophages playing the role of janitor cells.

Consequently, the significance of the HDLP as the most important unit of the antiatherosclerotic process was proved with the aid of numerous clinical and experimental data.

The level of HDLP in the blood is closely related to the nature of the diet and with the inclusion in it of polyunsaturated fatty acids [50, 70, 234, 310, 338, 351].

A growth in the apo-B level in the blood is one of the risk factors of IHD development. The pathogenetic significance of the increase in the level of apo-B is associated with disturbance of the reaction of lipoproteins with receptors because of the appearance of dyslipoproteinemias and also because of modification of the apo-B or under the influence of the insulin of the serum and esterified fatty acids. Of interest are the results obtained by M.Ruda who studied unstable and stable stenocardia, compared the change in the level of apo-B in these groups of patients, and showed that in patients with unstable stenocardia the apo-B level is higher, while the apo-B/apo-A-1 ratio exceeds unity. Analysis of the results of coronarography is very important. It was noted that in the patients of both groups the changes revealed in coronarography are approximately the same, there are no differences in atherosclerotic lesion, but there is a relation between the clinical course and the apolipoprotein level. With a severe course of the disease, the atherogenic shifts of apolipoproteins progressed, while with a favorable course they did not change.

When comparing in our studies the parameters of apo-B for healthy persons and IHD patients under lipid load conditions, we see clear distinctions: if in healthy persons, the apo-B level did not practically change (background – 100 mg/dl, in three hours – 100, in six – 86, and in nine – 96 mg/dl), in IHD patients the considerably increased background value was retained in the course of the lipid load. In group 2 the figures were 174, 168, 177, and 171 mg/dl. In group 1 the apo-B level was also high – 145, 133, 152, and 153 mg/dl.

Accordingly, the apo-B/apo-A-1 ratio in healthy persons remains at a low level. In group 1 IHD patients, the background value was 1.24, in three hours it was 1.1, in six it

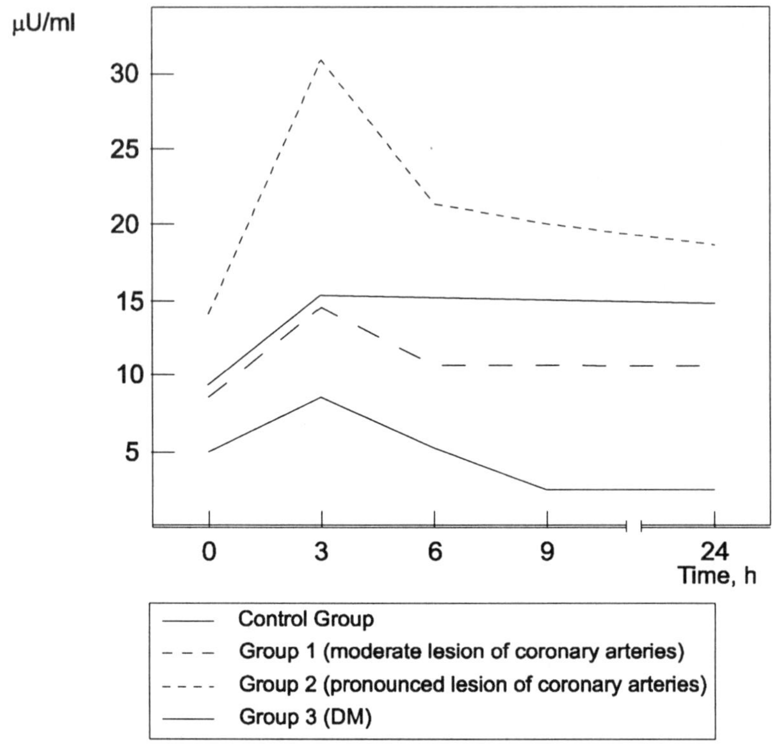

Fig. 20. Influence of Lipid Load on IRI Level in Blood in IHD and DM Patients.

was 1.28, and in nine it was 1.21. In group 2, the results were still more vivid: 1.50, 1.49, 1.45, and 1.51.

When analyzing the spectrum of lipids and apolipoproteins in the blood plasma in group 2 patients, we can note the following: the background parameters of the level of lipids and apoproteins in IHD group 2 patients were higher than in group 1. A fat load makes it possible to reveal the more pronounced atherogenic shifts of the lipid parameters and the level of apoproteins in the blood plasma in group 2 patients. This indicates the deeper disorders in the lipid transportation system of the blood plasma in these patients [89, 101, 115, 128, 148].

What are these disorders in lipid metabolism in IHD patients associated with? What hormones regulate these changes in metabolism?

We consider that insulin is the most significant regulating hormone in IHD patients. Our investigations (especially under conditions of a lipid load) conform this. We can see from Fig. 20 that in healthy persons (with the absence of disintegration of the regulating adaptive mechanisms) a lipid load produces a very moderate reaction of the insulin level in the blood: the background is 5.2 μU/ml, in three hours it is 10.8, in six it is 11.6, and in nine it is 5.0 μU/ml. In IHD patients, these changes are sharper: the figures for group 1 are 7.0, 17.5, 10.3, and 13.0 μU/ml, and for group 2 are 14.6, 34.9, 23.9, and 18.3 μU/ml. It follows from the above data that the evolution of the insulin level in group 2 of IHD patients points to the high significance of hyperinsulinemia in the pathogenesis of IHD development.

What pathophysiological mechanisms are triggered off in IHD patients with hyperinsulinemia and disturbance of lipid metabolism?

We agree with the opinion of many authors that in IHD patients hyperinsulinemia is noted (often in combination with hyperglycemia) caused by a diminished tolerance to glucose. The pathophysiological essence of this process consists in an increase in the amount of FFA in the blood that is associated with a disorder in the function of the liver. In IHD patients, the increased formation of lipids is due to the activation of lipogenesis of the adipose tissue, when lipids form from glucose and acetate.

Physiological studies showed that in pronounced hyperglycemia, an organism begins to actively perform a glycogen-forming function under the influence of insulin. In hypoglycemia, the glycogenolytic effect is activated. Insulin is a very important hormonal regulator of the formation of glycogen. It increases the permeability of the membranes to glucose, stimulates the activity of glucoso-6-phosphate facilitating the formation of lipids, carbohydrates, and acting on proteosynthesis and the inhibition of proteolysis [164, 166, 170, 222, 223].

Clinicians proved that the degree of utilization by the myocardium of a substrate strictly correlates with the level of metabolism of a number of hormones. The number and quality of the nutrient substances utilized by the myocardium are closely related to the processes of lipolysis and glycogenolysis in the liver and adipose tissue. The functioning of these organs and tissues determines the concentration in the blood and consumption by the myocardium of glucose and free fatty acids. The insulin concentration is a very important unit in the regulation of the level of glucose and FFA in the blood. Glucose-insulin mixtures have been finding an application in practical cardiology during many years, but the pathophysiological aspects of the action of these mixtures were not established. The role of carbohydrates in hypoxia of the myocardium (stenocardia, myocardium infarction, etc.) is especially important for supplying energy to the heart. Experimental studies proved the biological efficiency of insulin in the absence of glucose too (its influence on the transportation of amino acids, fatty acids, potassium, etc.). Insulin can stimulate anabolic processes in a cell, enhance reparative reactions, maintain equilibrium of protein metabolism, and raise the level of creatine phosphate in the myocardium.

Insulin is considered as the only universal anabolic hormone lowering the level of glycemia. Numerous clinical observations reveal that hypoinsulinemia produces deep disorganization of the metabolic processes in the myocardium, liver, and muscles. In experiments, hypoglycemia leads to the more rapid appearance of disturbances of the conductance in the myocardium. Hyperinsulinemia in the process of development of atherosclerosis may facilitate the change in the vessel permeability and infiltration of atherogenic lipids into the arteries. The inadequacy of the insulin present in an organism is expressed primarily in the lowering of the tolerance to carbohydrates. Disorders of the secretion of insulin and glucose are the main factors facilitating the development of coronary diseases [247, 251, 279, 294].

A very important matter in the insulin supply to an organism is the discrepancy between the absolute concentration of the hormone in the blood and its biological efficiency. Hyperinsulinemia in IHD patients can be considered as a compensating reaction of the beta cells aimed at normalization of carbohydrate metabolism.

The problem of the relative insulin insufficiency is quite involved. One of the hypotheses consists in that in IHD patients, the reactive sensitivity of the peripheral tissues to insulin lowers. Most clinicians consider that the pathophysiological basis of this phenomenon is the hyperactivity of hormonal and metabolic antagonists such an cortisol, the growth hormone, triglycerides, FFA, and uric acid. Lowering of the vessel permeability is another cause of the peripheral inefficiency of insulin [260, 309, 330, 397, 398, 421].

These comparisons show that the adipose tissue can use "bound insulin" for the normal functioning of metabolic processes in an organism. Lowering of the sensitivity of the adipose tissue to insulin in IHD patients was noted by many authors. The presence in IHD patients of a high level of FFA on an empty stomach and after a glucose load is one of the facts confirming these assumptions.

The membrane of a fat cell has two types of hormonal receptors: (i) receptors that react to stimulation by catecholamines, and owing to this the formation of cAMP increases with the participation of adenylatecyclase (which activates lipolysis), and (ii) insulin receptors that inhibit the activation of adenylatecyclase by these stimulating hormones.

Stimulation of insulin receptors is attended by lowering of the cAMP concentration and the suppression of lipolysis. These aspects are extremely important. Consequently, a fat cell membrane can react both to stimulation of lipolysis under the effect of catecholamines and to inhibition of lipolysis under the influence of insulin. The stimulation of the insulin receptors occurs very illustratively in DM, when the absence of insulin leads to uncontrolled stimulation of the intracellular lipase by lipolytic hormones. As a result, a clearly expressed mobilization of FFA from the fat cells is noted. A similar reaction is observed in emotional stress and a physical load, when FFA is mobilized from a depot or as a result of the increased secretion of adrenaline from the medullary substance of the adrenal glands into the general blood flow. In all cases of stress, the mobilization of fats, being a reply to an emotional tension, physical load, or to an increased demand for energy, increases the content of fatty acids needed to produce energy in various tissues of an organism [463, 469, 475, 522, 573, 621, 624].

During recent years, clinicians give great attention to the exchange of lipids between the blood plasma and erythrocytes. The interest is due to the fact that the metabolism of substances in cells (including in erythrocytes) depends on the structural and functional features of a cell membrane. The structural lipids, in addition to the mechanical function of separating the liquid spaces in cells, are also important for the metabolism of substances in a cell. Lipids play the main role in regulating cell permeability and transporting various substances through a membrane. In a number of cases, phospholipids are the main partners in enzymatic processes in cells. Cholesterol and phospholipids are the main components of the structure of membranes. In normal erythrocytes, the cholesterol/phospholipid ratio is 0.8. In atherosclerosis, this coefficient grows sharply to 1.2–1.4 as a result of an increase in the cholesterol content in a membrane that has the following duties: (i) it is a structural component of a plasmatic membrane ensuring its functioning, and (ii) it is a transportation form of free cholesterol that is rapidly exchanged with the lipoproteins of the plasma. The overloading of an erythrocyte membrane with cholesterol produces an increased rigidity of the erythrocyte, lowers its deformability, and changes its hemodynamic properties [495, 469, 621, 628].

In view of the major importance of the exchange of lipids between the blood plasma and erythrocytes, we studied the lipid and phospholipid spectra of the blood plasma and erythrocyte membranes in 12 IHD patients and 15 healthy persons.

In the patients, the cholesterol level in the blood plasma was increased considerably, on an average up to 233 mg/dl (in the healthy persons it was 184 mg/dl). We noted a growth in the content of triglycerides, LDLP cholesterol, VLDLP cholesterol, and a drop in the level of HDLP cholesterol. We considered the pathophysiological fundamentals of these disturbances and noted only a growth in the index of atherogenesis in the IHD patients to 4.25 (in the healthy persons it was 2.2).

Many clinicians and experimenters attach major importance to the state of the phospholipid spectrum of the blood plasma. It follows from Fig. 21 that especially sharp

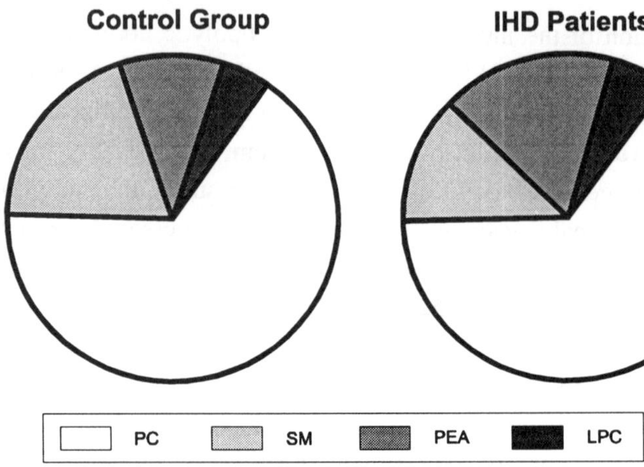

Fig. 21. Phospholipid Spectrum of Blood Plasma in IHD Patients.

disturbances are noted in the content of lysophosphatidylcholine (LPC) and sphingomyelin (SM).

It is considered that the phospholipids of the plasma mainly perform the function of biological detergents and together with the apoproteins increase the solubility of nonpolar lipids. The main part of the plasma phospholipids is formed by lecithins and SM. The high surface activity of the phospholipids is of major significance for building up a lipoprotein and its exchange with the plasma. It was proved in recent years by using tracer fatty acids that there is a close relation between the metabolism of polyunsaturated fatty acids and the phospholipids of the blood.

The content of LPC in the phospholipid spectrum in IHD patients (see Fig. 21) grows to 6.55% (in healthy persons it is 3.75%), and of SM drops to 10.59% (in healthy persons it is 19.23%).

In recent years, great attention is being given to the lipid composition of the erythrocytes in connection with their participation in the metabolism of lipids and to their influence on the hemodynamic functions of an organism. Special importance in these studies is attached to the composition of the lipid layer of an erythrocyte membrane [22, 36, 46, 123, 228].

The lipid layer of an erythrocyte membrane simultaneously has fluidity and order of its structure. The latter, on the one hand, is needed for the processes occurring between the membrane and environment (the reception of information, metabolism of energy and matter). On the other hand, the vital activities require that the components of a membrane

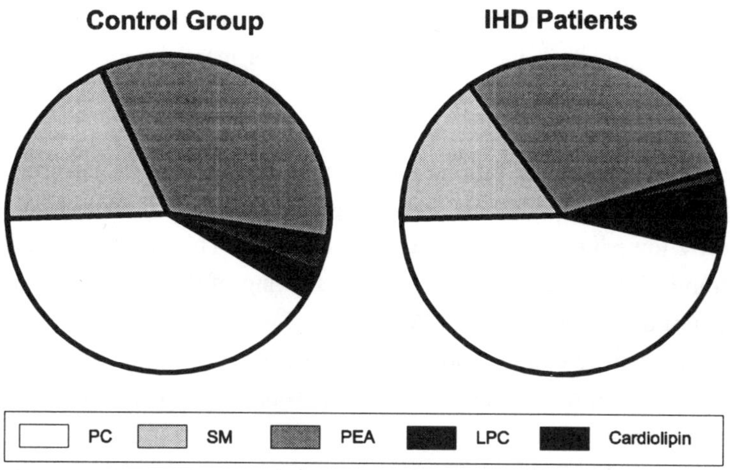

Fig. 22. Phospholipid Spectrum of Erythrocyte Membranes in IHD Patients.

could contact one another and form short-living complexes. The phospholipid composition of an erythrocyte membrane affects many of its functional properties. Among them are regulation of the active and passive transportation of organic substances and electrolytes through a cell membrane, sensitivity to the action of hormones, and participation in blood coagulation processes. The change in the phospholipid asymmetry affects the adhesion of an erythrocyte to a vessel endothelium, which results in disturbance of microcirculation.

In the membrane of erythrocytes in IHD patients, the fraction of cholesterol increases moderately to 52.8% (the norm is 44.6%), the content of phospholipids drops substantially to 28.5% (against 46.2% in healthy persons), the fraction of triglycerides does not virtually change, but the FFA content grows to 3.07%. Of special significance, from our standpoint, is the change in the cholesterol/phospholipid ratio. If in healthy persons it is 1.08, in the IHD patients it is 1.82. The SM fraction also drops sharply to 14.95% (against 19.6% in healthy persons).

Consequently, in IHD patients, considerable changes are observed in the phospholipid spectrum of the erythrocyte membranes (Fig. 22).

What do these changes indicate?

Phosphatidylcholine (PC) and phosphatidylethanolamine (PEA) together with cholesterol are the main structural components of a cell membrane. We attach great significance to the ratio PEA/PC for appraising the structural and functional state inside a membrane.

The cholesterol/phospholipid ratio reflects the rheological properties of erythrocytes. An isolated growth in the cholesterol level in a membrane without increasing the cholesterol/phospholipid ratio changes the viscosity of the blood and disturbs its rheological properties. A growth in the level of total cholesterol in the blood plasma and a change in the cholesterol/phospholipid ratio in a membrane are attended by lowering of the fluidity of the lipid phase of a membrane, increasing of the volume of an erythrocyte and of its rigidity, and by lowering of the deformability of a cell. This results in thickening of the basal membrane of the endothelium of vessels, slowing of the blood flow, and increasing of aggregation of the elements [121, 226, 353].

Moreover, a change in the phospholipid spectrum of the membranes of erythrocytes in IHD patients disturbs the function of the cytoplasmatic receptors. This relates first of all to the insulin receptors on the membranes of the erythrocytes. Experiments showed that the transportation and metabolic effects of insulin are a result of the binding of the hormone with membrane receptors. Changes in the phospholipid composition of the erythrocyte membranes affect both the number of these receptors and their functional activity.

It was noted that in IHD patients with an excess body weight the sensitivity of the fat cells to insulin is reduced. In connection with the fact that in obesity the activity of the LPL is disturbed, and the latter is controlled by the insulin-glucose system, changes in the sensitivity of the receptor apparatus and correspondingly of cAMP appear. Biochemical studies showed that the cAMP level regulates the activity of the LPL. The cAMP activity in a cell, in turn, depends on the action of the enzymes adenylatecyclase and phosphodiesterase that synthesize and decompose cAMP. The synthesis of cAMP is stimulated by hormones such as adrenaline, glucagon, and ACTH. The influence of insulin on the cAMP level in a cell can be characterized as an inhibiting one: it increases the activity of phosphodiesterase and enhances the decomposition of cAMP.

The enzyme systems ensuring the activity of a cell can be divided into several groups: (i) supporting the active transportation of nutritive substances in a cell from the environment, (ii) ensuring the energy reaction of a cell, (iii) regulating the synthesis of a protein and nucleic acids, (iv) ensuring cell mitosis, and (v) participating in the biosynthesis of proteins and polysaccharides.

Analysis of the state of the enzyme systems in the myocardium cells in cardiac insufficiency in IHD patients revealed various pathological changes: (i) the amount of FFA circulating in the blood may grow as a result of hydrolysis of the deposited triglycerides, (ii) the membrane phospholipids can be hydrolyzed by the activated phospholipases of the cytosol, which leads to the liberation of fatty acids, (iii) the intracellular content of fatty acids may grow as a result of their absorption by ischemic cells, inhibition of fatty acid oxidation, and condensing of the bicarbon fragments. All the above processes terminate in the accumulation of fatty acids and their metabolism products within and around the cells of an ischemic myocardium, The fatty acids have a complicated influence on the functioning of a membrane – they can either stabilize or disturb the membrane structure.

In summarizing, we can conclude that a major role in disorders of lipid metabolism in IHD patients is played by an increased level of cholesterol in the blood. The latter is a factor that considerably modifies the structural and functional properties of biological membranes in various organs and tissues, and it is this property that determines its leading role in the pathogenesis of atherosclerosis [83, 176, 292, 516].

The mismatch of the two cholesterol flows in an organism – the transported cholesterol in the lipoproteins circulating in the blood and the cholesterol in the cells, is the molecular prerequisite of atherosclerosis.

Experiments on animals showed that a cholesterol load on them alters the sensitivity of the thrombocytes to some aggregation inductors, in particular to adrenaline. Vessel walls enriched in cholesterol were distinguished by an increased sensitivity to the vasoconstrictive effect of potassium depolarization. We can thus conclude that cholesterol is an important factor in regulating the functional state of both the vessels and thrombocytes. An increased cholesterol content in the blood causes the development of vasomotor and thromboembolic disorders dangerous for an organism. Even transient hypercholesterolemia is the cause of development of regional dyscirculation. The regions of the blood channel where there are defects of the vascular endothelium are the weak spot.

In a postischemic syndrome, disturbance of the structural and functional properties of the membranes of erythrocytes reaches a critical level. When changes in the cell membranes lower their sorption properties, the erythrocytes become a source of thromboplastic substances and energy products that create favorable conditions for thrombus formation and blood coagulation.

We can conclude that in IHD patients, disturbance of lipid metabolism is based on the hormonal disintegration of the regulatory adaptive mechanisms, and the role of insulin is especially important.

In IHD, the structural and functional properties of the erythrocytes are altered drastically, the cholesterol/phospholipid ratio equals 1.2–1.4 instead of 0.8. This overloading of an erythrocyte membrane with cholesterol increases its rigidity, lowers its ability of deforming, and changes the hemodynamic properties.

1.5. Lipid Metabolism Disorders in Diabetes Mellitus Patients

Diabetes mellitus and atherosclerosis are diseases with similar disorders of lipid metabolism attended by hypercholesterolemia, hypertriglyceridemia, and hypoalphacholesterolemia.

A disorder in metabolism in DM occurs chiefly because of tissue insulin insufficiency. The metabolism of carbohydrates, and also of fats and proteins is disturbed. Not only insulin, but also the contrainsular hormones glucagon, somatostatine, glucocorticoids, and catecholamines are included in the chain reaction of the damaging hormonal units. In biochemical investigation of metabolism processes (under conditions of insulin insufficiency), there were noted a lower penetration of glucose into the muscle and fat cells, disturbance of oxidative phosphorylation, and lowering of glycogen synthesis. Hyperglycemia in DM ensures a better penetration of glucose into a cell. The disturbed participation of glucose in the supply of energy to an organism includes fats and proteins into energy metabolism as a compensation. As a result, an increased amount of FFA, ketone bodies, and amino acids appears in the blood. The increased supply of FFA into the liver causes the greater synthesis of triglycerides [155, 168, 171, 203, 231, 247].

The metabolic disorders in DM are thus attended by numerous functional disorders such as tissue hypoxia, dyslipoproteinemia, hormonal homeostasis, metabolic acidosis, angiopathies, and electrolytic disturbances. A special significance is acquired by the relation between disorders of the oxygen-transportation function of the blood, its coagulating system, and as a consequence, the development of microangiopathies. Tissue hypoxia in DM patients is realized according to the following scheme: erythrocyte membrane – plasma – capillary wall – tissue. An increased content of lipids in the blood plasma of DM patients impairs the transcapillary diffusion of oxygen.

In hypoxia of any genesis, acidosis appears in the myocardium as a result of anaerobic glycolysis. The utilization of pyruvic acid in the Krebs cycle is inhibited, and the increased transformation of pyruvate into lactic acid begins. Moreover, intracellular acidosis destabilizes the membranes of the lysosomes and causes hydrolytic enzymes to emerge from them, thus increasing their activity. All this lowers the energy potential of the cardiac myocytes. Great attention is attached to the peroxide oxidation of lipids in the myocardium. Activation of the free-radical peroxide oxidation of lipids (FPOL) may be associated with the accumulation of fatty acids under conditions of hypoxia, and also in DM and obesity. The growth in the FPOL is generally combined with activation of the sympathico-adrenal system [320, 399, 418, 429, 458].

It was shown under clinical conditions that in DM a growth in the production of the ACTH, STH, and also of glucocorticoids and catecholamines occurs.

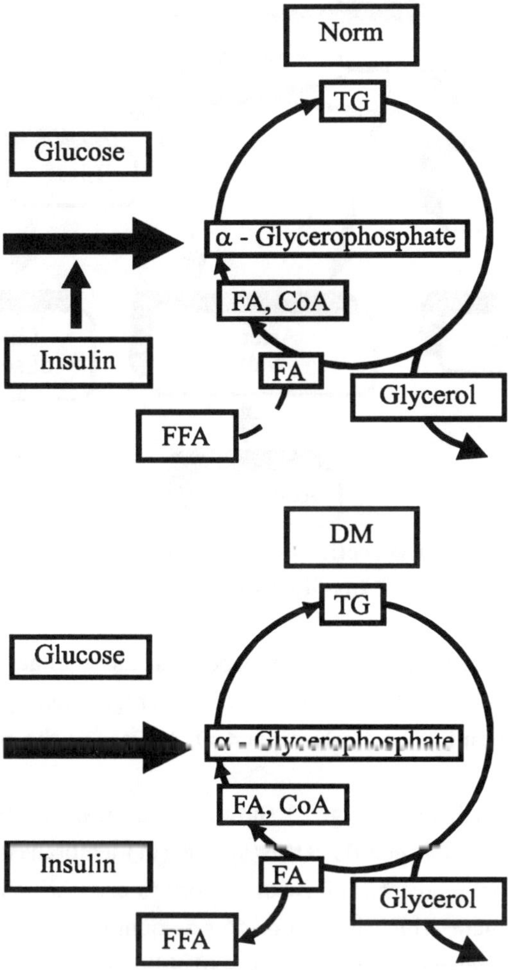

Fig. 23. Synthesis and Decomposition of Triglyceride in Adipose Tissue Cells Normally and in DM.

With an excessive physical load, destruction of the mitochondrias, expansion of the canaliculi of the endoplasmatic network, and fragmentation of the myofibrills may appear. An oxygen deficiency, disturbance of enzymatic processes, and an excess accumulation of catecholamines in the myocardium may be the pathophysiological bases of a physical overload. This leads to the appearance of disseminated necrosis.

Experimental studies showed a cell membrane to be the most important point of application of insulin action in an organism. The active mechanisms of a cell membrane

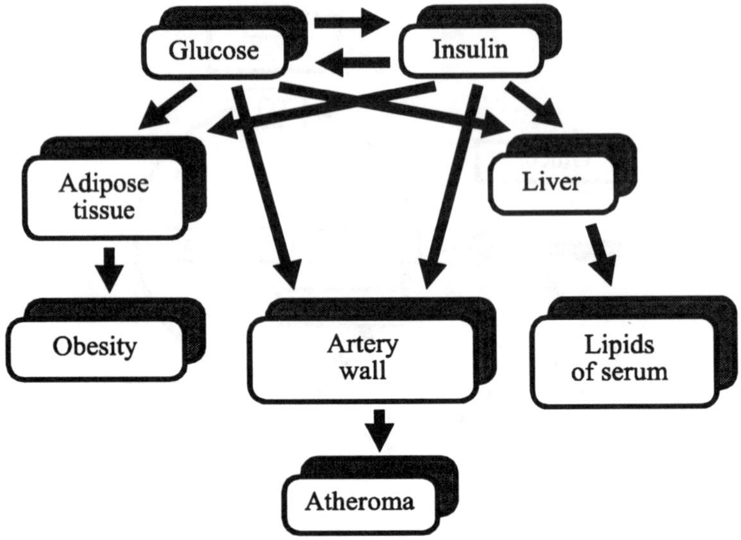

Fig. 24. Main Directions of Disturbed Relation between Insulin and Glucose.

pass over the influence of insulin to various aspects of the vital activities of a cell including its genetic function. Of special importance in the regulation by insulin of the permeability of a cell membrane. The Na-K-dependent ATPase is the point of application of insulin in a membrane.

A glance at the scheme in Fig. 23 shows that in healthy persons insulin regulates lipogenesis. With the aid of a specific carrier system (in membranes of both muscle and fat cells), insulin ensures an increased inflow of glucose to cells, which creates a greater pool of alpha-glycerophosphate. This causes a greater amount of fatty acids to be fixed on the cell acceptors, and lipogenesis increases [421, 438, 439, 463, 467].

In DM (and accordingly with a deficiency of insulin), glucose in accordance with the carrier mechanism is not assimilated in a sufficient amount, and the major part of the fatty acids enters the blood, and then the liver, thus creating a favorable medium for lipoprotein formation. This biochemical process often causes liver adiposis in DM patients.

Consequently, a deficiency of insulin underlies the inhibition of lipogenesis and activation of lipolysis.

We thus see that free fatty acids, like insulin, play a decisive role in the regulation of not only lipid, but also of carbohydrate metabolism. A deficiency of insulin (both absolute and relative) increases the level of the FFA and at the same time lowers the utilization of glucose, so that the muscle cells, as already noted, are switched over to FFA as an energy donor. Glucose accumulates in the fat cells and transforms into triglycerides if the amount of insulin is sufficient. Should hyperinsulinism appear for some reason or other, the lipogenesis becomes excessive, the fraction of triglycerides in the cell space grows, this leads to hypertrophy of the adipose tissue and, finally, to obesity (Fig. 24).

To date, the role of hyperinsulinemia in the development of atherosclerosis was proved. The atherosclerotic changes in the vessel walls in DM patients introduce complications into the course of angiopathies and, in particular, of cardiopathy, which leads to early invalidism and death [475, 488, 495, 522, 547].

An increased level of lipids in the blood and a high arterial pressure are the two main factors causing the development of IHD and atherosclerosis. But DM patients have features that are due to the greater frequency and combination of risk factors such as hyperlipoproteinemia, arterial hypertension, increased blood coagulation, and also the presence of hyperglycemia, microangiopathies, and changes in insulin secretion. In DM patients, three biochemical aspects of regulation of the transportation of lipoproteins in the plasma can be singled out [551, 570, 579, 585, 611]:

(1) an increased production of lipoproteins rich in triglycerides;

(2) lowering of the hydrolysis of triglycerides by the action of LPL;

(3) disturbance of the catabolism of lipoproteins rich in cholesterol.

Insulin affects both the synthesis of triglycerides and their withdrawal from circulation. Many investigators note the correlation between the content of insulin in the blood and the triglyceride level. This is sometimes attended by obesity which, in turn, is associated with hypertriglyceridemia. However, such a relation is also retained in people not suffering from obesity, but having resistance to insulin. Endogenic triglycerides are synthesized in the liver from fatty acids and glucose. Insulin, by acting on the liver, stimulates the synthesis of triglycerides. With a deficiency of insulin, the synthetic activities of the liver are suppressed. The synthesis and decomposition of lipoproteins rich in triglycerides are regulated in an organism by two enzymes: a hormonosensitive lipase and lipoproteinlipase.

Lipolysis in an adipose tissue occurs with the aid of a hormonosensitive lipase (HSL) whose activity is inhibited by insulin. With an insufficiency of insulin in the blood, the lipolysis of the adipose tissue grows and the level of circulating fatty acids increases. With an excess of insulin, the activity of the HSL lowers, and the level of FFA in the blood lowers accordingly. The supply of the FFA as a substrate is also regulated by insulin [45, 118, 133, 180, 201, 310, 584].

An important role in lipid metabolism is also played by LPL. It hydrolyzes the lipoproteins rich in triglycerides in the capillaries of adipose tissue and skeletal muscles, where it is bound to the glycosaminoglycanes localized on the surface of a capillary endothelium facing the lumen of a vessel. The enzyme is activated by apo-C-2 and is inhibited by sodium chloride and protamine sulfate. The LPL hydrolyzes lipoproteins rich in triglycerides mainly in the capillaries of vessel walls, while the liver lipase hydrolyzes them in the liver.

LPL is synthesized under the influence of insulin, when there in a deficiency of the latter there is also one of LPL. In extreme forms of insulin deficiency, the LPL level is very low, which leads to the accumulation in the blood of lipoproteins rich in triglycerides (diabetic lipemia). This process is compensated by the introduction of endogenic insulin. Consequently, depending on the form of DM, there are various mechanisms of development of hypertriglyceridemia: in DM of type I, direct suppression of triglyceride synthesis in the liver occurs. This is compensated by the excess supply of FFA from the adipose tissue, but here the removal of triglycerides from circulation is disrupted. In DM of type II that is attended by hyperinsulinemia or in obesity, the synthesis of endogenic triglycerides by the liver and the formation of triglycerides are enhanced [87, 96, 234, 381, 392, 537].

In DM of both types I and II, a correlation was revealed between the level of hypertriglyceridemia (HTG) and glycosylized hemoglobin. P.Janka and his co-workers in 1987 singled out hyperinsulinemia combined with HTG as a syndrome – a risk factor of the development of atherosclerosis in DM.

Apart from HTG, in DM patients a growth in the cholesterol level in the blood plasma was noted more frequently than in persons without disorders in carbohydrate metabolism. The influence of insulin on cholesterol metabolism has been studied insufficiently. Experiments showed the overwhelming influence of alloxan diabetes on the synthesis of cholesterol. On the other hand, the introduction of insulin via the portal vein leads to the enhanced inclusion of acetate into the cholesterol. Insulin stimulates the synthesis of cholesterol in the liver tissue, where it influences HMG-CoA-reductase (3-hydroxy-3-methylglutaryl-CoA-reductase) – an enzyme limiting the reaction rate in the metabolism of cholesterol.

The activity of this enzyme in the liver of rats with alloxan diabetes is very low, but insulin restores it to the normal value. According to P.Polumbo (1978), in DM patients there is both a growth in the synthesis of endogenic cholesterol and retardation of its excretion. On the basis of numerous observations, R.Stout in 1985 postulated the influence of uncontrolled hyperglycemia in DM on the increase in the production of cholesterol and also on the relation between the level of triglycerides and the excretion of bile acids – a product of cholesterol metabolism. The production of triglycerides rich in VLDLP and bile acids may be associated with the synthesis of cholesterol, which is presumably increased in the early, uncontrolled stages of DM. DM patients both with a deficiency of insulin and with hyperinsulinemia suffer from hyperphagia. The letter, in turn, leads to hypertrophy of the mucous membrane of the intestine and a growth in the activity of HMG-CoA-reductase. Consequently, hypercholesterolemia may be due to an increase in the synthesis of sterol in the small intestine, its excess supply with the food, and in increase in its absorption [76, 145, 159, 167, 214].

The path of penetration of cholesterol into a cell in LDLP by means of a receptor is the main factor for maintaining normocholesterolemia. Improper utilization of cholesterol may be a result of a deficiency or activity of the cell receptors, and also of changes in the structure of the LDLP themselves, e.g. the glycosylation of LDLP in DM patients, and the formation of autoimmunity complexes. R.Stout [555] considers that the mechanism of action of insulin is most likely associated with a growth in the number of LDLP receptors than with a growth in the affinity for them.

Investigations of the HDLP in DM patients have a contradictory nature. The protective role of the HDLP in DM remains unclear. It was shown that the level of HDLP cholesterol in type II DM patients is lowered. Some authors determine in these patients the presence of a negative correlation between the content of HDLP and the level of glycosylized hemoglobin, which was not observed in type I DM patients. The treatment of DM with the aid of a diet or sulfonyl-urea drugs normalizes the level of the HDLP. The restored activity of the LPL after the introduction of insulin increased the level of HDLP cholesterol, possibly in the catabolism of VLDLP. R.Eckel and his co-workers in 1981 found that in type I DM patients the level of HDLP cholesterol was slightly higher than in the control group, while the ratio of the high-density and low-density lipoprotein cholesterols is considerably higher in the patients. These changes correlated with the level of the $HDLP_2$ cholesterol.

In type II DM with hyperinsulinemia, there is a low level of the HDLP cholesterol. This is apparently associated with the fact that the point of application of insulin in type II DM is the liver, and not the peripheral tissues (as in type I DM). Here the activity of the liver lipase participating in the catabolism of the HDLP increases. This is what explains the low values of the HDLP cholesterol in type II DM patients and their high disposition to atherosclerosis. By one hypothesis, the protective influence of HDLP cholesterol on the risk of incidence of IHD consists in the participation of HDLP in the reverse transportation of cholesterol from the tissues that lowers its accumulation in an arterial wall. This is what we call a cholesterol-acceptor function. By another hypothesis, HDLP or one of their components prevent the formation of thrombi on the surface of the atherosclerotic plaques, e.g. by stabilizing the prostacyclin in a vessel wall as a result of prolonging the duration of its functioning. This process occurs via apo-A-1 [203, 214, 239, 275, 288, 342].

Experiments with a culture of fibroblasts and endothelial cells revealed that HDLP can block the receptors of LDLP cells, after which the accumulation of cholesterol in the cells lowers.

Consequently, the content of HDLP cholesterol in DM depends on the type of the disease, the presence or absence of obesity, the degree of control of the course of the disease, and the kind of treatment.

The disorders of lipid metabolism in DM depend directly on the form and severity of the diabetes. A change in the lipid content in the blood may be a result of the increase in the delivery of substrates for the synthesis of lipids, their increased synthesis, and the incomplete excretion of their metabolism products.

The question of the role of HTG in the development of IHD and atherosclerosis to date is under discussion. Some authors give HTG the leading place among the risk factors of IHD development. The Framingham investigation showed that HTG is an independent IHD risk factor (men with a normal level of HDLP cholesterol are an exception). Individual epidemiological studies failed to confirm the leading role of HTG among the risk factors [393, 415, 422, 450, 487].

An increased triglyceride level in the blood plasma of DM patients may underlie the accelerated development of atherosclerosis indirectly via other lipid fractions, but the triglycerides may also give a direct atherogenic effect. It was found that HTG has a toxic effect on the endothelium of vessels, increases the aggregation of thrombocytes, and lowers the fibrinolytic activity of the blood. A direct correlation between the level of glycemia and the triglyceride content in the blood was noted.

Currently, morphologists studying atherosclerosis do not single out separately atherosclerosis typical of DM. They consider that in IHD and in DM the transformation of the smooth muscle cells or macrophages is the key link in the formation of an atherosclerotic plaque. A major significance in this transformation is attached to disorders in the metabolism of lipids and lipoproteins. The most important for the development of atherosclerosis is a combination of an increased level of cholesterol, LDLP, VLDLP, and apo-B with a reduced content of HDLP, and all this is clearly connected with the development of IHD.

The Framingham studies showed the presence of a clear inverse correlation between the level of HDLP cholesterol and the frequency of development of IHD independent of the sex. A protective effect of high concentrations of HDLP cholesterol was manifest in patients up to 80 years old, whereas low values of HDLP cholesterol were associ-

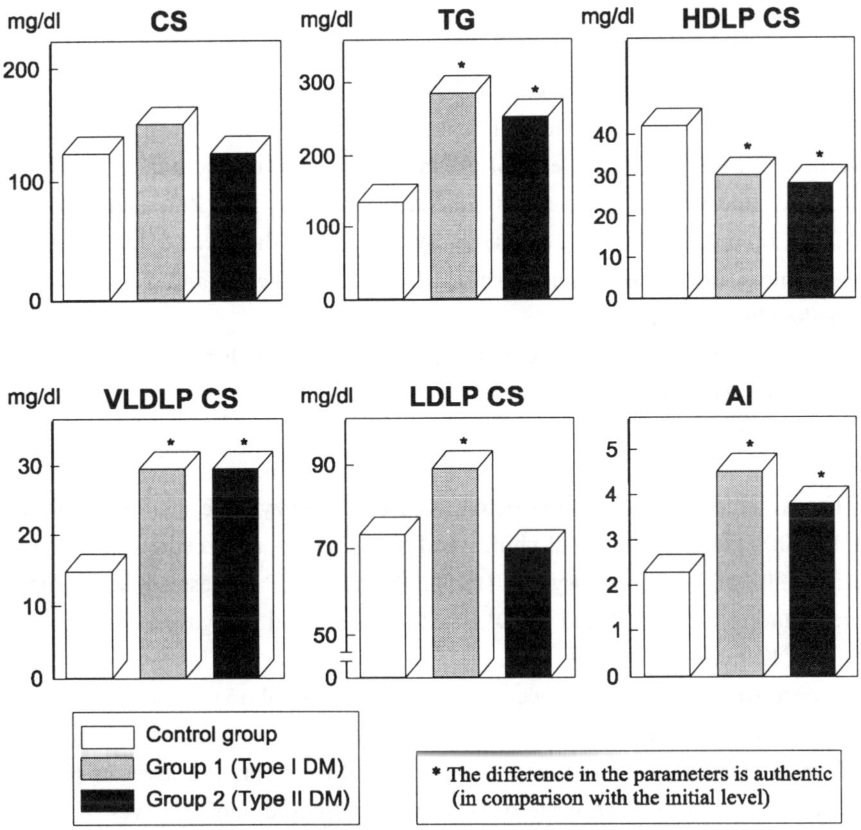

Fig. 25. Level of Lipids in Blood in DM Patients.

ated with an increased risk of atherosclerosis with any level of total cholesterol in the blood.

We studied the lipid and phospholipid spectra of the blood plasma and membranes of the erythrocytes in all persons under observation. Blood was taken from on ulnar vein after 12 hours of fasting and before the taking of sugar-containing preparations. A 0.6% solution of ethylenediameine tetraacetic acid was used as an anticoagulant. The erythrocyte material was obtained by threefold washing with a 0.89% solution of sodium chloride, and next 0.5 ml were taken from the deep layers of the precipitate. The suspension of erythrocytes and plasma was frozen and stored at -18°C not over one month. The lipids of the plasma and erythrocytes were extracted with a mixture of chloroform and methane in a ratio of 2:1. Separation into fractions was performed by thin-layer chromatography on Silufol plates. A system consisting of hexane, diethyl ether, and glacial acetic acid in a proportion of 80:20:2 was used to separate the lipids, and one consisting of methanol, chloroform, and bidistilled water in a proportion of 60:20:4 to separate the phospholipids. The lipid spots were determined with a 10% alcohol solution of phosphoromolybdenic acid. The following fractions of the lipids were determined: total phospholipids (TP), free cholesterol (FC), FFA, TG, cholesterol esters (CSE) and the spectrum of phospholipids: LPC, PC, SM, and PEA. The fractions were registered in a Karl Zeiss Jena densitometer EP1-65. The lipid fractions were estimated by calculating the areas of the relevant peaks of the substances, using the formula $S=h/\alpha$, where h is the height and α is the width taken at half the height of a peak.

Each lipid fraction was expressed in percent. When analyzing the results of the investigations, we used the relation:

$$\frac{FC+CSE}{PL}$$

The triglyceride content was determined by the enzymatic method in an autoanalyzer Olly 300 (Finland), the total cholesterol by the enzymatic method in a biochemical analyzer FP-901 of the firm Labsystem (Finland), and HDLP cholesterol in the supernatant after precipitating the LDLP and VLDLP with heparin and manganese dichloride by the procedure of Klimov *et al.* [292].

In performing all the work, we constantly used intralaboratory control of the quality with the aid of control serums.

Inspection of Fig. 25 shows that in type I DM patients the level of triglycerides, LDLP cholesterol, VLDLP cholesterol, and AI is increased and the level of HDLP cholesterol is lowered in comparison with the control group, while in type II DM patients the level or triglycerides, VLDLP cholesterol, and AI is increased and that of HDLP cholesterol is lowered. The content of total cholesterol in the plasma for the type I DM (148 mg/dl) and type II DM (123 mg/dl) patients did not differ from that in the control group and was within the normal values of this parameter established for the given age group.

The high level of triglycerides in the blood plasma in type I DM patients did not correlate with the body weight index and the glucose level in the blood, although an increased glucose level is known to stimulate triglyceride synthesis. In the labile course of DM with phenomena of dysinsulinemia, processes can be observed that inhibit and activate the system of lipases. The inhibiting effect on the HSL of the adipose tissue and the activating one on the LPL and liver triglyceridelipase result in lowering of the mobilization of free fatty acids from the adipose tissue, but accelerate the catabolism of the VLDLP formed in the liver with the production of free fatty acids. Lowering of the insulin level In the blood causes the intense supply of FFA from the adipose tissue into the liver and the development of HTG. The type I DM patients we studied showed a considerable lowering of the level of HDLP cholesterol (29.0 mg/dl).

When analyzing the correlations between the level of total cholesterol and the content of glucose in the blood, triglycerides, LDLP cholesterol, VLDLP cholesterol, and HDLP cholesterol in type I and II DM patients, no correlation was found between the content of cholesterol and glucose in the blood, triglycerides, VLDLP cholesterol and LDLP cholesterol, although in type I and II DM the content of VLDLP cholesterol is authentically increased (in type I DE patients it is 29±5.4 mg/dI, and in type II ones it is 29±6.2 mg/dl). In type I DM, the content of LDLP cholesterol is authentically increased (89±5.4 mg/dl against 72±12.9 mg/dl in the control group).

What is the pathophysiology of the process of the increase in the content of LDLP and VLDLP in DM?

We consider that there are three pathophysiological factors, namely, (i) lowering of the HDLP level, (ii) nonenzymatic glycosylation, and (iii) the number of LDLP receptors.

Most clinicians attach special importance to the first factor – lowering of the HDLP level in DM. This is due to the fact that the HDLP stimulates the catabolism of LDLP and VLDLP. If the level of the HDLP drops (in DM), the catabolism of the LDLP and VLDLP is disturbed accordingly. Our results indicate that the HDLP level in type I DM patients is lowered to 29±3.8 mg/dl, and in type II – to 27±3.1 mg/dl.

It was shown in experiments that the HDLP also regulate the level of apo-B. With a high level of the HDLP, the activity of the apo-B receptors on the cells of the peripheral tissues grows, and the catabolism of the apo-B is accelerated accordingly, i.e. the atherosclerotic process is inhibited [508, 547, 551, 561].

Moreover, the high level of the VLDLP (especially in combination with HTG) is due to biochemical disturbances in DM: the increased supply of FFA from the fat depots to the liver, activation of triglyceride synthetase and HMG-CoA-reductase – specific enzymes of the liver and endothelial cells.

It was established in biochemical investigations that in diabetes mellitus, the VLDLP contain more cholesterol than is normal, while the LDLP contain an increased amount of triglycerides. All these factors facilitate the development of atherogenesis in DM patients.

These enzymes play a key role in lipid metabolism, and their activity is regulated by the endocrine system. It is very important that the LPL is controlled by the "insulin-glucose" system. Experiments showed the activity of LPL to drop when the insulin level lowers. The coordinating role in the regulation of the LPL is played by cAMP. It is just via cAMP that hormones affect the activity of these enzymes.

How is the level of cAMP regulated in a cell?

As indicated earlier, the cAMP content is determined by the reaction of two enzymes – adenylatecyclase and phosphodiesterase. These enzymes participate in the synthesis and decomposition of cAMP. The synthesis of cAMP is stimulated by adrenaline, glucagon, and ACTH. Insulin increases the activity of phosphodiesterase and increases the decomposition of cAMP. The thyroid-stimulating hormone (TSH) inhibits the activity of LPL, while thyroxin increases the sensitivity of lipases to the effect of adrenaline and thus stimulates lipolysis [54, 280, 301, 310, 597].

The influence of the thyroid hormones increases the release of FFA from the adipose tissue and their utilization.

The second pathophysiological factor of disturbing lipid metabolism in DM is nonenzymatic glycosylation. Clinicians noted that the atherogenic type of dyslipoproteinemia in DM patients is encountered very often. An important role in disturbance of the biochemical properties of the lipoproteins in DM patients is played by their nonenzymatic glycosylation. The attachment of glucose to HDLP with the formation of the glycosylized form of the HDLP alters the functional properties of these lipoproteins, prevents the transportation of cholesterol from the tissues into the liver and the participation of the apoproteins of HDLP in the synthesis of LPL, and lowers the antiatherogenic properties of the HDLP.

Matters are similar with LDLP and VLDLP. DM is accompanied by the hyperproduction of glycosylized LDLP and VLDLP with a change in their functional properties and with inhibition of the reaction with specific receptors. This transformation of the LDLP and VLDLP increases their capture by fibroblasts and endotheliocytes. This is attended by the development of hypercholesterolemia and lipid infiltration of a vessel wall with the appearance of sources of atherosclerosis [351, 365, 381, 516].

Many authors consider that the crossing binding of plasma proteins to the matrix proteins of vessel walls (to the collagen part) by means of the products of pronounced glycosylation and binding of the insoluble matrix components of the basal membrane with one another by means of the same products is the main pathogenetic unit of diabetic micro- and macroangiopathies. This relation results in the formation of complexes that are the site of attachment of other proteins of the plasma, and this results in the cascade growth of the process.

An important place in the normal homeostasis of a vessel wall is relegated to macrophages that "purify" the vessel wall from denaturized protein complexes. In DM patients, an excess accumulation of the products of pronounced glycosylation occurs, equilibrium of homeostasis shifts in the direction of a drop in the processes of removal of the denaturized protein complexes, and proliferation thus grows. Two main pathogenetic units of vascular complications in DM are considered: glycosylation, and the polyol path of glucose metabolism [58, 163, 176, 567].

The third cause of the growth in the level of LDLP and VLDLP is the number of LDLP receptors. Clinical observations of children with family hereditary hypercholesterolemia proved that a genetic defect in the LDLP receptors causes the formation of "atherogenic dyslipoproteinemia" with a clinically severe manifestation of atherosclerosis. With a genetic defect in the LDLP receptors, the cholesterol content in the blood cells, fibroblasts, and liver cells is not increased, but its level varies. This is why more importance is attached to the number of LDLP receptors rather than to the total cholesterol content. In the works of D.Goldstein and M.Brown, who were awarded the Nobel Prize (1985), it was noted that in healthy cells the synthesis of cholesterol stops as soon as LDLP is added to the medium because 25-oxycholesterol – a very strong inhibitor of cholesterol synthesis from acetate – forms from a part of the LDLP cholesterol captured via the receptors [73, 76, 83, 88, 106].

In the opinion of these authors, the mismatch of two cholesterol flows, namely, the transportation cholesterol in the lipoproteins circulating in the blood and the cholesterol in cells, is the main molecular prerequisite for the development of atherosclerosis in an organism. The authors evaluated the pathogenesis of atherosclerosis from the physicochemical, biochemical, and physiological viewpoints. They noted that under conditions of hypercholesterolemia, the LDLP stop synthesizing cholesterol: receptor endosytosis of the LDLP sets in and the formation of cholesterol ethers is activated, i.e. a balance in cholesterol metabolism is achieved as a result of coordination between its "inflow" and "deposition" in the cells. In the hepatocytes that synthesize and absorb the major part of the lipoproteins, the intense inflow of the cholesterol into the cells is associated with activation of the synthesis and secretion of the lipoproteins.

Since the development of IHD in DM patients is closely related to the development of disorders in lipid metabolism, the progressing of atherosclerotic changes is generally combined with a growth in the level of triglycerides and cholesterol, especially of the latter contained in the low-density lipoproteins [126, 136, 168, 192, 214].

On the other hand, the HDLP cholesterol has an antisclerotic effect. A number of investigations revealed a negative correlation between the content of HDLP cholesterol and the level of triglycerides. It was shown that the content of HDLP cholesterol depends on the body weight (its level is diminished in persons with obesity). In DM patients, a negative correlation was found between the content of alpha-cholesterol and the triglyceride level in the blood. The latter correlation was found only in type II DM patients, and was not discovered in type I patients.

The growth in type I DM patients of the LDLP level to 89.0 mg/dl (against a normal value of 72.0) and of the VLDLP level to 29.0 mg/dl (against 14.0 in healthy persons) that we revealed is exceedingly important for the progressing of IHD. The growth in the LDLP level is closely related to the lowering of the HDLP cholesterol one. Biochemical studies showed that the supply of free and esterified cholesterol to cells is regulated by receptors of LDLP, whereas its emergence from the cells is controlled by the HDLP receptors (mainly by apo-A-1). In DM, the level of HDLP cholesterol is reduced drastically, it being 29.0 mg/dl in type I DM patients and 27.0 in type II ones (against 41.0 in healthy persons).

In connection with the sharp drop in the content of HDLP cholesterol (with the relevant drop in the level of apo-A-1). the emergence of the cholesterol from the cells is retarded, and it accumulates in the cells.

Of special significance in the metabolic disorders in DM, in our opinion, is the disintegration of the cyclic nucleotides in the cells under the influence of a deficiency of insulin.

An important role in the metabolic processes in the blood plasma is played by the phospholipids, which play (chiefly) the role of biological detergents. Together with the apoproteins, they facilitate an increase in the solubility of nonpolar lipids. In combination with proteins, they can retain cholesterol, its esters, and neutral fats in the micellar form in an aqueous solution. The increased surface activity of the phospholipids is very important for the construction of lipoproteins and their exchange with the plasma. Between individual lipoproteins, the exchange of phospholipids occurs, and simultaneously exchange between the lipoproteins and cell membranes takes place. Studies with the use of tracer fatty acids showed that there is a close relation between the metabolism of polyunsaturated fatty acids and phospholipids of the blood [177, 221, 237, 247, 257].

Of the phospholipids, PC (lecithin) and sphingomyelin are constantly present in the blood plasma. The phospholipids are synthesized in virtually all the tissues, but chiefly in the liver. One of the phospholipids – lecithin – enters the plasma from the chylomicrons.

Phospholipids are an essential and very important component of all cell membranes. In rest and under a physical load, a constant exchange of PC and SM occurs between the plasma and erythrocytes. These phospholipids are constantly present in the blood plasma and maintain the nonpolar lipids (triglycerides and cholesterol esters) in a soluble state.

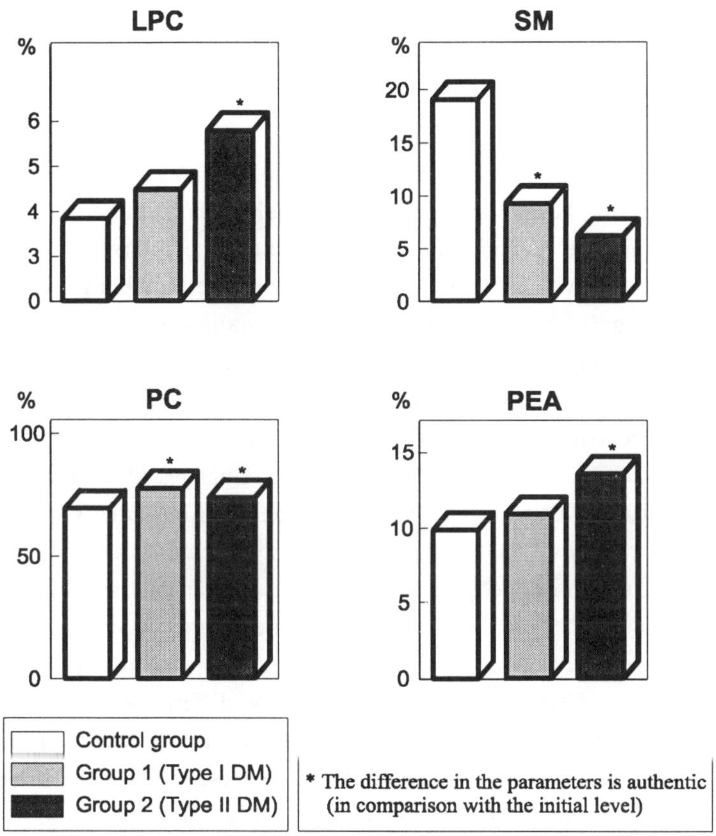

Fig. 26. Phospholipid Spectrum of Blood Plasma in DM Patients.

A comparison of the phospholipid content in the blood plasma of type I and II DM patients and of the control group revealed considerable distinctions. The fraction of LPC in type I DM patients was 4.46%, in type II patients was 5.77%, and in the control group was 3.75%. The SM content in DM patients of both type I (8.91%) and type II (6.59%) is lowered considerably (in the control group it is 19.23%). The PC content differs insignificantly: it is 73.9% in type I DM patients, 75.4% in type II, and 68.6 % in the control group (Fig. 26).

In the involved process of incidence and progressing of atherosclerosis in patients with IHD and DM, the exchange of lipoproteins between the blood and cell elements, in particular erythrocytes, is of a major significance. Studying of the lipid composition of the membrane of erythrocytes is interesting not only from the viewpoints of the quantitative migration of lipids from the blood plasma into the membranes, but also from the viewpoints of disorder of the rheological properties of the blood in IHD and DM patients. The lipids contained in the membrane of an erythrocyte can be divided into polar ones (phospholipids, glycolipids, sphingolipids) – 60% and neutral ones (sterols) – 30%. The membranes also contain triglycerides and FFA (about 10%).

The cholesterol/phospholipid ratio in all biological membranes is 1. The phospholipids in a membrane are present in approximately the following proportion: PC – 30%, PEA – 28%, and SM – 25%. As already indicated, the lipid bilayer of a membrane is its basic structural unit and maintains its liquid-crystalline state. Physiological studies showed that a lipid bilayer of an erythrocyte membrane simultaneously has fluidity and order of the structure. Owing to the order of its structure, an erythrocyte carries out a number of processes occurring between the membrane and its surroundings (the reception of information, an exchange of energy and substances). Changes in the surroundings affect the composition of an erythrocyte membrane, the degree of its "adhesion" to the endothelium of the vessels changes, and the aggregative ability is disturbed.

Cholesterol and phospholipids are the main components of biomembranes. The former, responsible for 23% of the total mass of a membrane's lipids, is a very important regulator of the membrane in an erythrocyte. It acts like a shuttle transferring glucose, amino acids, and salts. Cholesterol is what ensures the difference in the physicochemical state of a membrane, and it is in two forms – free and esterified. If an erythrocyte membrane is too dense, the cholesterol "liquefies" it, as it were; conversely, if a membrane is too "liquid", the cholesterol compacts it. Consequently, the state of aggregation of the membrane of erythrocytes is ensured to a considerable extent with the aid of cholesterol. There is a constant exchange of cholesterol between the erythrocytes and blood plasma [282, 362, 378, 417].

The cholesterol of erythrocytes, like that of the plasma, relates to the readily exchanged pool of this substance in an organism.

Cholesterol is transferred between the erythrocytes and lipoproteins of the plasma in both direction. The process is not enzymatic, requires no energy, and terminates in 6–20 h. It was noted in experiments that a cholesterol molecule when incorporated into a membrane may have a liquefying effect and increase the mobility of the fatty acid chains, or a condensing one and compact them. A surplus of cholesterol in a membrane increases the microviscosity of the bilayer and lowers the rate of many reactions. The amount of membrane cholesterol depends on the activity and amount of the specific membrane receptors and the ratio in the blood plasma of the LDLP (as cholesterol donors) and HDLP (as acceptors of it).

The level of phospholipids in. the plasma allows us to judge on the cholesterol-acceptor ability of the HDLP, because the esterification of cholesterol is the duty of the phospholipids (lecithin) of the HDLP. An erythrocyte membrane is protected from a surplus of cholesterol by the oxidation and esterification of the cholesterol. The lipoprotein complexes removing the cholesterol from the membranes first esterify it with the aid of LCAT. It was noted that if phospholipids containing large amounts of polyunsaturated fatty acids are added to food, the activity of the cell enzymes and the microviscosity of the membrane of an erythrocyte are restored [419, 473, 487].

An important function in the structure of a membrane is performed by its phospholipid composition: the latter regulate the active and passive transportation of organic substances and electrolytes through a cell membrane, ensure sensitivity to the action of hormones, and regulate the activity of the enzyme systems of succinate oxidase, adenosinetriphosphatase by changing the viscosity of the enzymes, i.e. phospholipids are of a major importance for the formation of multienzyme complexes.

In an erythrocyte membrane, the phospholipid molecules are distinguished according to the degree of saturation of their fatty acid "tails". The PC and SM are in the outer layer of the membrane, while the PEA and phosphatidylserine are in the inner one.

Analysis of the phospholipid composition of various membranes performing various functions revealed that they contain the same set of lipids, but in different combinations.

An unfavorable factor in IHD and DM is the increase in the content of cholesterol in an erythrocyte membrane and the change in the cholesterol/phospholipid ratio.

A drop in the amount of polyunsaturated fatty acids and a growth in that of saturated ones is also important. Under the conditions of insulin and energy insufficiency in DM, conditions are created for the accumulation of products of FPOL, and their biological effect on the cell membranes and receptors deteriorates. This is accompanied by disturbance of the mobility of the fatty acid chains, the passive transportation of ions and nonelectrolytes is inhibited, as is active transportation. The growth in the area of an erythrocyte membrane, and an increase in its rigidity (under conditions of an increased amount of cholesterol in a membrane) detract from the deformability of the erythrocytes, disturb microcirculation, and alter the activity of the membrane enzymes. These biochemical shifts facilitate the proliferation of the smooth muscle cells of a vessel wall. The retarding of the blood flow and increase in the aggregation properties of the forming elements facilitate the development of microangiopathies [395, 501, 524, 605].

In many recent studies, changes in the structural and functional properties of the erythrocytes in DM and disorders in the enzyme composition of a membrane were noted.

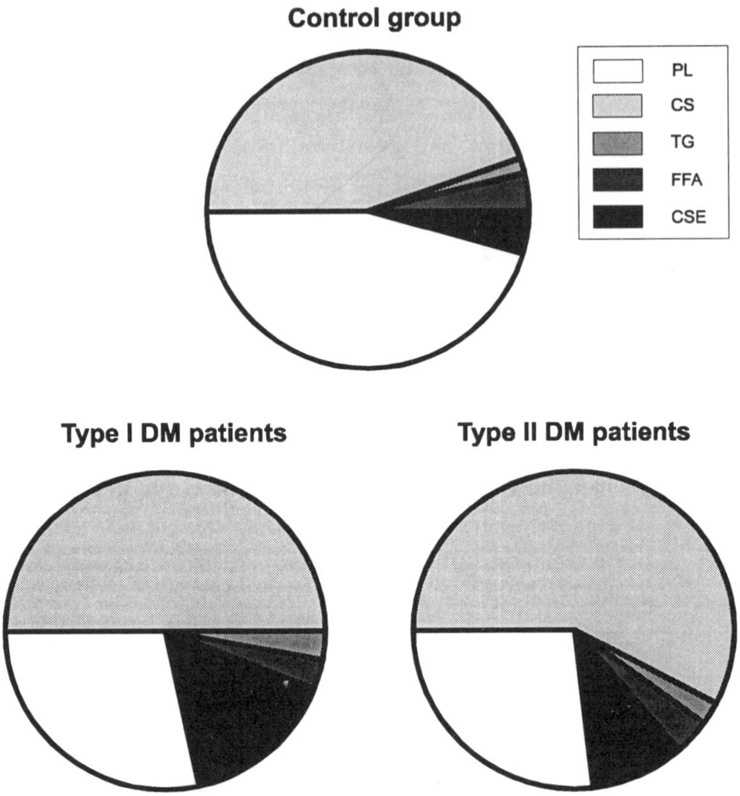

Fig. 27. Lipid Spectrum of Erythrocyte Membranes in DM Patients.

Insulin was found to participate in the metabolism of the lipid peroxides and facilitate their utilization. With insulin insufficiency (in DM), conditions are created for the accumulation of products of FPOL and their pathological effect on the cell membranes and receptors. In DM, a growth in the processes of FPOL is noted in the erythrocytes, and lowering of the activity of the antioxidant protection. Consequently, in DM, a change in the fractional composition of the lipids and their fatty acids occurs in an erythrocyte membrane. These shifts facilitate disturbance of the rheological properties of the erythrocytes and, in the long run, the development and progressing of diabetic microangiopathies.

Examination of Figs. 27 and 28 shows that the lipid and phospholipid spectra of erythrocyte membranes in type I DM patients are characterized by a growth in the content of triglycerides to 3.8% (1.23% in the control group), of fatty acids to 4.34%, (2.85%), of esterified cholesterol to 15.30% (4.87%), of the cholesterol/phospholipid ratio of 2.04

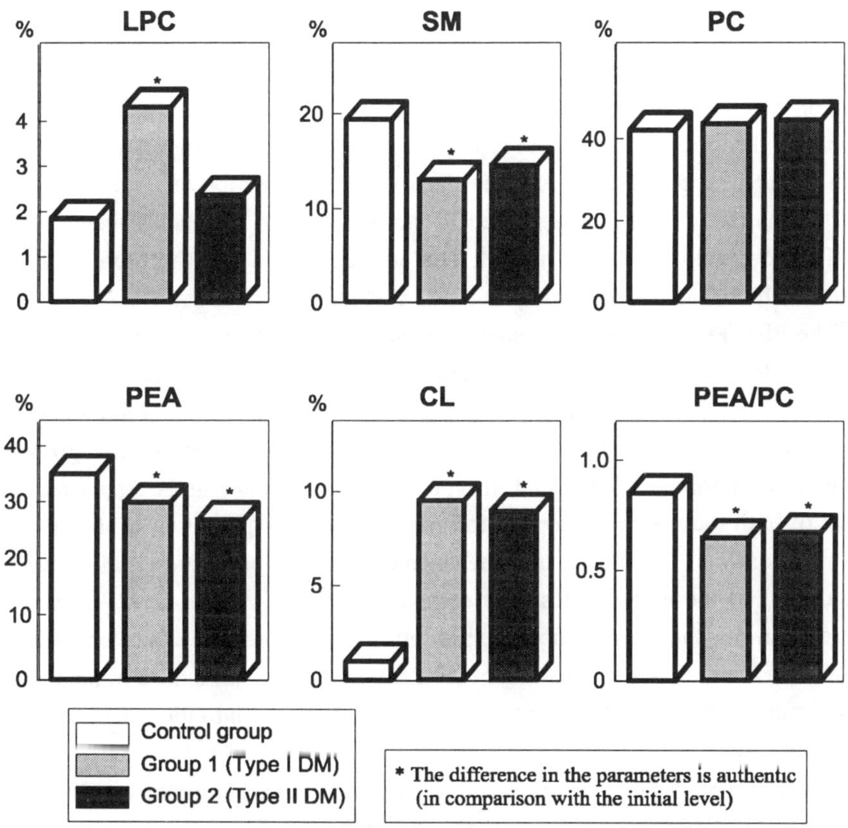

Fig. 28. Phospholipid Spectrum of Erythrocyte Membranes in DM Patients.

(1.08), and of the content of phospholipids to 30.5% (46.2%). This indicates a growth in the density of the lipid packing of an erythrocyte membrane, lowering of its deformability, and a growth in the aggregation properties of an erythrocyte. The proportion of the phospholipid fractions changes substantially: the fraction of the SM is lowered to 13.01% (the normal value is 19.6%), which possibly reflects disorder of the receptor sensitivity of a membrane, while the PC level is increased to 4.24% (1.79% in the control group).

The changes in the lipid spectrum of an erythrocyte membrane in type II DM patients was characterized by an authentic increase in the cholesterol fraction to 55.8% (44.6% in the control group) and in the esterified cholesterol to 9.84% (4.87%), lowering of the phospholipid fraction to 29.1% (46.2%), and a growth in the cholesterol/phospholipid ratio. Moreover, a drop in the fraction of SM to 14.3% (19.7%), PEA to 26.8% (34.9%), and a drop in the PEA/PC ratio were noted.

Consequently, the shifts we discovered in the lipid and phospholipid spectra of the blood and erythrocyte membranes in type I and II DM patients differed authentically from those in the control group of persons in their identical direction and standing out of the changes in the parameters being studied in the direction of their atherogenic nature, but they did not depend on the type of DM. The only exception was the content of LPC in a membrane, which was increased only in type I DM patients.

The PEA/PC ratio, with the same authentically low value in type I and II DM patients, characterizes the change in the fluidity of the lipid phase of an erythrocyte membrane. This results in an increase in the coagulating properties of the blood, disturbance of the trophism of a vessel wall, and in progressing of microangiopathy. The growth in the cholesterol/phospholipid ratio found in type I and II DM patients is associated with impairment of both the macrorheological (an increase in the aggregation of the erythrocytes, a higher viscosity of the blood) and the microrheological properties (diminishing of the deformability) of the erythrocytes. The compacting of the surface layers of a membrane (an increase in the cholesterol/phospholipid ratio) and disturbance of the physicochemical processes in a membrane (diminishing of the PEA/PC ratio) with a drop in the SM fraction in a membrane may possibly hinder the penetration of insulin into the cells of DM patients [60, 130, 180, 424].

The disorder in microcirculation and oxygenation of the tissues in DM affects the physicochemical properties of an erythrocyte membrane. The growth in the cholesterol level in the plasma increases the amount of cholesterol in a membrane. As a result of hyperglycemia and disturbance of microcirculation, the amount of cholesterol in a membrane grows. In hyperlipidemia, the permeability of an erythrocyte membrane drops, and the diffusion of oxygen is disturbed. In DM, molecules of glucose react with the residues of lisine of the membrane proteins, and the amount of glycolisines doubles the normal values. This lowers the mechanical and osmotic stability and reduces the duration of circulation of an erythrocyte.

Opinions diverge with respect to the amount of phospholipids in the erythrocyte membranes in DM patients. Some scientists note a drop in the content of all the phospholipid fractions. A growth in the amount of fatty acids in the membranes points to the activation of phospholipases, which selectively destroy the lipids of a membrane. A special activity of the phospholipases was noted in the period of decompensation of DM [40, 121, 180, 569, 594].

It is the common opinion of all clinicians and experimentators that in DM patients, the level of total cholesterol and the cholesterol/phospholipid ratio increase. These changes lower the fluidity of the lipid phase of a membrane, increase the volume of an erythrocyte, increase the rigidity, and lower the deformability of a cell. This results in thickening of the basal membrane of the endothelium of vessels, slowing down of the blood flow, and a greater aggregation of erythrocytes.

The introduction of insulin to DM patients facilitates the emergence of FPOL products from a cell membrane. These products accumulate as a result of insulin insufficiency and weakening of the antioxidant protection.

Some authors do not exclude the possibility of the participation of FPOL products in the modification of the fatty acid composition of the erythrocyte membranes in DM, especially under conditions of its decompensation.

The accumulation in an erythrocyte membrane of the metabolism products (FFA, FPOL) combined with the accumulation of cholesterol in a membrane disturb the rheological properties of the blood and lead to the development of diabetic angiopathies.

The functioning of the lipid transportation system in DM patients is disordered quite considerably; the principles of the biochemical disturbances of carbohydrate metabolism are similar to those of patients with atherosclerosis [22, 178, 576].

The results of examining two patients in whom the course of DM was made more complicated by hypertension or obesity can serve as examples.

Clinical Examples

Example 1.

Patient B. When 62 years old, she was admitted to the endocrinological department with complaints of general weakness, headaches, poor sleep, and thirst.

DM was established 15 years ago, when the patient began to lose weight, and thirst and general weakness appeared. Examination on an empty stomach revealed hyperglycemia. Her state improved after being given sugar-lowering drugs, and the glucose content in the blood became normal. During 20 years, she has been suffering from hypertension, and is constantly receiving hypotensive therapy.

Objectively: the general state is satisfactory. Height 160 cm, body weight 68 kg. Integuments of normal moistness. Vesicular respiration in the lungs. The heart tones are muffled. The pulse is rhythmic, 72 bpm. The arterial pressure is 160/80 mmHg. The stomach is soft, painless, the liver is not palpated. The general blood and urine tests are without special features. The glucose level in the blood when admitted varied from 12.0 to 14.0 mmol/l. After the increase of the maninil (glibenclamide) dose to three pills a day and selecting an adequate diet, the glucose content in the blood was normalized, and DM was compensated completely.

To determine the state of the patient's lipid transportation system, a fat load was conducted on the basis of 130 g of fat per 2 m^2.

Parameter	CS, mg/dl	TG, mg/dl	α-CS, mg/dl	apo-A-1, mg/dl	apo-B, mg/dl	IRI, μU/ml
Value						
after 0 h	224	142	37	121	106	9.1
after 3 h	207	311	35	121	142	15.8
after 9 h	201	359	32	121	157	16.8

Example 2.

Patient K.

When 56 years old, she was admitted to the endocrinological department with complaints of general weakness, fatigue, thirst, sweating, irritation.

She has been suffering from DM 10 years. The disease was incurred after influenza: thirst appeared, as well as general weakness. During several years, DM was compensated by a strict diet, and then by taking peroral sugar-lowering drugs.

Objectively: the general state is satisfactory, increased nutrition. Height 168 cm, body weight 93 kg. The obesity is proportional. Integuments of normal moistness. Vesicular respiration in the lungs, the heart tones are clear and muffled. The arterial pressure is 140/85 mmHg. The stomach is soft, painless, the liver is not palpated.

The general blood and urine tests are without special features. When taking two pills of maninil (glibenclamide) a day and lowering the caloricity of the food to 1200 kcal a day, the glucose content in the blood became 6.0 mmol/l on an empty stomach and up to 9.0 mmol/1 during the day, which indicates compensation of the DM. The patient was also subjected to a fat load.

Parameter	CS, mg/dl	TG, mg/dl	α-CS, mg/dl	apo-A-1, mg/dl	apo-B, mg/dl	IRI, μU/ml
Value						
after 0 h	302	319	38	130	220	12.0
after 3 h	292	426	36	122	261	15.3
after 9 h	331	716	33	145	308	14.8

The results indicate that both patients have been suffering from type II DM over 10 years. Their age is almost identical. The background parameters of the lipids and the hormonal profile of these patients is different. The cholesterol level in patient B. is 224 mg/dl, and in K. it was 302 mg/dl. A considerable difference was noted in the background values of TG: in B. it was 142 and in K. – 319 mg/dl. The level of HDLP cholesterol in both patients is the same – 37 and 38 mg/dl. The apo-A-1 content in B. was 121 and in K. – 130 mg/dl. A very significant difference was noted in the parameter apo-B. It is 106 mg/dl in B., while it is increased to 220 mg/dl in K. The insulin level in these patients also varies; it is 9.1 μU/ml in B. and 12.0 in K.

Why are there substantial distinctions in the background parameters of these type II DM patients who took the same drug (maninil) with duration of the disease of 10 and 15 years?

We consider that the pronounced disorders in lipid metabolism in patient K. are due to three factors: disorder in the function of the liver, hyperinsulinemia, and hypercortisolemia. Moreover, she suffers from obesity, which facilitates the development of dyslipoproteinemia.

After administering a lipid load, our assumptions were confirmed. It follows from the above data that the reaction to a lipid load differed. The cholesterol level in both cases did not change substantially. The dynamics of the triglyceride content was more pronounced in K. If in B. after three hours the triglyceride level increased to 311, and after nine hours to 359 mg/dl, in K. the background value was 319, in three hours it was 426, and in nine hours it was 716 mg/dl.

Such a dynamics of the triglyceride level confirms the assumption that the triglycerides in the chylomicrons are subjected insufficiently to lipoproteinlipolysis. As a result, the triglyceride level after nine hours increased to 716 mg/dl; here the dynamics of the apo-B content is very interesting. In B., the background level of apo-B is 106 in three hours it was 142, and in nine hours it was 157 mg/dl. In K., the dynamics was more pronounced: 220, 261, and 308 mg/dl.

The apo-A-1 level is also important. Its background level in B. was lowered somewhat – 121, and in K. it was 130 mg/dl. In B. the apo-A-1 content did not change under the influence of a lipid load, whereas in K. after three hours it dropped somewhat to 122 mg/dl, and then grew to 145. In other words, the protective function of apo-A-1 was stronger in K., and the atherogenesis process tends to normalization, as it were.

The insulin level after three hours increased identically in both patients to 15.3, while after nine hours it was 16.8 in B. and 14.8 μU/ml in K.

Considering the growth in the content of triglycerides, apo-B, the background hyperinsulinemia, and the presence of obesity in K., we can state that hyperinsulinemia combined with obesity lowers lipoproteinlipolysis (with an increase in the triglyceride level after a lipid load) and accordingly leads to progressing of the atherosclerotic changes.

Great significance with HTG in obesity patients is attached by clinicians to damage of a vessel wall (stimulation of thrombus formation).

Apolipoproteins play an important role in lipid metabolism and ensure three very important biochemical reactions: (i) they help, by reacting with phospholipids, to solubilize the esters of cholesterol and triglycerides, (ii) they regulate the biochemical reactions of lipoproteins with enzymes such as LCAT, LPL, and liver lipase, and (iii) they determine the sites of capture and the rate of degradation of lipoproteins by binding with cholesterol receptors on the surface of cells.

We attach special significance to the growth in the apo-B level in K. as an indication of the disintegration of the regulatory metabolic processes in lipid metabolism.

The pathogenesis of this phenomenon can be explained by disturbance of the process of VLDLP decomposition. As noted, VLDLP in DM contains more cholesterol than is normal. This ensures their increased atherogenic properties. LDLP contain an increased amount of triglycerides. The atherogenic effect of LDLP in DM is due to the circumstance that they are bound by glycosylized collagenic fibers at the places of damage to the endothelium. This leads to disturbance of an aggregation of thrombocytes, the formation of an atheromatous plaque, and disturbance of the prostacyclin/thromboxan ratio in the direction of an increase In the thromboxan level.

A clearly noted difference with a fat load is noted in the content of apo-B: if in healthy persons its background value of 153 mg/dl does not virtually change under the influence of a lipid load, in patient K. with type II DM the background value of 220 mg/dl grows after nine hours to 308. Such a growth in the apo-B level may also affect the formation of an atherosclerotic process and lesion of the endothelial membrane,

Disorder of the lipid metabolism in patient K. is also due to the presence of obesity and insulin resistance. It is exactly the latter that most often appears in DM patients of type II.

When studying the pathophysiological mechanisms of the disorders appearing in DM, most authors consider that insufficiency in the supply of energy to the erythrocytes is the most important link. Under conditions of a deficiency of insulin, it weakens the antioxidant systems and steps up the FPOL processes. This lowers the content in erythrocytes of the most readily oxidized phospholipid fractions and increases the content of the PC and SM fractions difficult to oxidize. These biochemical processes increase the renewal and exchange of lipids between the erythrocytes and plasma. A result is a higher content of phospholipids in the plasma and a lower one in the erythrocytes.

Studying of the lipid composition of an erythrocyte membrane points to the important role of the lipids in its functioning in vascular lesions. In view of the above, correction of the disorders in the lipid spectrum of the blood plasma and erythrocyte membranes is currently an urgent matter.

1.6 Conclusion

Consequently, dyslipidemia and the progressing of atherosclerotic changes are links of one chain. The development of an atherosclerotic process in a person's organism depends on the relation between the intensity of two nonenzymatic processes, namely, the transportation of cholesterol esters in the composition of HDLP into the liver (for the excretion of cholesterol from the organism) and the getting of the esters into an endothelium of a vessel wall. The prevailing of one of the processes determines the degree of activity of atherosclerosis development.

To determine the physiological significance of lipid metabolism, it is necessary to conduct a lipid load. It allows us to appraise the physiological mechanisms and predict the biochemical relations between the various links of lipid metabolism.

We noted that in a lipid load, the content of triglycerides, LDLP and VLDLP change the most clearly. Of special significance in the progress of atherosclerosis is the increase in the apo-B level under the influence of a lipid load. If this dynamics is combined with lowering of the level of apo-A-1, from the prognosis viewpoint we consider this to be an unfavorable symptom of atherosclerotic lesion.

We attach special significance under the conditions of a lipid load to the hormonal regulating factors, especially to a growth in the insulin level. The presence of hyperinsulinemia combined with dyslipidemia is a negative factor witnessing the progress of IHD.

The deep studying of the risk factors of IHD by clinicians, morphologists, and biochemists during recent years brought to light one of the important disorders of metabolism in the pathogenesis of atherosclerosis, namely, syndrome X – a combination of hyperinsulinemia and insulin resistance. Syndrome X is a pathogenetically integrated disorder of metabolism including hyperinsulinemia, DM, and obesity, on the one hand, and disturbance of the tolerance to glucose and hypertension on the other. Degradation of the hormonal-mediator units of metabolism in syndrome X leads to disorder of the physiological regulator mechanisms and to the progressing of IHD.

An important role in the development of syndrome X is played by insulin resistance, which is accompanied not only by hyperinsulinemia, but also by disturbance of the tolerance to glucose, and basal hyperglycemia. The syndrome of insulin resistance forms in an integrative way, including both genetically determined defects of metabolism (disorder of both carbohydrate and lipid metabolism) and acquired pathological states – obesity and hyperinsulinemia.

When studying the biochemical fundamentals of formation of syndrome X, most authors give major attention to glyconeogenesis, which is one of the chief adaptive mechanisms of carbohydrate homeostasis and a very important factor in supplying the energy cycle in an organism with glucose. We agree with the assumption that glyconeogenesis is one of the main regulatory contrivances in the integration of the syndrome X. It facilitates the formation of hyperglycemia and hyperinsulinemia.

When analyzing the hormonal regulation of syndrome X, we arrived at the conclusion that this disease is attended by complete degradation of the regulatory hormonal processes with disturbance of the physiological relation between the content of insulin, STH, and catecholamines. Hypertrophy of the myocardium appearing as a result of activation of the synthesis of nucleic acids and proteins is an important clinical symptom of syndrome X. The clinical manifestation of hypertrophy of the myocardium in syndrome X is its hyperfunction. The increased function of the myocardium in syndrome X is the cause of the increased mobilization of the energy and structural reserves in the myofibrills. This is attended by a growth in their lesion. Ischemia of the myocardium leads to disturbance of the functioning of the membrane phospholipases. Electrical instability of the lipid bilayer of the membranes appears, and they lose their barrier functions. All these morphobiochemical changes in the heart in syndrome X lead to the development of pathological processes in the myocardium that grow progressively and manifest themselves clinically in the form of IHD and cardiac insufficiency.

Disturbance of fat metabolism is one of the important pathogenetic factors of IHD development. The key factor in the metabolic disturbances is an increase in the glucose level that is accompanied by inhibition of lipolysis of adipose tissue. These relations stand out especially clearly in DM. If in DM glucose is not assimilated in a cell, the major part of the fatty acids is removed into the blood via the extracellular space. DM is attended by distortion of carbohydrate-lipid metabolism: lowering of the utilization of glucose by the muscles is attended by switching over of energy metabolism to the use of free fatty acids. The freed glucose accumulates in the fat cells, transforming into triglycerides and disturbing lipid metabolism. An important part in these metabolism disturbances is played by insulin. On the one hand, it stimulates lipogenesis and, accordingly, facilitates the accumulation of cholesterol in such vessels, while on the other, it inhibits the lipolysis processes and lowers the mobilization of lipids from the walls of vessels.

Clinicians and morphologists proved in numerous investigations that the main role in atherogenesis is played by the reaction of lipoproteins with the smooth muscle cells of an arterial wall. It is cholesterol (unesterified) that affects the macromolecular structure, permeability, and electrical potential of a membrane. Great importance in the development of atherosclerosis of the coronary vessels is attached to activation of the peroxide oxidation of lipids in the cells and tissues of an organism that occurs in the development of DM. The key link in disturbance of the peroxide oxidation of lipids in IHD and DM patients is the "lipid triad" that integrates a combination of functioning of carbohydrate and lipid metabolism. The disturbance of various relations (biochemical) in the lipid triad causes lesion of the cardiac myocyte membranes because of potentiation of the catecholamines and activation of lipases and phospholipases.

Maintaining of the balance between the inflow into the cells of vessels of cholesterol in the composition of the LDLP (especially of cholesterol esters) and its outflow from the peripheral cells into the liver with the acid of HDLP is a significant principle underlying the metabolic changes in atherosclerosis. We noted in our studies that in IHD and DM patients the ability of the HDLP to accept cholesterol from the smooth muscle cells of the aorta is drastically lowered. The value of the ratio apo-B/apo-A-1 is the laboratory biochemical criterion of the atherogenic state in IHD and DM patients. We noted a considerable growth in this ratio for patients with coronarographic registration of IHD.

In studying the regulatory mechanisms of lipid metabolism in IHD patients, we concluded that all the disorders in lipid metabolism in these patients are due to the hormonal disintegration of the regulatory adaptive mechanisms, and the role of insulin is especially important. It is exactly insulin, being absorbed by adipose tissue, and also regulating the pentose and glycolytic cycles of glucose exchange, that stimulates in IHD patients the increased synthesis of fatty acids and triglycerides from the products of carbohydrate metabolism.

In DM, the incidence of insulin insufficiency entails disturbance of the assimilation of glucose, an increase in the FFA level in the blood, and the development of dyslipidemia.

In DM, the development of angiopathies and IHD occurs on the background of an integrative combination of risk factors such as hyperlipoproteinemia, arterial hypertension, increased blood coagulation, and hyperglycemia. An important role in the regulation of lipid metabolism is played by LPL, whose level in DM (on the background of a deficiency of insulin) is lowered considerably. A deficiency of LPL in the blood is attended by a growth in the level of lipoproteins rich in triglycerides (diabetic lipemia).

Morphological studies of the vessels in DM patients allowed authors to conclude that in these patients, the transformation of the smooth muscle cells or macrophages is the key factor in the formation of an atheromatous plaque. Biochemical studies here showed the reaction of lipid metabolism units such as an increased level of cholesterol, LDLP, VLDLP, diminishing of the level of HDLP, and a greater content of apo-B in the blood.

Lesion of the vessels in DM of both type I and II is manifest in the form of diabetic micro- and macroangiopathies. Most authors find that the crossed binding of the plasma proteins to the matrix proteins of vessel walls by means of the products of pronounced glycosylation plays a role in the pathogenesis of these lesions.

The deficiency of insulin in DM is attended by degradation of the level of cyclic nucleotides (cAMP and cGMP) that play an important regulatory function in the metabolism of lipoproteins. We proved that in DM patients, the function of the erythrocytes is disturbed because of the appearance in their membranes of a surplus of cholesterol. It is this state that increases the microviscosity of a membrane bilayer and lowers the rate of many enzymatic reactions. Insulin insufficiency in DM favors the accumulation of the products of peroxide oxidation of the lipids in the membranes of erythrocytes. This causes disturbance of the mobility of the fatty acid chains, inhibition of the passive transportation of ions and nonelectrolytes, and also disturbance of the receptor sensitivity of the erythrocytes to HDLP. Lowering of the protection processes in the blood cells of DM patients is an important pathogenetic factor in the development and progressing of diabetic microangiopathies.

CHAPTER 2.
DISORDERS IN THE HEMOSTASIS SYSTEM AND CHANGES IN THE RHEOLOGICAL PROPERTIES OF THE BLOOD IN ISCHEMIC HEART DISEASE AND DIABETES MELLITUS PATIENTS

2.1. Hemostasis System and Rheological Properties of the Blood in Healthy Persons

2.1.1. Vessel Wall as a Regulator of the Functional State of the Hemostasis System

Hemostasis is an evolutionally developed and genetically determined adaptation system of an organism. It includes an integrated unity of the various physiological constants of an organism that ensure the functions of the organism under conditions of stress.

One of the important components of preserving the constancy of the internal medium is maintaining of the intactness of the wall of vessels and the liquid state of the blood in a vascular channel. Clinical and experimental studies of recent years showed the presence of functional reaction of the coagulating system of the blood with both the fibrinolytic and kinin systems of an organism [42, 158, 163, 322, 411, 454]. An important stage is the determination of the role of prostaglandins, their cyclic endoperoxides, thromboxans (synthesized by thrombocytes) in the aggregation of thrombocytes and studying of the functions of prostacyclin (PrC) that inhibits the process of aggregation of thrombocytes [105, 216, 219, 528, 529].

In a healthy organism, all the hemostasis links are in dynamic equilibrium and, like other biological systems, have a high elasticity and reliability with various changes in the external and internal medium of an organism. The hemostasis system not only adapts itself successfully to many physiological and pathological effects, but within broad limits corrects disorders of the vital activities of an organism, namely, it maintains intactness and a closed state of the vascular system, a constant volume, liquid state, and a number of other properties of the circulating blood. All the shifts of a protective nature in the functional state of the hemostasis system appearing when various factors act on a healthy organism are usually rapid, therefore a state of relative constancy is restored rapidly.

The regulation of the hemostasis system is a closed dynamic organization of central and peripheral mechanisms. The biochemical processes occur according to the self-regulation principle when each level of the system has its own functional role.

Table 11. Vascular Factors of Hemostasis System.

Factor	Function
Tissue thromboplastin	Participates in the formation of prothrombinase by an "external" mechanism
Factor VIII (FW)	Participates in the formation of prothrombinase (by an "internal" mechanism). Willebrand's factor (FW) participates in the adhesion and aggregation of thrombocytes
Tissue fibrin-stabilizing factor	Transforms soluble fibrin S into insoluble fibrin I
Collagen	Activates factor XII, induces thrombocyte aggregation
ADP	Induces thrombocyte aggregation
Thromboxan A_2, formed in the activation of thromboxan-synthetase of the endothelial cells	Induces thrombocyte aggregation, is a powerful vasoconstrictor
Anticoagulants:	
antithrombin III	Inhibits thrombin, factors Xa, IXa, XIIa, kallikrein
alpha-2-macroglobulin	Inhibits thrombin, kallikrein
heparin sulphate	Inhibits thrombin
protein C, protein S	Inactivates factors Va, VIIIa, increases fibrinolytic activity of the blood
thrombomodulin	Is a cofactor of the activation of protein C by thrombin
Prostacyclin, formed in the activity of prostacyclin-synthetase	Inhibits the aggregation of thrombocytes, is a vasodilator
Activator of plasminogen	Activates the fibrinolytic system (plasminogen)
Inhibitor of plasminogen activator	Inhibits the effect of the plasminogen activator

The blood vessels are not only a structural component of the cardiovascular system, but also one of the links of the hemostasis system [97, 98, 202, 410]. For the blood to be in its customary state, intactness of the vessel endothelium is required. This plays the main role in maintaining "blood-tissue" hemostatic balance and performs the following functions:

(1) active transportation of metabolic substances between the blood and tissues;

(2) the formation of a relative barrier for the blood macromolecules;

(3) the synthesis of mediators regulating the reaction between a vessel wall and the blood and ensuring hemostatic homeostasis;

(4) removal from the blood of highly active substances and complement (the phagocytaric function):

(5) maintaing the thromboresistance of the vessel walls (antithrombogenic properties);

(6) regulation of the tonus of the vessels;

(7) participation in immune reactions.

A vessel endothelium has an involved balanced system of synthesizing the factors ensuring hemostatic homeostasis (Table 11). It prevents the entry into the blood of thromboplastin, activation of the factor XII (contact factor), and supplies natural anticoagulants (antithrombin III, a plasminogen activator) into the blood flow. Prostacyclin and the enzyme apirase (adenosinediphosphatase) preventing thromboaggregation are synthesized in the endotheliocytes. The glycosaminoglycanes contained in the endothelium of a vessel have a negative surface charge identical to that of the blood cells, which ensures its intactness [67, 150, 406, 413, 498, 610].

However, the absence of adhesion of circulating thrombocytes on a normal endothelium cannot be explained only by the negative charge of the intima of the vessels.

Normally neither adhesion nor aggregation of thrombocytes occurs on the endothelial lining of vessels because of the liberation of PrC from the endothelium [215, 372, 607]. The main place of its synthesis is the wall of a vessel, especially the endothelium. The intima, which forms only 5% of the mass of the arterial tissue, produces about 40% of the total amount of PrC [405]. The synthesis of PrC in a wall of a vein in humans is higher than in arteries, and in this connection moat likely sclerotic lesions of the veins do not appear in them [533]. Apart from this, the lungs are a generator of PrC, whence it is discharged into the blood [217, 218]. Great significance is attached to the local synthesis and secretion of PrC in the stimulation of the endothelium by definite factors [361].

The stimulation of the synthesis and secretion of PrC by the endothelium cells is the responsibility of vasoactive agents including bradikinin, thrombin, histamine, and also kallikrein, peptide derivatives of fibrin-fibrinogen, interferon, and the growth factor separated from alpha granules of activated thrombocytes [7, 13, 25, 43, 55, 135, 202, 240, 407, 603].

Prostacyclin forms from arachidonic and eicosapentaenic (series ω-3) acids, the latter having the most pronounced antiaggregation effect.

Arachidonic acid is freed from the phospholipids of the membrane of endothelial cells under the influence of stimulating agents and phospholipase. The enzyme cyclooxygenase acts on the acid to form the endoperoxides of prostaglandins G_2 and H_2. In the next stage, PrC forms under the influence of the enzyme prostacyclin-synthetase and enters the blood.

PrC can also be synthesized from the prostaglandins G_2 and H_2 that form in thrombocytes when they are activated. Part of them under the influence of thromboxan-synthetase transform into thromboxan A_2, while part is secreted, is captured by the endothelium, and again under the influence of prostacyclin-synthetase PrC forms from them.

Prostacyclin is a short-living physiological substance with a half-life of about three minutes. The stable product of its decomposition was identified as 6-keto-PGF_1-α [52].

Thrombin stimulates the synthesis of PrC by direct influence on the endothelium and activation of prostacyclin-synthetase and indirect by activation of the thrombocytes with the subsequent formation of endoperoxides of prostaglandins (PGH_2 and PGG_2), aggregation of the thrombocytes, and the freeing of biologically active substances.

Prostacyclin is not inactivated in the lungs, which is of a fundamental significance in maintaining hemostatic homeostasis.

The main effect of PrC is inhibition of the aggregation of thrombocytes, and to a smaller extent it suppresses adhesion. It also has a vasodilating effect, weakens the smooth muscles of vessels, including the coronary arteries, causing dilatation, lowers the arterial pressure, and has antiarrhythmic, antisclerotic, and antiulcerogenic properties.

The antiaggregation effect of PrC is associated with its ability to increase the effect of the adenylatecyclase of the thrombocytes leading to an increase in the synthesis of cAMP and a decrease in the ability of the thrombocytes to aggregation.

Under the influence of various inductors, PrC is secreted from an endothelium and ensures the effect of the mechanism of limiting the formation of a thrombus at a site of injury of a vessel. The PrC binds to the specific receptors of thrombocytes, this resulting in activation of adenylatecyclase, an increase in the synthesis of cytoplasmatic cAMP, and lowering of the level of cytoplasmatic calcium. All this facilitates the preservation of the discoid shape of the thrombocytes, which makes them less active when the cells interact.

PrC deactivates thrombocytes, preventing the too intense aggregation of them outside a site of injury to a vessel and the intravascular aggregation of circulating thrombocytes [104, 233].

The required concentrations of the PrC on the surface of the endothelial calls, its constant decomposition and transformation into an inactive form do not prevent the temporary physiological adhesion of the thrombocytes to a normal endothelium; here its angiotrophic function is manifest. Every day about 15% of the circulating thrombocytes participate in this process.

Under physiological conditions, the endothelial cells regulate the local concentration of the adenosine derivatives having thrombocyte-activating properties owing to the effect of ectonucleotidases.

The ATP secreted by the endothelium and the blood cells is catabolized to adenosine, which is a powerful inhibitor of thrombocyte activation.

Other vasoactive substances are also catabolized on the surface of the endothelium [498]. Normally, the participation of endothelial cells in the catabolism of nucleotides and vasoactive substances greatly exceeds that in the catabolism of the formed elements of the blood.

The biologically active substance thrombomodulin, which is a membrane glycoprotein, when binding with thrombin, inhibits the process of conversion of the factor VIII:C into the active form VIII:Ca, the process of activation of the factor V, and also the transformation of fibrinogen into fibrin and the activation of the thrombocytes occurring under the influence of thrombin [146, 147].

The vascular endothelium actively binds the thrombin formed in a process of increasing the functioning of the coagulating system of the blood and the active factors Xa, IXa, XIa, Va, VIIIa, and kallikrein.

The thrombin-binding activity of the endothelium is due to antithrombin III, heparin, glycocalyx, α-2-macroglobulin, protein C and protein S.

The antithrombogenic properties of a vessel wall are ensured by the following factors:

(1) a negative charge of the internal surface of an endothelium;
(2) glycocalyx,
(3) prostacyclin:
(4) anticoagulants: antithrombin III, α-2-macroglobulin, protein C, protein S;
(5) thrombomodulin;
(6) ATPase;
(7) activator of plasminogen;
(8) intraendothelial fibrin layer,

Antithrombin III is the most powerful inhibitor of blood coagulation, it is synthesized by the endothelium and hepatocytes, is on the surface of the endothelial cells and in the blood [78], and is responsible for 75–90% of the spontaneous anticoagulant activity of the blood. It neutralizes the enzymatic activity of many blood coagulation factors, and also of plasmin [348, 349].

The anticoagulating effect of antithrombin III is due to its ability to form equimolecular stable "enzyme-inhibitor" complexes, which inactivates thrombin [348, 349].

The anticoagulating effect of heparin is aimed at inactivating thrombin, which is manifest only in the presence of antithrombin III. Heparin plays the role of a catalyst and accelerates the reaction of antithrombin III with thrombin from 2000 to 3000 times.

The anticoagulant α-2-macroglobulin, localized on the surface of the endothelium, inhibits thrombin and kallikrein [228] and thus participates in maintaining the thromboresistance of vessels.

The anticoagulant activity of the endothelium is due to the effect of the heparin and dermatan sulfates synthesized in it [202]. The latter considerably potentiate the reaction of thrombin inactivation and act as cofactors of antithrombin III [564]. The endothelial cells can also bind exogenic heparin on their surface.

Anticoagulant activity is also associated with protein C and protein S that inhibit factors V and VIII. The rate of inhibition of the factors of coagulation by protein C is regulated by its cofactor – protein S [146, 147]. Protein C is not only an anticoagulant, it raises the fibrinolytic activity by increasing the level of the circulating vessel activator of plasminogen, thus ensuring lysis of a clot, but not causing fibrinogenolysis [61, 93, 94, 207, 208, 209, 514, 560].

The activator of plasminogen is also synthesized in the endothelium of the blood vessels. But there is also a tissue activator of plasminogen. It is secreted from the endothelium under the influence of various vasoactive substances (adrenaline, noradrenaline, nicotinic acid), under a physical load, in occlusion of the vessels [577].

An inhibitor of the activator of plasminogen is also synthesized in the endothelial cells [352]. The basal membrane of the vessels produces glycoprotein with a molecular weight of 440,000 that is present in two forms – soluble (fibronectin monomer) and insoluble (tissue fibrillar fibronectin) [230].

The tissue fibronectin cements the main substance of the connecting tissue, and participates in the formation of intercellular contacts and in the attachment of the cellular layer to a basal membrane [241, 408]. The fibronectin, having free amino groups, forms a covalent bond with the amino groups of various substances, especially with soluble fibrin, collagen, i.e. performs an antithrombogenic function [20, 40, 140, 230, 408, 627].

It also inhibits the "thrombin-fibrinogen" reaction, the formation of a fibrin clot, and lowers its hemostatic functions; by binding to fibrillar collagen of the subendothelium, it can impart antithrombogenic functions to it.

D.Zubairov in 1966 [626] advanced the concept of the continuous functioning of the coagulating system of the blood and the continuous latent microcoagulation of fibrinogen in an organism. The formed profibrin may facilitate the formation of an external and internal endothelial profibrin layer that lyses with the simultaneous activation of fibrinolysis. The absence of adequate fibrinolysis at the level of a microcirculatory channel leads to disturbance of the function of the organs and tissues. Zubairov considers that the tissue thromboplastin entering the blood flow in small amounts participates in the continuous maintenance of the blood coagulation process. This is confirmed by the detection in the blood of blood coagulation products and thrombin.

A.Copley [97, 98] advanced the concept of the intraendothelial fibrin coat according to which this layer functions like a critical interface between the two parts of the vessel and blood organ – the blood and the vessels, and has anticoagulant and antithrombogenic properties. He considers the wall of a capillary as a surface consisting of three barriers. The first is the fibrin layer covering the endothelium lining and extending into the interendothelium and subendothelium spaces. The second barrier is the endothelium. The transendothelial canals and the plasmalemma pores in the endothelial cells play the main role in capillary permeability. The basal membrane is the third barrier. Copley called the internal endothelial fibrin layer "cement fibrin" that fills the interendothelial and subendothelial spaces [98]. The cement fibrin consists of three components, namely, (i) fibrin particles in the initial stage of polymerization, (ii) calcium-containing fibrin, and (iii) fibrin particles split by the accompanying fibrinolysis.

The fibrin layer diminishes the viscosity and has antithrombotic and anticoagulation action because of its enormous surface area covering the intima of the vessels [97, 98].

Consequently, the thromboresistance of the vessel wall is determined by a number of factors, some of which prevent the adhesion and aggregation of the thrombocytes, while others inactivate thrombin and the active factors of the blood coagulating system, thus preventing the formation of thrombocytes, and third ones activate the lysis of the forming fibrin

The maintaining of an integral, healthy vessel wall and hemostatic homeostasis is also due to procoagulant and proaggregate factors of the wells of vessels. The main ones of them are the Willebrand factor, the factors of adhesiveness and aggregation of the thrombocytes, tissue thromboplastin, the convertin-like, fibrin-stabilizing, and antiheparin factors.

The Willebrand factor is synthesized by the endothelial cells, and it is in the circulating blood combined with factor VIII: it participates in the adhesion of thrombocytes to the subendothelium and in aggregation of the thrombocytes in the presence of ristocetin. It was indicated that ADP induces the reaction of the Willebrand factor with thrombocytes.

Collagen, a basal membrane, microfibrills, and elastin are factors of the adhesiveness and aggregation of thrombocytes. They react with thrombocytes by means of the Willebrand factor when the wall of a vessel is damaged [34, 100, 177, 245, 246, 459, 499, 530, 531].

An intact endothelium has a weak procoagulant activity. The intima of vessels has the greatest, and the adventitia – the least thromboplastin activity. The synthesis of thromboplastin is sharply increased by lymphocytes and granulocytes. The walls of vessels con-

Table 12. Hemostasis System in Practically Healthy Persons.

Parameter	Average value
Fibrinogen, g/l	3.35±0.12
Tolerance of plasma to heparin, s	831.0±15.0
Duration of plasma recalcification, s	250.0±13.0
Antithrombin III, s	28.2±1.7
Fibrinolytic activity of blood, %	15.6±1.1
Index of intravascular aggregation of thrombocytes, %	1.04±0.03
Spontaneous aggregation of thrombocytes, %	0.2±0.01
Aggregation of thrombocytes induced by:	
ADP 10^{-5} M, %	60.9±2.1
ADP 10^{-6} M, %	4.1±0.15
collagen, %	58.0±1.5
thrombocyte activation factor, %	14.8±1.2
adrenaline 10^{-4} M:	
1st wave, %	13.6±1.3
2nd wave, %	59.2±2.4
aggristine, %	59.6±2.2
Beta-thromboglobulin, ng/ml	31.4±1.8
Thrombocyte factor IV, s	15.1±0.8
Thromboxan B_2 in plasma, pg/ml	32±2.9
Thromboxan B_2 in thrombocytes, pg/3·10^8 cells	281.0±28.0
cAMP in thrombocytes, pmol/3·10^8 cells	4.2±0.3
6-Keto-PGF$_1$-alpha in plasma, pg/ml	53.0±3.3
6-Keto-PGF$_1$-alpha / thromboxan B_2	1.7±0.2
Willebrand factor:	
un/ml	1.12±0.08
%	110±8.9

tain an antiheparin factor – an enzyme transferring fibrinogen into fibrin in the absence of thrombin [47, 322, 483, 626].

The wall of a vessel also contains a fibrin-stabilizing factor that participates in the formation of stable blood clots dissolving with difficulty. In the presence of calcium ions, it forms covalent bonds between two adjacent monomers of fibrin and transforms it into insoluble fibrin [41, 322, 323].

Consequently, a vessel wall contains a large number of factors participating in the coagulation of blood, fibrinolysis, adhesion and aggregation of thrombocytes, and the formation of a hemostatic thrombus. A vessel wall is an efferent regulator of hemostatic homeostasis keeping the blood in the liquid state in the vessel channel and ensuring reliable hemostasis when vessels are injured.

In connection with the establishment of the significance of the vessel wells as a regulator of the functional state of the hemostasis system, investigators were confronted with the problem of how to determine the antithrombogenic (fibrinolytic, antiaggregation and anticoagulation) activity of a vessel wall.

However, determination of the state of the vascular link of the hemostasis system turned out to be the most difficult procedure in hemostasiology from the methodical standpoint, because direct studies are associated with the obtaining of a bioptic material of the vessel walls of a patient being examined, which does not allow this method to be used widely.

Oivin *et al.* [433] showed that brief occlusion of forearm vessels caused by applying the cuff of a sphygmomanometer to the forearm causes an activator of plasminogen to be freed from a vessel wall and enter the blood. This raises the fibrinolytic activity of the blood. On the basis of these studies, Oivin *et al.* developed a way of determining it.

Published information and our own studies which showed that the brief occlusion of vessels of healthy persons leads to the freeing and delivery into the blood of an antiaggregation substance and antithrombin III underlied the ways of diagnosis of the antiaggregation and anticoagulation activity of vessel walls we developed.

To study the antithrombogenic properties of vessel walls, we studied 168 healthy persons. Information on the state of the procoagulant, anticoagulant, and fibrinolytic activity of the thrombocytes is presented in Table 12.

Thus, the level of fibrinogen, the tolerance of the plasma to heparin, and the siliconized duration of plasma recalcification were within the limits of the physiological norm described in the literature,

The activity of antithrombin III fluctuated from 20.1 to 37.2 s (the norm is 25–35 s), while the fibrinolytic activity of the blood fluctuated within 13–22% (the norm is 15–17%). The intravascular aggregation of thrombocytes was absent in most of the healthy persons, as was the spontaneous aggregation of thrombocytes.

The aggregation of thrombocytes induced by ADP 10^{-5} M was within 45 to 72%, by ADP 10^{-6} M – within 2.1 to 6.1%; the collagen-induced aggregation of thrombocytes was 44–74%, while in three of the tested persons it exceeded 80%; the adrenaline-induced aggregation of thrombocytes (the second wave) was 55–70%.

The content of thrombocyte proteins in the blood plasma reflecting the state of the intravascular activation of the thrombocytes (beta-thromboglobulin) was 31.4±1.8 ng/ml, the thrombocyte factor IV was 15.1±0.8 s.

The content of PrC and thromboxan, the main regulators of the functional activity of the thrombocytes, determined from the level of their stable metabolites, was 53.0±3.3 and 32.0±2.9 pg/ml, respectively.

Table 13. Antithrombogenic Activity of Vessel Walls in Practically Healthy Persons.

Parameter	Average value
Antiaggregation activity of vessel wall	
Aggregation of thrombocytes induced by ADP 10^{-5} M, %:	
before compression	60.9±2.4
after compression	58.1±2.0
index	0.95±0.1
Aggregation of thrombocytes induced by ADP 10^{-6} M, %:	
before compression	4.1±0.18
after compression	4.0±0.19
index	1.0±0.11
Aggregation of thrombocytes induced by collagen, %:	
before compression	58.0±1.5
after compression	55.0±1.6
index	0.92±0.06
Aggregation of thrombocytes induced by TAF, %:	
before compression	14.8±1.1
after compression	13.9±1.0
index	0.94±0.08
Aggregation of thrombocytes induced by adrenaline 10^{-4} M,%:	
before compression	59.2±2.2
after compression	63.0±2.3
index	1.1±0.09
Thrombocyte factor IV, s:	
before compression	15.1±0.7
after compression	14.6±0.9
index	0.98±0.07
6-Keto-PGF_1-alpha in plasma, pg/ml:	
before compression	53.0±3.8
after compression	57.1±4.0
index	1.08±0.09
Anticoagulant activity of vessel wall	
Activity of antithrombin III, s	
before compression	28.2±1.5
after compression	29.2±1.6
index	1.03±0.09
Fibrinolytic activity of vessel wall	
Fibrinolytic activity of blood, %:	
before compression	15.6±1.1
after compression	17.9±1.3
index	1.2±0.11

* The index is determined by dividing the parameter being studied after a cuff test by its value before compression.

An important parameter showing the antiaggregation activity of the thrombocytes is the 6-keto-PGF_1-α/thromboxan B_2 ratio, which in the studied persons averaged 1.7 with individual fluctuations from 1.2 to 1.9.

Table 14. Index of Antithrombogenic Activity of a Vessel Wall in Various Groups of Practically Healthy Persons.

Group	Activity		
	antiaggregation	anticoagulant	fibrinolytic
All practically healthy persons	0.95±0.07	1.03±0.07	1.2±0.06
1st – with functionally active antithrombogenic properties af a vessel wall	0.8±0.06	1.31±0.07	1.61±0.08
2nd – with lowering of one component of the antithrombogenic activity of a vessel wall:			
- antiaggregation	1.23±0.05*	1.1±0.07	1.19±0.08
- anticoagulant	0.79±0.04	0.95±0.06*	1.13±0.09
- fibrinolytic	0.83±0.05	1.07±0.08	0.8±0.08*
3rd – with lowering of two components of the antithrombogenic activity of a vessel wall:			
- antiaggregation and anticoagulant	1.17±0.05*	0.96±0.06*	1.25±0.08
- antiaggregation and fibrinolytic	1.21±0.06*	1.04±0.07	0.9±0.05*
- anticoagulant and fibrinolytic	0.79±0.07	0.87±0.07*	0.78±0.05*

* The difference of the parameters is authentic (in comparison with the level in the 1st group).

The cAMP content in the thrombocytes averaged 4.2±0.3 pmol per $3 \cdot 10^8$ cells and fluctuated from 2.6 to 5.6 pmol per $3 \cdot 10^8$ cells.

The content of the Willebrand factor playing a substantial role in the adhesion of the thrombocytes to the subendothelium averaged 1.12±0.08 un/ml.

Table 13 presents data on the state of the antithrombic activity of a vessel wall. The indexes of antiaggregation and anticoagulation activity were within unity, while that of the fibrinolytic activity averaged 1.2.

At first sight, analysis of the average statistical data obtained with the aid of a cuff test showed a good physiological reaction of a vessel wall directed at maintaining hemostatic homeostasis under extremal conditions in the entire group of healthy persons. But when analyzing the individual parameters, we found that in 52% of the studied persons, the antithrombogenic activity of a vessel wall was high: after brief occlusion of the forearm vessels, a considerable secretion of prostacyclin, antithrombin III, and the plasminogen activator was observed in this group. This corresponds completely to the physiological

Table 15. Content of Fatty Acids in Thrombocytes in Practically Healthy Persons.

Series	Fatty acids (symbol)	Content, %
ω-6	$C_{16:0}$	16.83±0.61
	$C_{16:1}$	1.14±0.06
	$C_{17:0}$	0.46±0.04
	$C_{17:1}$	2.37±0.19
	$C_{18:0}$	21.31±0.76
ω-6	$C_{18:2}$	5.77±0.29
	$C_{20:3}$	1.43±0.07
	$C_{20:4}$	20.92±0.79
	$C_{20:5}$	0.3±0.02
	$C_{22:4}$	0.51±0.05
ω-3	$C_{18:3}$	1.64±0.14
	$C_{20:5}$	0.35±0.04
	$C_{22:5}$	1.60±0.09
	$C_{22:6}$	1.73±0.05
ω-9	$C_{18:1}$	17.57±0.54
	$C_{20:1}$	0.98±0.16
	$C_{20:3}$	0.88±0.14

state of the antithrombogenic properties of a vessel wall ensuring stable hemostatic homeostasis of an organism.

In 48% of the examined persons, one or two components of the antithrombogenic activity of a vessel wall were altered, which was indicated either by a growth in the index of antiaggregation activity, or a drop in the index of the anticoagulant or fibrinolytic activity of a vessel well.

The results indicate that in 48% of the healthy persons, the antithrombogenic activity of a vessel wall is lowered, which may be due to its functional or structural injury. Examples are persons who smoke a large number of cigarettes and use animal fats in their food, or persons in whom a deeper examination revealed a concealed cardiovascular pathology.

Depending on the reaction to the cuff test, i.e. on the state of the antithrombogenic activity of a vessel wall, we divided the studied persons into three groups and compared these parameters for persons of the different groups with the results of clinical instrumen-

Table 16. Content of Fatty Acids in Thrombocytes in Practically Healthy Persons according to the State of Antithrombogenic Activity of Vessel Walls.

Group	Fatty acid, %				Coefficient K_{ac}
	$C_{18:2}$, ω-6	$C_{20:4}$, ω-6	$C_{20:5}$, ω-3	$C_{22:6}$, ω-3	
All practically healthy persons	5.77±0.29	20.92±0.04	0.35±0.04	1.73±0.05	10.06±0.71
1st – with functionally active antithrombogenic properties af a vessel wall	5.70±0.28	19.10±0.76	0.40±0.03	2.05±0.06	7.80±0.61
2nd – with lowering of one component of the antithrombogenic activity of a vessel wall:					
- antiaggregation	6.36±0.28	19.77±0.76	0.38±0.03	1.32±0.07*	11.63±0.7*
- anticoagulant	6.06±0.26	23.32±0.75*	0.34±0.04	1.30±0.07*	14.22±0.61*
- fibrinolytic	5.99±0.31	20.17±0.82	0.36±0.04	1.75±0.06	9.56±0.56
3rd – with lowering of two components of the antithrombogenic activity of a vessel wall:					
- antiaggregation and anticoagulant	5.79±0.4	22.21±0.74*	0.15±0.04*	1.03±0.08*	18.82±0.82*
- antiaggregation and fibrinolytic	5.76±0.36	26.10±0.8	0.39±0.04	1.55±0.07*	13.45±0.73*
- anticoagulant and fibrinolytic	6.15±0.31	26.94±0.81*	0.19±0.05*	1.59±0.07*	15.22±0.68*

* The difference of the parameters is authentic (in comparison with the level in the 1st group).

tal examination, complex investigation of the system of hemostasis, and the content of fatty acids in the lipids of the thrombocytes (Tables 14, 15, 16).

In the group of persons with lowered antiaggregation and fibrinolytic activity of a vessel wall, the indexes were on an average 1.5 times higher and 1.8 times lower than in persons with retained antithrombogenic properties of the vessel walls.

In the group of persons with a lowered anticoagulant and fibrinogenic activity of the vessel walls, the indexes were on an average 1.5–2 times lower than in persons with a pronounced antithrombogenic activity of the walls.

The thrombocytes and endothelial cells of a vessel wall have an autonomous system of synthesizing fatty acids [428] and their derivatives: prostaglandins, thromboxan, and prostacyclin, the main substrate for whose building up is arachidonic acid (Table 17).

Eicosapentaenic and docosahexaenic ($C_{20:5}$, ω-3; $C_{22:6}$, ω-3) acids inhibit the synthesis of arachidonic acid from linolic. The series ω-3 polyunsaturated fatty acids compete with arachidonic acid for the second position in the membrane phospholipids. This results

Table 17. Atherogenity and antiatherogenity of fatty acids.

Position of terminal connection	Fatty acid	Atherogenity, ++ Antiatherogenity, 00
$C_{16:0}$	Palmitic	+++
$C_{18:0}$	Stearic	+++
$C_{18:0}$, ω-9	Oleic	+++
$C_{18:2}$, ω-6	Linolic	0
$C_{20:4}$, ω-6	Arachidonic	+
$C_{18:3}$, ω-3	Alpha-linolic	00
$C_{20:5}$, ω-3	Eicosapentaenic	000
$C_{22:6}$, ω-3	Docosahexaenic	000

in diminishing of the amount of arachidonic acid in the plasma and cells. Eicosapentaenic acid has a high affinity for cyclooxygenase and competes with arachidonic acid, inhibiting its transformation into thromboxan A_2 [296, 506, 546, 575, 594]. In the endothelial cells of a vessel wall, eicosapentaenic acid does not inhibit the production of prostacyclin noticeably. When analyzing the fatty acid composition of the phospholipids of thrombocytes, special attention was given to the polyunsaturated fatty acids of the series ω-6 and ω-3 that play a key role in the formation of the functional state of the thrombocytes and a vessel wall [242, 243, 428].

Since these acids lower the synthesis or arachidonic acid and thus lower the level of thromboxan A_2 in the thrombocytes without affecting the synthesis of prostacyclin in a vessel wall, we introduced a coefficient K_{ac}:

$$\frac{C_{20:4}, \omega - 6}{C_{20:5}, \omega - 3 + C_{22:6}, \omega - 3}$$

that characterizes the competition between arachidonic acid on the one hand and eicosapentaenic and docosahexaenic acids on the other. On an average, for the entire group of studied persons, the coefficient was 10.06. Our results showed that the deeper changes in the antithrombogenic activity of a vessel wall were a result of greater disturbances of the ratio of polyunsaturated fatty acids in the thrombocytes. This pointed to predomination of the level of arachidonic acid and, accordingly, to an increase in the activity of the "arachidonic cascade" in the thrombocytes that shifted thromboxan-prostacyclin equilibrium in the direction of a greater synthesis of thromboxan. The latter creates a risk of development of the intravascular aggregation of thrombocytes and corresponds to a growth in the disturbances of the antithrombogenic activity of a vessel wall.

In the group of persons with disturbance of the antithrombogenic activity of the vessel walls, in comparison with the group having functionally active antithrombogenic properties, the number of smokers was greater, the consumption of animal fats was higher, the incidence of IHD among the parents of the examined persons and the number of subjective complaints, changes revealed on an ECG in a veloergometric test, echocardiography, and zonal rheography were considerably higher.

In the group of persons with disturbances of the antithrombogenic activity of a vessel wall, complex clinical instrumental examination revealed a pathology in 92% of the cases. Consequently, to interpret correctly the physiological, biochemical, and functional parameters characterizing the state of the organs and systems in various diseases and compare them with the physiological norm for practically healthy persons, it was necessary to select the persons for the control group on the basis of data obtained by means of clinical instrumental and laboratory methods of examination making it possible to reveal persons without a latent pathology and the presence of a predisease.

2.1.2. Rheological Properties of the Blood in Healthy Persons

The process of motion of a liquid is shear of the layers in it. The viscosity (η) is the ratio between the shear stress and the gradient of the velocity along an axis at right angles to the stream:

$$\eta = \tau_x \left(\frac{\partial v_x}{\partial y} \right)^{-1}$$

The property of a liquid that is the reciprocal of the viscosity is called the fluidity and characterizes the ability of the substance to be deformed. In the blood, a change in the rate of shear leads to a change in the viscosity. This relation is an inversely proportional one. The moment when a substance begins to flow under the action of a shear stress applied to it is called the yield point.

Currently, biophysics has no clear scheme of parameters that could characterize the rheological properties of blood and erythrocyte suspensions. The most acceptable is the Kesson model on whose basis the rheological parameters of the blood are characterized by the following parameters: (i) the yield point; (ii) the Kesson viscosity, and (iii) the apparent viscosity (the viscosity of the blood determined with a viscometer) at a shear rate of 1 s^{-1}.

We determined the rheological properties of the blood with the aid of a rotary viscometer with a freely floating cylinder designed by V.Zakharchenko. The obtained curve

Table 18. Rheological Parameters of Blood in Practically Healthy Persons.

Parameter	Average value
K^2, cP	5.55±0.13
Hematocrit (Ht)	0.41±0.005
τ_0, dyn/cm^2	0.055±0.002
A, dyn/cm^2	0.75±0.028
T_1, s	9.2±0.28
T_2, s	41.95±0.81
B, s^{-1}	34.48±0.85
Er, 10^{12}/l	4.39±0.017
Thr, 10^9/l	245.9±0.96
V_{av}, μm^3	89.9±0.56
Fibrinogen (F), g/l	3.16±0.18

of blood motion (τ vs. $\partial v_x / \partial y$) was replotted in Kesson coordinates ($\sqrt{\tau}$ vs. $\sqrt{\partial v_x / \partial y}$), which made it possible to determine the yield point (τ_0), the Kesson viscosity (K^2). and the coefficient of erythrocyte cohesion (A).

Microrheological studies of the suspension of erythrocytes were conducted using an aggregometer designed on the basis of a Couette viscometer. When processing the curves, two important processes were singled out that differed in time. The first, characterized by the time T_1, reflects the formation of small aggregates, while the second, characterized by the time T_2, characterizes their merging into larger structures.

Analysis of the dependence of aggregation of the erythrocytes on the rate of shear allows us to determine the hydrodynamic strength of the aggregates B. The number of thrombocytes (Thr), erythrocytes (Er), and the average volume of the erythrocytes (V_{av}) were calculated in a coeloioscope "Analus instrument Bicellcounter-134" (Sweden).

To study the blood flow in healthy persons, we ran rheological tests of the blood of 15 healthy persons from 16 to 49 years old (Table 18).

The leading integral parameter determining the fluidity of the blood is its viscosity. In healthy persons, K^2 was 5.55±0.13 cP. According to present concepts, the hematocrit, degree and reversibility of erythrocyte aggregation, and the viscosity of the blood plasma are the main factors affecting the viscosity of the blood.

We found the hematocrit in healthy persons to be 0.41 (with fluctuations from 0.38 to 0.45), and the number of erythrocytes to be $4.39 \cdot 10^{12}$ per liter. We established that the number of erythrocytes and the level of the hematocrit in healthy persons correlate directly

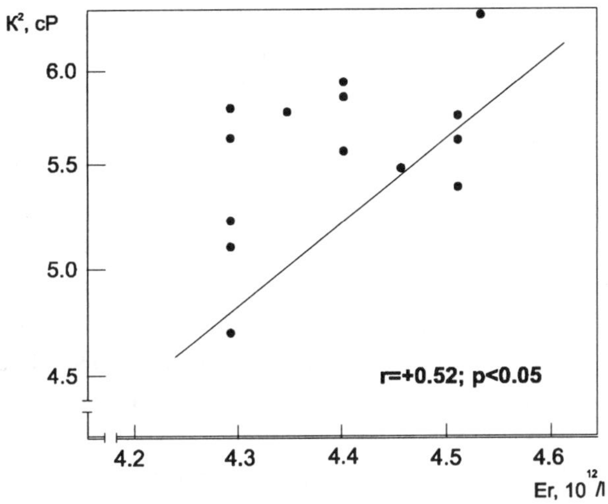

Fig. 29. Correlation between Kesson Viscosity of the Blood (K^2) and the Number of Erythrocytes (Er) in Healthy Persons.

with the value of K^2 ($r=+0.52$, $p<0.05$; $r=+0.42$, $p<0.05$) because with an increase in the number of red cells in the blood the density of the "coin piles" of the erythrocytes increases (Fig. 29).

At an identical rate of shear of the blood, the viscosity changes exponentially with a change in the hematocrit if it is below 50% and as a greater function of the hematocrit if it is above 50% [75, 265].

The processes of regulation of the blood viscosity depending on the level of the hematocrit are associated with the nature of the blood flow along vessels varying in diameter [162, 451].

When blood flows along small-diameter vessels, its fluidity properties are affected by phenomena such as the Fareus-Lingquist effect, the axial flow of the cells, and "slipping away" of the plasma.

According to the Fareus-Lingquist effect, the hematocrit lowers with a decrease in the diameter of the arteries and becomes very low in the zone of microcirculation. This is associated with the appearance of sections of plasma without cells [160, 161, 265], and the fluidity of the blood grows. Moreover, under conditions of a physiological norm with a high hematocrit, the processes of self-regulation of the blood viscosity are engaged. They are due to the growth in the deformability of the erythrocytes and lowering of their aggregation [75, 81, 161, 162, 265, 340, 451].

The axial flow of blood is characterized by the predominating transportation of erythrocytes along the central part of a vessel.

The phenomenon of slipping away of the plasma is explained by the diminishing of the diameter of the microvessels being attended by a drop in the erythrocyte concentration in them in such a way that the hematocrit is much lower in the capillaries than in the supplying arterioles and the removing venules [75, 340, 451].

The high shear rate and shear stress are typical features of the blood flow in the microvessels. This underlies the insignificant influence of the viscosity of the blood in the microcirculation system on the blood flow. With a growth in the shearing force, more and more blood cells acquire an ellipsoidal shape and become oriented with their long axis along the flow. This lowers the resistance to motion and, consequently, reduces the viscosity of the blood [79, 81].

The viscosity of the blood grows especially sharply at a shear rate of the order of magnitude of 1 s^{-1} [81, 161, 162, 520]. This anomalous behavior of the blood is associated with the formation of large erythrocyte aggregates and their ability to interact with one another.

The viscosity of the blood depends substantially on the shape and volume of the erythrocytes [252, 359, 427]. The stability of the structure of the erythrocytes, the constancy of their composition, and their function are ensured to a considerable extent by the external plasmatic membrane.

We found that under conditions of a physiological norm, the Kesson viscosity increases with a growth in the average volume of the erythrocytes ($r=+0.60$, $p<0.05$) whose average volume was 89.8±0.56 μm^3.

A considerable growth in the volume of the erythrocytes may disturb the hemodynamics in a microcirculation channel and cause a growth in the viscous dissipation of energy in the fine vessels.

The internal viscosity of the contents of an erythrocyte, the elasticity of a membrane, and the ratio of its surface area to the volume of a cell determine the deformability of the erythrocytes [60, 510, 511].

A change in the fluidity of a suspension of erythrocytes connected with the transformation of discocytes into spherulated cells (stomatocytes, echinocytes) is an important factor affecting the blood viscosity. Insignificant echinocytosis is a natural process accompanied by a growth in the blood viscosity [53, 181, 252, 451, 513]. The explanation is that the erythrocytes shaped like a pin rotate and capture the adjoining layer of plasma. With an increase in the rate of shear, the blood viscosity lowers in connection with the transformation of part of the echinocytes into discocytes [75, 337, 451].

With aging of the red cells of the blood, their deformability gradually lowers, which impairs their passing through narrow capillaries up to 3 μm in diameter. It is assumed that this is why the "old" erythrocytes are identified and eliminated from the blood flow [226, 445, 497]. The ability to deform is noticeably lowered in the old erythrocytes because of the growth in them of the concentration of hemoglobin, which affects considerably the fluidity of the cell contents [226, 445, 497, 512].

The flow of blood in a microcirculation channel is affected the most by rheological factors. Since the size of the microvessels is compatible with the diameter of an erythrocyte, of significance are also the mechanical properties of individual cells, primarily the deformability of the erythrocytes. Its lowering is important for the blood flow in microvessels. The rigid cells that cannot enter a capillary rapidly cause the formation of cell clusters in the vessels (stasis), emptying of the capillaries, and, in the long run, the formation of regions with local hypoxia [24, 53, 81, 181, 440, 504, 540].

The deformability of the erythrocytes is affected greatly by fluctuations of the osmotic pressure as a result of the change in the concentration of the electrolytes and the protein composition of the plasma [127, 151, 359, 390, 403, 448, 504]. When the pH of the blood diminishes, the rigidity of the membrane of an erythrocyte grows. Other causes of the change in the deformability of the erythrocytes is the accumulation in them of cholesterol and a change in the concentration of ATP and the level of ATPase activity [60, 121, 181, 351, 557].

The process of accumulation of intracellular cholesterol is determined mainly by the lipoprotein profile of the plasma, and not by the total cholesterol level in it [351]. The intense exchange of cholesterol between the erythrocytes and blood plasma ensures a large and rapidly exchanged erythrocyte-plasma pool of cholesterol, The large total adsorption area of the erythrocytes in the peripheral blood, especially in hypercholesterolemia, is capable of excluding the circulating atherogenic lipoproteins from the plasma. Young cells have this property to a greater extent.

The increase in the microviscosity of the erythrocyte membranes in DM patients is associated with deep metabolic disorders and a growth in the peroxide oxidation of the lipids that leads to a higher rigidity of the lipid bilayer of the membranes. Such a state of the surface of the erythrocyte membranes is most pronounced with a severe course of DM [22, 46, 151, 181, 182, 362, 481].

The rheological properties of the blood are also determined to a considerable extent by the aggregating ability of the erythrocytes. It consists of the forces of cohesion (aggregation) between them, the duration of forming small and large aggregates, and also of the ability of these aggregates to disintegrate hydrodynamically.

A growth in erythrocyte aggregation disturbs the protein and lipid spectra of the blood plasma, in particular it leads to hypoalbuminemia, hypergammaglobulinemia, hyperlipidemia (dyslipidemia) and hyperfibrinogenemia [181, 485, 563]. It was proved that low-molecular compounds diminish, and high-molecular ones increase the aggregation of the erythrocytes as a result of a sharp drop in the electrokinetic charge of the latter, especially when spherulized forms are present.

The fluidity of the blood in large vessels where the velocity of the blood flow is high depends only slightly on the properties of the formed elements and is due entirely to the pumping function of the heart. At the same time, with a low velocity of the blood in the vessels, i.e. in the venules and arterioles, the viscosity of the blood is determined by the aggregation of the erythrocytes [75, 161, 162, 265, 340].

Erythrocytes aggregate by two basic mechanisms, a bridge and an electrostatic one [160, 512].

The main condition for the former is the approaching of erythrocytes to a distance not exceeding the length of a molecule forming a bridge. A growth in the erythrocyte concentration brings the cells closer together, while the presence of forces of electrostatic repulsion produced by the zeta potential prevents this.

Normally, the erythrocytes carry a large negative charge ensuring their repulsion from one another at very low shear rates of the blood. Lowering of the surface charge is associated with a growth in the plasma of the concentration of fibrinogen and other high-molecular proteins. They are adsorbed on the membranes of the erythrocytes and lower the zeta potential considerably [478, 485].

An increase in the formation of large molecular filaments of fibrinogen, soluble complexes of fibrin monomers, and the products of degradation of fibrinogen not only affects the aggregation of the erythrocytes, but also increases the viscosity of the blood and plasma.

Pronounced intravascular aggregation causes stasis and deposition of part of the blood in the microcirculation channel. This results in lowering of the number of circulating erythrocytes and is a cause of ischemia of the tissues and organs.

The blood moves in a laminar flow in the blood vessels of a healthy person. With the considerable and widespread aggregation of the erythrocytes observed in a sharp change in the conditions of hemodynamics, a pathological type of blood flow may occur, e.g. a gravitationally stratified one. It consists of a layer of settled or stationary cell aggregates having different physical properties. With this type of blood flow, the erythrocyte aggregates cause fragmentation of the blood flow, which increases the peripheral resistance of the vessels and the viscosity of the blood [6, 24, 79, 81, 130, 178, 517].

The aggregation of the erythrocytes is a physiological phenomenon. The ability of the erythrocytes to form aggregates is by itself not yet a decisive factor determining the nature of blood flow in the vessels. The erythrocyte aggregates under the influence of the shear forces due to the difference in the pressure at the vessel ends may break up, which can be determined from the dynamics of the parameter B.

In healthy people, the parameter B was found to diminish with a growth in the hematocrit. The aggregates with a high hematocrit have the lowest strength (the coefficient of correlation between the hematocrit and the parameter B is 0.37, $p<0.05$). With a drop in the number of erythrocytes, the number of macromolecules (fibrinogen, immunoglobulins G, etc.) per erythrocyte grows, which increases the cohesion of the cells and the strength of the forming aggregates.

A negative influence of the erythrocyte aggregates on the microhemodynamics is possible when the forces of cohesion between the erythrocytes in the aggregates are increased and the forming aggregates do not break up under the effect of the shear stress, or when the contracting ability of the myocardium ensuring hydrodynamic breaking up of the cell aggregates lowers.

A major role in the formation of the hemorheological properties belongs to plasma factors. The blood plasma should be considered as a Newtonian liquid with a constant viscosity. The latter fluctuates within comparatively small limits and mainly depends on the content in the plasma of proteins and lipids that increase its viscosity. The plasma layer near a wall deprived of formed blood elements, but having a high viscosity and force of friction against the inner surface of a vessel, may cause peeling off of the endothelium over a great length.

Experiments revealed a clear relation between the viscosity of the blood (plasma), aggregating ability of the erythrocytes, and the content in it of fibrinogen, gammaglobulins, and plasma lipids (especially triglycerides) [181, 478, 485, 580].

Numerous investigations [323, 504] showed erythrocytes to have an independent coagulating and fibrinolytic activity, which in many ways determines the fluid properties of the blood. The role of erythrocytes in the pathogenesis of disturbances in the functional activity of thrombocytes was clarified [255, 335, 485, 504]. It was noted that in the microgenesis of the red blood cells, procoagulant factors are singled out that induce the aggregation of the thrombocytes. Moreover, a direct correlation was found between the viscosity of the blood and all the parameters of aggregation of the plates.

The erythrocytes of healthy persons have a high thromboplastic, heparin and antiheparin, fibrinolytic and antifibrinolytic activity. These parameters are in dynamic equilibrium and are hemostasis regulators. With an increased activity of blood coagulation, the coagulation activity of the erythrocytes lowers, and vice versa [142, 211, 255, 322, 335, 485].

Of importance for the suspension stability of the blood is the functional state of the thrombocytes.

The aggregates of the thrombocytes, on the one hand, can facilitate the intravascular coagulation of the blood, and on the other, by clogging up small capillaries and arterioles, may cause stasis in the system of microcirculation [440, 619].

Both phenomena, in turn, increase the aggregation of the erythrocytes and underlie the phenomenon of inversion – a sharp growth in the viscosity of the blood at a definite critical diameter of a capillary. Thrombocytes acquire a special significance in the rheological stability of the blood at high shear rates because (in contrast to erythrocytes), the thrombocytes experience irreversible aggregation in the velocity gradient [36].

2.2. Hemostasis System and Rheological Properties of the Blood in Ischemic Heart Disease Patients

During recent years, many authors advanced the concept of the continuous functioning of the coagulating and anticoagulating systems of the blood, continuous latent microcoagulation, and the destruction of fibrinogen in an organism [40, 162, 367, 402, 433]. This was proved, in particular, by the detection of blood coagulation products in the blood, e.g. thrombin. In the development of atherosclerosis, especially at an elder and senile age, microthrombi form. Although they lyse, they prevent the normal diffusion of oxygen through a capillary wall, which leads to hypoxemia.

The thromboresistance of a vessel wall is determined by the contribution of not one, but a number of factors, some of which prevent the adhesion and aggregation of the thrombocytes, others inactivate thrombin and active factors of the blood coagulating system and thus prevent thrombus formation, still others activate fibrinolysis, which leads to destruction of the formed fibrin. Lowering of the antithrombogenic activity of the walls of the vessels contributes substantially to thrombogenesis.

Atherosclerosis and IHD are based on a tendency of thrombus formation and of coronarospastic phenomena [11, 57, 81, 158, 178, 277]. Both processes are affected by prostanoids. The most interesting among them are thromboxan A_2 (TX) and prostacyclin (PrC). The latter provides powerful vasodilative and coronary dilating effects, and has an antiaggregation, fibrinolytic and antiatherosclerotic, antioxidant, energy-saving, antiarrhythmic, anticatecholemic, and membrane-stimulating action. In healthy persons, the synthesis of prostacyclins by a vessel endothelium predominates over the level of thromboxan secreted by thrombocytes, which prevents the aggregation of the latter and excludes spontaneous thrombus formation.

IHD is characterized by the increased formation of thromboxan and a reduced one of prostacyclin. However, of importance here is what is primary and what is secondary. It is still unclear whether disturbance of these relations is primary with subsequent development of IHD, or whether the incidence of IHD results in a change in these relations. It is very important to study the properties of prostanoids. Experiments on animals showed that the formation of prostacyclin diminishes with age [35]. We concluded that this is probably due to the lowering in the phospholipid membranes of vessels of their content of polyunsaturated fatty acids that are precursors of the prostanoids, in particular, arachidonic acid.

We consider it to be proved that with age the diminishing of the synthesis of prostacyclin and the growth in the formation of thromboxan are primary, therefore the disturbance of these relations is a factor facilitating the incidence of IHD. In atherosclerosis patients, the ability of the vessels, including the coronary ones, to synthesize prostacyclin drops sharply. This is associated both with the development of atherosclerosis and, probably, with a growth in the peroxide oxidation of lipids and a growth in the content of LDLP inhibiting the activity of PGI_2-synthetase unlike HDLP, which activate this enzyme and inhibit the synthesis of thromboxan.

It should be noted that the formation of prostacyclin is inhibited more noticeably in men than in women, and after 70 the content of PiC in the blood of women is higher then in men. This explains the more rapid development of atherosclerosis in men.

That LDLP acts differently on the activity of PGI_2-synthetase in men and women is very interesting. In men, the inhibiting effect of the LDLP is more pronounced than in women. At the same time, in the latter, HDLP stimulate the formation of PGI_2-synthetase more than in men. The same occurs in the biosynthesis of thromboxan – it forms more actively in men than in women. It was proved that estrogens increase, and androgens lower the level of the HDLP, while they have the opposite effect on the LDLP [273, 301, 342].

2.2.1. Rheological Properties of the Blood in Ischemic Heart Disease

The following hemodynamic, microcirculation, and rheological factors play a major role in the pathogenesis of coronary atherosclerosis and its complications: a high turbulence of the flow, side pressure, shear stress, intravascular aggregation of the erythrocytes and thrombocytes, shunting of the blood. These factors may facilitate the formation of atheromatous plaques on sections of the aorta and coronary arteries having bends, bifurcation, or branches, i.e. at spots that are the most complicated in a hydrodynamic respect for the blood flow.

Table 19. Rheological Parameters of Blood in IHD Patients.

Parameter	Control group	IHD patients
K^2, cP	5.55±0.13	6.48±0.2*
Hematocrit (Ht)	0.41±0.005	0.47±0.007*
τ_0, dyn/cm^2	0.055±0.002	0.097±0.006*
A, dyn/cm^2	0.75±0.028	1.14±0.04*
T_1, c	9.2±0.28	6.91±0.22*
T_2, c	41.95±0.81	37.93±0.98
B_1, c^{-1}	34.48±0.85	33.05±2.54
B_2, c^{-1}	-	55.2±2.32
Er, 10^{12}/l	4.39±0.017	5.0±0.12*
F, g/l	3.16±0.18	4.25±0.22*
V_{av}, μm^3	89.9±0.56	91.2±1.27

* The difference of the parameters is authentic (in comparison with the level in the control group).

In a number of observations based on the results of selective coronarography, morphological changes in the coronary arteries were absent even in patients with pronounced stenocardia [130, 138, 160, 178, 265, 300, 357, 370, 417]. On these grounds, we singled out a group of persons in whom the leading role in the origin of the diseases belonged to microcirculation disorders whose causes of development were not established. We can assume that in these cases in addition to other factors, a major role is played by disorders in the rheological properties of the blood that facilitate the formation of thrombi in the fine vessels. This enabled some authors to consider stenocardia as a rheological syndrome. The rheological parameters of the blood in IHD patients are presented in Table 19.

The Kesson viscosity of the blood in IHD patients was higher than in healthy persons. A positive correlation of K^2 with the level of fibrinogen was established ($r=+0.41$, $p<0.01$). The concentration of the latter exceeded the values in the control group by 34.6% ($p<0.05$) (Fig. 30).

Apart from fibrinogen, the blood viscosity in IHD patients is affected by a change in the lipid composition of the plasma. It was shown [121], that the highest viscosity of the blood was observed in hyperlipidemia of types II and IV. A direct correlation was established between the blood viscosity and the triglyceride level. Here the blood viscosity in persons with normal ratios of the lipids approached the values for healthy persons.

In hyperfibrinogenemia and dyslipoproteinemia different effects are noted, in particular a growth in the viscosity of the plasma [130, 226, 336, 390, 403, 448, 485]. This is associated with the appearance in the blood plasma of particles with a high molecular

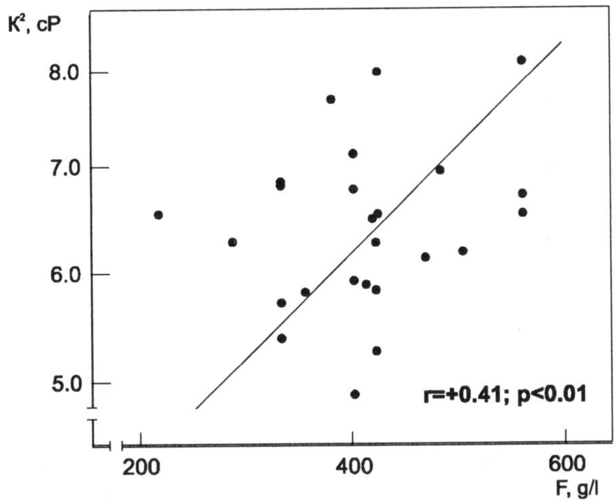

Fig. 30. Correlation between the Kesson Viscosity of the Blood (K^2) and the Fibrinogen Level (F) in IHD Patients.

weight, i.e. LDLP and VLDLP. The HDLP does not affect substantially the rheological properties of the blood in connection with the small size of a molecule.

A growth in the viscosity of the blood was noted in IHD patients by a number of authors [40, 79, 130, 226, 480]. According to Belousov [40], in stenocardia patients an authentic growth in the blood viscosity was found with different shear stresses. However, in patients with blood circulation insufficiency of stages IIB and III in comparison with patients without it or with stage IIA, a high blood viscosity was found much more rarely, and it was more often within the limits of normal values or lowered.

In IHD patients, the cohesion of the erythrocytes (A) is 1.5 times ($p<0.001$) higher. A growth in the concentration of fibrinogen (Fig. 31) is accompanied by a growth in A ($r=+0.68$, $p<0.1$). The fibrinogen settles on the erythrocytes and plays the role of an inter-cellular bridge that facilitates their aggregation. The LDLP and VLDLP have a similar property. The role of fibrinogen and its complex compounds in the aggregation of erythrocytes becomes more significant if we consider that even with its normal content in the blood, but with a high concentration of the decomposition products, pathological aggregation of the cells can be observed [11]. It was shown that with a growth in the cholesterol content in the erythrocyte membranes (observed in IHD patients), the degree of adsorption of the fibrinogen by the membrane increases, which facilitates a growth in the aggregation and a drop in the degree of disaggregation of the blood cells.

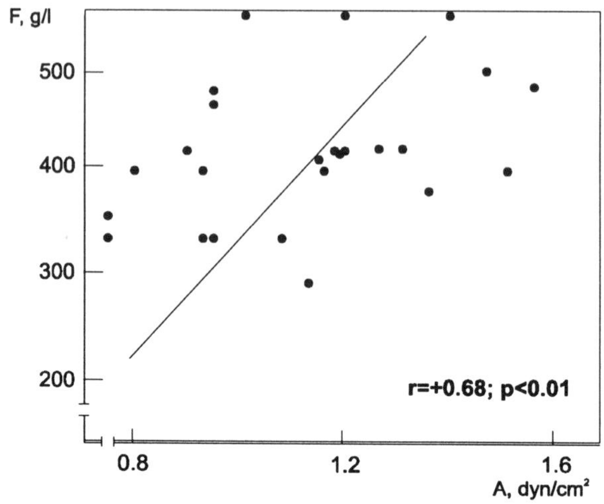

Fig. 31. Correlation between the Fibrinogen Level (F) and the Cohesion of Erythrocytes (A) in IHD Patients.

With a growth in the cohesive forces of interaction, in IHD patients the duration of formation of small erythrocyte aggregates diminished authentically, and small aggregates resistant to disaggregation appeared with the absence of appreciable changes in the dynamics of the large ones. The discovered changes may facilitate the blocking of the terminal vessels as a result of the sludge phenomenon and disturbance of microcirculation. This results in diminishing of the exchange surface of the blood cells, which cannot fail to affect the transportation of oxygen to the tissues. In our opinion, the integral strength of the aggregates, which depends both on the number of cells in the aggregates and on the cohesion between the cells, affects the blood flow in the microcirculation channel as a whole more strongly than only the aggregation of the erythrocytes.

According to Schmid-Schonbein [509, 510, 511, 512, 513], in the clumpy aggregation of erythrocytes observed in vessel pathology, the blood transforms from an emulsion into a suspension because aggregation remaining at high shear rates prevents the rotation of an erythrocyte membrane about their contents, as is observed in complete disaggregation under normal conditions. The clumpy aggregates with a growth in the shear rate do not break up into "coin stacks", but retain their structure even at very high shear rates.

In addition to the plasma components of the blood, the cohesive forces acting between erythrocytes are affected greatly by pathological alterations in the erythrocytes themselves revealed in IHD [40, 60, 130, 161, 162].

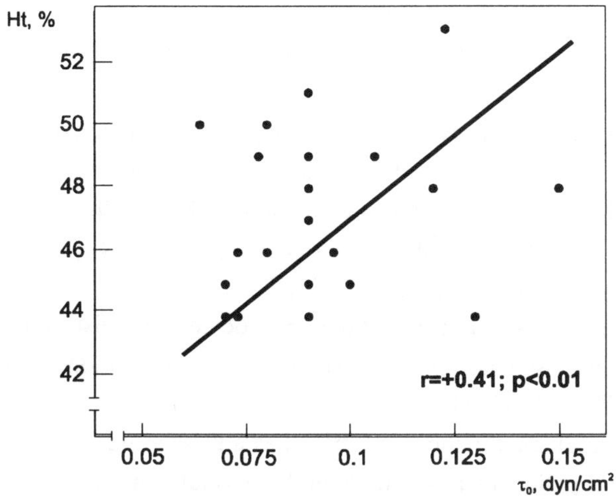

Fig. 32. Correlation between the Hematocrit (Ht) and the Ultimate Shear Stress of Blood (τ_0) in IHD Patients.

We failed to obtain authentic data on the change in the average volume of the erythrocytes in IHD patients (91.2±1.27 μm^3) in comparison with healthy persons (89.9±0.56 μm^3), but according to published information [188, 351, 590, 598], in IHD, a growth in the size of the erythrocytes and the transformation of discocytes into spherocytes and echinocytes are observed [60]. This is connected with a growth in the cholesterol/phospholipid index of a membrane. Aggregation of the cells was found to depend on their shape. For example, a negative correlation was observed between the rate of aggregation of the erythrocytes and discocytes, while a positive one was observed between the rate of aggregation of the erythrocytes and the number of pathological forms, which is due to the growth in the degree of their cellular interaction.

It is typical that a change in the morphological parameters is accompanied by disturbances of the electrophoretic mobility of the erythrocytes [60, 130, 508]. Some authors noted that the morphofunctional composition of the erythrocytes affects the viscosity of the blood more than their volume concentration, and the degree of membrane disorders correlates with the magnitude of the clinical manifestations of the disease.

In the morphological studying of blood smears in IHD patients [60, 130, 359, 504, 569], a growth in the number of large erythrocytes over eight micrometers in diameter was found. It was attended by lowering of the index of deformability of the erythrocytes. It was noted [485, 517], that in progressing stenocardia and acute myocardial infarction, the blood flow contains noticeably more large erythrocytes then in the stable form of IHD. It is assumed that in IHD and DM of type II, there is a separate pool of erythrocytes having an increased resistance to deformation [181].

The rheological properties of the blood depend to a great extent on the deformability of the erythrocytes, which determines to a considerable extent the viscosity of the blood. For example, at given values of the hematocrit Ht and the shear rate, the viscosity of a suspension of erythrocytes with an increased rigidity exceeds the viscosity of a suspension of ordinary erythrocytes [75, 181, 451, 512].

A change in the shape and size of the erythrocytes may disturb their filterability and retard the blood flow, especially in the system of microcirculation, development of stasis, and aggregation of the blood cells.

An important characteristic of the blood as a Kesson liquid is the value of the yield point, which in IHD patients was higher than the normal value 1.8 times ($p<0.001$). Under the conditions of the physiological norm, the value of τ_0 is regulated by changing the cohesion of the erythrocytes. In IHD patients, this relation is violated: an increase in A and Ht was noted on the background of high values of τ_0. A positive correlation between τ_0 and Ht appears ($r=+0.41$, $p<0.01$) that is absent in healthy persons, which points to disruption of the compensating rheological mechanisms of the organism (Fig. 32).

Whenever an increased value of τ_0 was found, a growth in the number of pin-shaped discocytes was observed [130, 160]. The presence of pins or tongues considerably increased the area of interaction between the cells and increased the forces of cohesion between them.

Clinical Exaμple.

Patient M. 60 years old was admitted with complaints of frequent pains in his chest under a small physical load.

He previously had a myocardium infarction. Under outpatient conditions, he took prolonged nitrates, calcium antagonists, and up to 8–10 nitroglycerin tablets a day with a slight effect. His state when admitted was satisfactory. There were no peripheral edemas. Over the lungs, respiration was vascular with no rales, the frequency of respiration was 16 a minute. The boundaries of cardiac dullness were expanded to the left by 1 cm. The heart rhythm was regular, the heart rate was 80 bpm, pathological noises were absent. The arterial pressure was 130/80 mmHg. The liver was not increased, at the edge of the rib arch.

The blood rheology was: K^2=7.13 cP, Ht=0.48, τ_0=0.12 dyn/cm^2, T_1=6.78 s, A=1.51 dyn/cm , T_2=38.8 s, B_1=26.17 s, B_2=51.7 s^{-1}, and F=4.0 mg/dm^3.

Veloergometry was conducted: at a load of 25 W after one minute the patient had a seizure of stenocardia attended by lowering of the segment ST. Coronarography revealed stenosis up to 90% with prevailing lesion of the left circumflex artery.

Consequently, the patient suffered from stenosis disorder of the coronary vessels combined with pronounced disorders of the fluid properties of the blood in the form of a growth in the blood viscosity, aggregation ability of the erythrocytes, acceleration of the formation of small and large aggregates, and disturbance of the disaggregation of the erythrocytes, which facilitated the course of IHD. In the given case, there was a combination of morphological changes in the coronary vessels and a rheological anomaly, disturbance of the blood flow in both large vessels and in the microcirculation channel. This resulted in the development of myocardial infarction and a drastic drop in the tolerance to a physical load.

An analysis of the level of the myocardial blood flow in IHD patients showed that it is lower at high values of the viscosity of the blood and plasma [357, 440]; here no distinctions in the rheological parameters in patients with injured and intact (in coronarography) vessels of the heart were found [11, 160, 370, 417].

In ischemia of the myocardium, a local accumulation of lactic acid and other incompletely oxidized metabolism products in the system of microcirculation is a factor leading to the pathological intravascular aggregation of erythrocytes, injury of their membrane, and the appearance of a large amount of adenine nucleotides that are the most powerful stimulator of blood plaque aggregation.

According to Firsov [160], the change in the rheological parameters of the blood in myocardial infarction depends directly on the depth and extent of lesion of the myocardium. An inverse correlation was found between the stroke index and τ_0. It was noted that with a drop in the tolerance to a physical load in IHD patients, the blood viscosity increased authentically [130, 485, 517]. Information was published [40, 520] indicating that the strength of the erythrocyte aggregates in myocardial infarction patients with blood circulation insufficiency and disturbances of the cardiac ejection was higher than in patients with no complications in the course of the disease. It was presumed that the drop in the propulsive ability of the myocardium makes erythrocyte aggregation possible pathogenetically as a result of lowering of the blood flow shear stress.

The change in the rheological properties of the blood in severe manifestations of IHD may be of a phase nature. It was shown [40, 162, 520] that with acute myocardial infarction, a drop in the viscosity of the blood and the yield point was observed in the terminal period of the cardiogenic shock on the background of a drop in the hematocrit, the content of fibrinogen, and a growth in the aggregation of the erythrocytes (in comparison with the initial level of these parameters when patients are admitted to an inpatient ward). This is due to sequestration of the formed blood elements and the emergence from active blood circulation of compact, undeformed aggregates of erythrocytes with blocking of the microcirculation channel and development of a plasma flow.

In patients with IHD (in comparison with healthy persons), an authentic increase in Ht was revealed as a result of a growth in the number of erythrocytes. The increase in Ht on the background of high cohesive forces between erythrocytes is an unfavorable factor facilitating an increase in both the volume of the erythrocyte conglomerates and their hydrodynamic strength. This is also witnessed by the appearance of a slight, although opposite in sign (in comparison with healthy persons) correlation between the strength of the aggregates and the Ht level ($r=+0.18$ and $r=-0.37$, respectively).

Consequently, in severe manifestations of IHD, disturbances of the rheological properties of the blood are due to changes in the composition of the plasma and morphological and functional changes in the formed elements of the blood that impair microcirculation in the vitally important organs.

2.2.2. Hemostasis System in Ischemic Heart Disease

The pathophysiological process of thrombosis consists of several units. The initial one of thrombus formation is the activation of the thrombocytes when they interact with a damaged surface of a vessel wall and the shape of the thrombocytes alters. Unique protuberances appear. Next thrombocytes adhere to the damaged zone and their aggregation begins. This leads to the secretion of thromboxan, factor III of the thrombocytes, beta-thromboglobulin, and other physiologically active substances that are very strong aggregation promoters and boost the beginning process of aggregation and destruction of the thrombocytes.

Thrombocytes are not only the structural basis of a thrombus, but also the source of the factors activating all the hemostasis units. Activation of the thrombocytes triggers off a whole cascade of biochemical reactions and leads to the formation of a thrombocyte thrombus with its subsequent overgrowth by other formed elements, especially erythrocytes. Under the influence of the substances secreted from the thrombocytes, the inactive prothrombin transforms into thrombin, acts on fibrinogen, and transforms it into the insoluble compound fibrin whose filaments form a thrombus and make it immobile.

A vessel wall, whose morphological and functional properties ensure the liquid state of the circulating blood, is an important unit in the hemostasis system. Of the three layers of a vessel wall, the intima (subendothelial layer and internal elastic membrane) has the greatest significance. The normal functioning of the blood vessels and their antithrombogenic activity are closely related to the intactness of the endothelium and its functional activity. The endothelium performs numerous functions for maintaining hemostasis:

(1) transports metabolic substances actively between the blood and tissues;

(2) creates an obstacle to the penetration of blood macromolecules, being a sort of a barrier;

(3) synthesizes various mediators regulating the reaction between a vessel wall and blood for preserving hemostatic homeostasis;

(4) supplements the phagocyte function by removing highly active biological substances, active factors of the hemostasis system, etc. from the blood;

(5) ensures the antithrombogenic properties of the walls of vessels and their thrombus resistance;

(6) participates in regulation of the vascular tonus and facilitates the secretion of prostacyclins;

(7) ensures proliferation of the cells, metabolism, and functioning of the mechanisms of cell migration;

(8) participates actively in immune reactions.

A number of experimental studies revealed that the functions of thrombocytes in the regulation of atherogenesis are diverse and depend functionally on the state of metabolism. The modulating effect of thrombocytes on metabolism and the functions of the endothelium can be regulated both as a result of direct contact of the cells and the secretion into the zone of wall damage of a mitogenic factor with the subsequent proliferation of the smooth muscle cells and as a result of chemical modification of the LDLP by the malonic dialdehyde formed in the biosynthesis of thromboxan: modified lipoproteins can penetrate into a vessel wall in a nonreceptor, i.e. nonregulated way.

Hyperadrenalinemia, hypercortisolemia, hypercholesterolemia, an increase in the content of LDLP and VLDLP, the formation of free radicals, and the peroxide oxidation of lipids found in patients with IHD [29] disturb the antithrombogenic activity of the vessel walls.

It is generally acknowledged that an increase in the concentration of LDLP and VLDLP leading to damage of vessel walls and disturbance of the lipoprotein spectrum of the plasma called dyslipoproteinemia increasing the supply of cholesterol to an artery wall are risk factors of atherosclerosis [453, 536, 613].

The hypercholesterolemia appearing chiefly as a result of an increase in the content in the blood of cholesterol in the composition of LDLP, which are the main carrier of cholesterol from the liver and intestine to the cells of organs and tissues, is one of the manifestations of such DLP [453, 455].

When speaking of the participation of cholesterol in atherogenesis, we must have in view two aspects of the problem, namely, the pathogenetic and etiological ones.

N.N.Anichkov back in 1915 determined the pathogenetic binder of atherosclerosis when he declared that there is no atherosclerosis without cholesterol.

The highest atherogenic potential is noted in dyslipoproteinemia of types IIA, IIB, III, and IV [82].

The accumulation of lipids in the cells of the intima of a vessel wall is one of the typical manifestations of spontaneous atherosclerosis in a person [267]. Kacharova *et al.* [267] also proved that the serum of IHD patients has an atherogenic potential. It is revealed by the ability to cause the accumulation of intracellular lipids in a culture of smooth muscle cells in an uninjured intima of the aorta.

Baumgartner [35] and Austin [23] revealed a direct relation between the frequency of the incidence of IHD and the level of triglycerides in the blood.

It was assumed that changes in lipid metabolism by leading to stenosis of the vessels have a major significance in the incidence of stenocardia of stress, while the activation of thrombocyte hemostasis may play the main role in the development of myocardial ischemia at rest [593].

What is the relation between disorders in the system of lipids, the state of hemostasis, and lesions of a vessel wall?

Bokarev [57] considers that LDLP stimulate the activation of the factor VIIIa and inhibit the synthesis of prostacyclin; an excess of saturated fatty acids increases the aggregation of thrombocytes, while a lack of lipoic acid prevents the formation of prostacyclin.

An increased consumption of fats of an animal origin and hypertriglyceridemia cause hypercoagulation and increase the aggregation of thrombocytes [243, 428]. At the same time, the incidence of IHD in the inhabitants of northwestern Greenland (Eskimos) is extremely low, which is associated [30, 31] with the low content of cholesterol, triglycerides, LDLP, and a high level of HDLP. This is explained not by the genetic factors of the Eskimos, but by the nature of their food [30, 31, 128]. One of its features is the high content of eicosapentaenic acid that does not cause aggregation of thrombocytes *in vitro*. The reason is that the acid is a source of formation of thromboxan A_3 that is not similar to thromboxan A_2 and does not affect the functional activity of thrombocytes. Moreover, the cyclooxygenase of a vessel wall can use the eicosapentaenic acid for the synthesis of the thromboxan antagonist PrC I_3 [101].

Consequently, in people in whose ration eicosapentaenic acid predominates, the balance between the pro- and antiaggregation components in the plasma is shifted in the direction of a growth in the synthesis of PrC I_3 and a drop in the synthesis of thromboxan A_2. This lowers the risk of appearance of thrombosis and is probably the cause preventing the early development of atherosclerosis and IHD.

The novel domestic food products (eiconol, dietic protein caviar and an emulsion protein-fat product distinguished by its content of polyunsaturated fatty acids (PUFA) can be used for the prophylaxis and treatment of atherogenic dyslipoproteinemia by lowering the hemostatic potential and aggregation activity of the thrombocytes [296, 506, 575, 594].

Koudo [311] concluded that in IHD patients, the ratio of the level of the eicosapentaenic acid to the content of the arachidonic acid in the FFA of the plasma is authentically lower than in healthy persons. He singled out the low level of the eicosapentaenic acid as a new risk factor of cardiovascular diseases.

In IHD patients, a risk of thrombus formation appears because of the lowering in the blood of the relative content of the fatty acids $C_{20:3}$ and $C_{20:5}$ (the substrates of the synthesis of thrombolytics and vasodilators) and the greater inhibition of prostacyclin synthesis in comparison with the inhibition of thromboxan with a growth in the concentration of $C_{22:6}$. Saturated fatty acids facilitate the formation of thrombi and the development of IHD [243, 428].

Srivastava [546] considers that the PUFA are an antithrombogenic factor as a result of the partial inhibition of thrombocyte cyclooxygenase, which is indicated by the lower formation of thrombocyte endoperoxides and malonic dialdehyde. Here the degree of inhibition depends directly on the degree of unsaturation of the fatty acids.

These investigations point to the necessity of further investigation of the relation between dyslipidemia and the system of hemostasis and their influence on the process of local blood coagulation under conditions of atherogenesis.

The pathogenesis of thromboses is a multifactor and multicomponent process based on injury of a vessel wall, a change in the functional state of the hemostasis system and blood properties, disturbance of blood circulation, and a change in the lipid spectrum of the blood.

The initial cause of the formation of a hemostatic thrombus is injury of a vessel wall. A damaged wall is the trigger and functionally active field on which the adhesion and aggregation of thrombocytes occur with the formation of a primary thrombocyte thrombus and activation of the coagulation unit or hemostasis, which leads to the formation of a hemostatic and pathological thrombus.

Ross [491, 494] advances a hypothesis of the pathogenesis of atherosclerosis according to which the following processes occur at spots of injury of an endothelium:

(1) adhesion, aggregation, and the reaction of release of thrombocytes, the secretion of thrombocyte factors; thromboxan A_2 and ADP induce further aggregation of the thrombocytes and the formation of a thrombocyte thrombus, while the thrombocyte growth factor, by penetrating into the subendothelium layer, induces the migration of smooth muscle cells through an opening in the internal elastic membrane and their proliferation in the intima;

(2) synthesis of collagen, proteins of elastic fibers, and proteoglycans of the smooth muscle cells of the intima;

(3) injury associated with the formation of a thrombus;

(4) intra- and extracellular accumulation of lipids;

(5) subsequent adiposis of the cells and the formation of an atheromatous plaque.

This hypothesis of the pathogenesis of atherosclerosis attaches special significance to the proliferation of the smooth muscle cells in the intima of the vessels and thrombogenesis. The existence of a thrombocyte growth factor was confirmed by Kohler [299], who showed that in the destruction of thrombocytes (*in vitro*), the supernatant has a mitogenic effect in the cell culture. The glia cells in the culture can stimulate the synthesis of the thrombocyte growth factor [229].

In addition to the growth factor, the Willebrand factor, ADP, a number of proteins, thrombocyte factors III, IV, and V, thrombocyte fibrinogen, serotonin, and thromboxan A_2 are released from the thrombocytes in aggregation.

However, morphological injuries are not always found in artery walls. Numerous functional disturbances of the endothelium are possible. They are manifest in lowering of the antithrombic activity (antiaggregation, anticoagulant, fibrinolytic), which is the initial unit in the pathogenesis of atherosclerosis. The reason is that here favorable conditions are created for the adhesion and aggregation of thrombocytes on the wall of vessels and the local deposition of fibrin. The progress of atherosclerosis may occur with the participation of several mechanisms, including the thrombocyte growth factor [507].

Atherosclerotic disorders are also accompanied by qualitative changes in the membranes of the erythrocytes. The peroxidation syndrome was studied in experiments with animals [226]. In this process, the content of acylhydroperoxides in the blood serum increased, the resistance of the erythrocyte membranes to spontaneous hemolysis lowered, and the rate of accumulation of malonic dialdehyde in the liver tissue increased. A growth in the level of LDLP and VLDLP in the blood was noted, and a drop in the amount of prostacyclin in the aorta wall. The presence of lipid infiltration was discovered morphologically. Moreover, a growth in the sensitivity of the injured sections of the aorta to vasoactive substances – adrenaline and histamine – was established. Numerous studies showed the importance of the peroxidation syndrome in the development of atherosclerotic changes.

To study the antithrombic activity of vessel walls and the homeostatic balance between a vessel wall and the blood factors of the hemostasis system, we studied 320 patients with different forms of IHD from 34 to 65 years old, with an average age of 52 ± 2.8 years (Table 20).

We studied only men with a view to the greater occurrence of IHD among them.

Table 20. Hemostasis in IHD Patients.

Parameter	Control group**	IHD patients
Fibrinogen, g/l	2.36±0.10	3.74±0.11*
Tolerance of plasma to heparin, s	948.0±14.0	724±13.0*
Duration of plasma recalcification, s	286.0±11.0	268±12.0
Antithrombin III, s	37.6±1.5	27.3±0.9*
Fibrinolytic activity of blood, %	20.3±0.9	11.2±0.8*
Index of intravascular aggregation of thrombocytes, %	1.04±0.03	2.6±0.1*
Spontaneous aggregation of thrombocytes, %	0	6.9±0.3*
Aggregation of thrombocytes induced by:		
ADP 10^{-5} M, %	61.0±2.3	79.1±2.4*
ADP 10^{-6} M, %	4.2±0.17	10.3±0.2*
collagen, %	57.0±1.4	69.0±2.2*
factor thrombocyte activation, %	14.5±1.0	22.3±1.3*
adrenaline 10^{-4} M:		
1st wave, %	13.4±1.1	30.2±1.8*
2nd wave, %	59.3±2.1	79.2±2.4*
aggristine, %	57.2±2.0	77.4±2.1*
Beta-thromboglobulin, ng/ml	29.2±1.9	38.0±2.2*
Thrombocyte factor IV, s	14.9±0.9	18.9±0.9*
Thromboxan B_2 in plasma, pg/ml	31.0±3.1	37.0±2.9
cAMP in thrombocytes, pmol/$3 \cdot 10^8$ cells	4.1±0.4	3.0±0.3*
6-Keto-PGF_1-alpha in plasma, pg/ml	65.0±2.2	51.0±2.9*
6-Keto-PGF_1-alpha/thromboxan B_2	2.1±0.2	1.4±0.2*
Willebrand factor:		
un/ml	1.04±0.08	2.18±0.09*
%	104±8.0	221±8.0*

* The difference of the parameters is authentic (in comparison with the level in the control group).

** The control group includes persons with functionally active antithrombogenic properties of the vessel walls.

tients with different forms of IHD from 34 to 65 years old, with an average age of 52±2.8 years (Table 20).

We studied only men with a view to the greater occurrence of IHD among them.

We selected IHD patients without improper cardiac rhythm, not suffering from DM, severe forms of obesity, arterial hypertension, chronic pneumonia, diseases of the kidneys or other accompanying diseases in the aggravated stage. Among the IHD patients 115 (36%) previously had myocardial infarctions, while 13 (4%) had it twice.

All the patients were subjected to selective coronarography to determine the characteristics and propagation of the lesion of the coronary vessels.

The case histories of the IHD patients showed that the overwhelming majority of them (85%) smoke; 67% of the men with stable stenocardia of tension and who had had myocardial infarction have been smoking for a long time (20–30 years) over 20 cigarettes

By comparing our information on smoking, the use of fats, and information on the incidence of atherosclerosis in the parents of the patients with published information, we can conclude that the above factors underlie the causal relations of the earlier incidence and progress of IHD in the studied group of patients.

One of the major factors of the risk of development of IHD and arterial thrombosis is smoking. Numerous investigations in our and other countries showed that the regular smoking of up to 20 cigarettes a day increases the risk of the incidence of myocardial infarction 1.5 times in men over 40 [158, 402]. There is a clear correlation between tobacco smoking and the incidence of IHD and myocardial infarction.

The influence of smoking on a human organism and the development of IHD does not depend on the level of cholesterol in the blood and the arterial pressure [402]. It was shown in experiments on animals that nicotine causes injury of the endothelial cells, lowers the synthesis of prostacyclin, and increases the aggregation and adhesive properties of thrombocytes [461]. The thrombocytes in the blood of smokers have an increased sensitivity to the inductors of aggregation.

Nicotine produced changes in the luminal surface of membranes of the aorta in rats with multifold lesion of the endothelial cells rising in the lumen of the aorta like protruding bubbles, and also the formation of thrombocyte microthrombi [461, 612]. The thrombocytes of smokers, as indicated, have an increased sensitivity to inductors of aggregation, and a growth in the sensitivity was noted even if one cigarette was smoked [339]. This is why the probability of the appearance of an arterial thrombosis in states when an inductor of thrombocyte aggregation gets into a vessel channel, e.g. in an emotional stress, is higher in smokers than in nonsmokers.

Nicotine inhibits the synthesis of prostacyclin, but does not affect the formation of thromboxan [354, 414, 461, 527, 609]. On the background of a drop in the synthesis of prostacyclin and an increased sensitivity of the thrombocytes to aggregation inductors, the probability of the appearance of arterial thrombosis and the progress of atherosclerosis grows, even passive smoking lowers the sensitivity of thrombocytes to prostaglandins, which increases their aggregation [527].

Smokers have a higher content of serum cholesterol, triglycerides, LDLP, and a lower level of HDLP. They have a higher degree of lipolysis produced by catecholamines and an increased activity of the LPL of the adipose tissue [393].

Consequently, smoking is one of the leading risk factors of the development and progress of atherosclerosis and arterial thrombosis.

When comparing the hemostasis system in healthy persons and IHD patients, we noted clear distinctions: first of all, the fibrinogen level in IHD was higher and was 3.74 g/dm^3 (in healthy persons it was about 2.36 g/dm^3).

The coagulation of blood can be represented in the form of the phases: (i) activation of prothrombin and the formation of thrombin, and (ii) under the influence of the thrombin, the fibrinogen first transforms into fibrin monomers, and then into a polymer. In clinical practice, both the amount of thrombin and its influence on the fibrin monomers are very important. It was shown in experiments that in a healthy person, fibrinogen transforms into fibrin at very low thrombin concentrations in the blood [36, 97, 349, 403].

In the appearance of pathological processes such as DIC and IHD, both qualitative and quantitative changes occur in the system of blood coagulation. First of all, the sensitivity to thrombin lowers and the process of polymerization of the fibrin monomers is disturbed. They partly form complex compounds with fibrinogen, probably with early products of the decomposition of fibrinogen by plastin. Products of paracoagulation appear – soluble fibrin-monomer complexes (SFMC). These large and medium-size molecular SFMC coagulate poorly under the influence of thrombin. This thrombin resistance of the complexes disturbs the normal transformation of fibrinogen into fibrin and may produce the phenomenon of the DIC syndrome (syndrome of disseminated intravascular coagulation).

The anticoagulant activity of the blood is ensured by numerous biological factors. One of the most active anticoagulants is the plasma cofactor of heparin – antithrombin III (AT III). It is a universal inhibitor of almost all the enzyme factors of blood coagulation (especially of thrombin and the factor Xa). In the hereditary absence of this factor in an organism, severe thromboses of various arteries and veins occur with the relevant infarctions and embolisms.

The fibrinolytic system in an organism breaks up fibrin-fibrinogen asymmetrically into fine fragments. The main component of the system is fibrinolysis (plasmin). Fibrinolysis is activated by a protein activator of the tissue type that is synthesized in a vascular endothelium. When conducting a cuff test, this compound is excreted intensely from the endothelium.

The products formed as a result of fibrinolysis are biologically active and affect the permeability and tonus of the blood vessels, the properties of the endothelium, inhibit the aggregation of thrombocytes, retard fibrinolysis, and have an activating influence on the macrophage system.

Our data indicate that the tolerance of the plasma to heparin is increased and averages 724 s (in healthy persons it is 948 s). The activity of AT III changed significantly: if in healthy persons it is 37.6 s, in IHD patients it is 27.3 s.

AT III is a universal inhibitor of all the enzyme coagulation factors, especially of thrombin and the factor Xa. It is the basic plasma cofactor of heparin. It was noted in clinical observations that an acquired or congenital lowering of the AT III level entails thrombosis of arteries and veins, and thrombophilia.

The fibrinolytic activity of the blood experiences considerable changes. In healthy persons, it is 20.3%, while in IHD patients it lowers to 11.2%. Fibrinolysis is a very mosaic biochemical process in which an important role is played by plasmin. A drop in the level of the fibrinolytic activity of the blood is an important risk factor of thrombosis appearance, which is due to the insufficient destruction of fibrin.

Local injuring factors play an important role in the development of atherosclerotic changes in a vessel wall [29, 491, 507]. The lesions of vessel walls may be either structural (traumatic, atheromatous, an "antigen-antibody" reaction, etc.) or functional due to disturbance of cellular metabolism, exhaustion of energy stores, inhibition and disturbance of the function of individual systems in a vessel wall (enzymes). Especially great significance in the pathogenesis of atherosclerosis is attached to functional disturbances of vessel walls such as lowering of the antiaggregation, anticoagulation, and fibrinolytic activity. Favorable conditions for the aggregation and adhesion of thrombocytes and local deposition of fibrin appear when injuring factors act on a vessel wall.

The endothelium of vessels is a very sensitive morphological substrate that reacts very delicately to the modulating effect of the thrombocytes. The latter act on the endothelium by the direct contact of cells, and by the release and entering into the zone of lesion of a vessel well of a mitogenic factor with the subsequent proliferation of the smooth muscle cells [287].

Lesion of a vessel wall creates conditions for the embodiment of the adhesive and aggregation function of the thrombocytes [124, 489, 491, 493, 494, 495]. Experiments showed the process to begin with the contact of the thrombocytes with the subendothelium, in particular with the main stimulator of adhesion collagen. In the thrombocytes, distinctive processes swell. Cofactors of this occurrence are the calcium ions in the blood plasma and the protein synthesized in the endothelium – the Willebrand factor, and in the thrombocytes – the thrombocyte aggregation factor interacting with it. This leads to adhering of the blood platelets to one another and the deposition of aggregates in the region of injury of a vessel, In addition to collagen, ADP, catecholamine, and serotonin secreted from a vessel wall are stimulators of thrombocyte aggregation.

The next stage of aggregation is determined by the substances separated from the thrombocytes: beta-thromboglobulin, antiheparin factor IV, etc. When the thrombocyte plasma factors interact, thrombin forms in the hemostasis zone. It accelerates drastically the aggregation and thrombus formation.

The physicochemical aspects of thrombocyte aggregation are a process consisting of many units appearing under the influence of various inductors that are secreted when the intimas of vessels are injured. The receptors of the thrombocyte membranes interact with collagen, ADP, thrombin, fibrinogen, and the Willebrand factor. When an inductor interacts with a receptor, a signal is fed into a thrombocyte that activates phospholipase A_2 and phospholipase C, and Ca^{2+} is mobilized from the thrombocytes into the cytoplasm. The further development of the biochemical processes is associated with arachidonic (eicosatetraenic) acid in whose conversion process thromboxan A_2 appears. It is just what facilitates the appearance of thrombocyte aggregates as the basis of the physiological formation of a hemostatic thrombus at the site of lesion and in the blood flow itself.

When comparing the quantitative parameters of thrombocyte aggregation in healthy persons and in IHD patients, we noted significant distinctions. First of all, spontaneous aggregation of the thrombocytes (6.9%) was discovered in the patients that is absent in healthy persons. The aggregation of the thrombocytes induced by ADP is considerably increased – up to 79.1% (61.0% in healthy parsons). Aggregation stimulated by collagen increases very clearly – up to 69.0% against 57.0% in healthy persons.

The collagen-induced aggregation is very important because with an insignificant morphological change in a vessel wall, conditions appear for the formation of an intravascular thrombocyte thrombus. Injury of the endothelium with the uncovering of collagen is an important starting factor of thrombocyte aggregation.

The aggregation of thrombocytes also changes considerably in IHD under the influence of adrenaline. If in healthy persons it is 13.4%, in IHD it increases to 30.2%. This points to the sharp growth in the sensitivity of the receptor apparatus of thrombocytes to adrenaline. The conclusion follows that insignificant emotional stresses or a moderate physical load may provoke the activation of thrombocyte aggregation with possible progress up to incidence of the DIC syndrome.

The pathophysiological process of the relation between thrombocytes and the endothelium begins with contact with collagen. This stimulates the swelling of the thrombocytes and the formation of processes with whose aid attachment to the subendothelium occurs. Calcium ions and the Willebrand factor synthesized in the endothelium are important cofactors of this process. A deficiency or absence of the Willebrand factor is attended by sharp disturbance of the adhesiveness of the thrombocytes.

The discharge of catecholamine, serotonin, and ADP from the damaged part of a vessel are the biochemical basis of thrombocyte aggregation (with the active participation of collagen).

An exceedingly important role in the regulation of thrombocyte hemostasis is played by the products of arachidonic acid. Prostaglandins form under the influence of phospholipases from a vessel wall and under the influence of cyclooxygenase from the membrane phospholipids of the thrombocytes. A powerful aggregating agent – thromboxan A_2 – forms upon the further transformation of the prostaglandins. If the activity of the cyclooxygenase lowers, the function of the thrombocytes is disturbed, and bleeding develops. Another variant of disturbance of blood coagulation (an increased coagulant activity) appears in the disturbance of the synthesis and secretion of prostacyclin into the blood from a vessel wall. This results in an increased tendency of the thrombocytes to aggregation and the appearance of thromboses.

We determined a growth in the content of beta-thromboglobulin up to 38.7 ng/ml (in healthy persons 29.2 ng/ml) and the activity of the thrombocyte factor IV to 18.9 s (up to 14.9 s in healthy persons). Beta-thromboglobulin and thrombocyte factor IV are released from the alpha granules of the thrombocytes and are markers of the reaction of freeing of the thrombocytes. A growth in their content in IHD indicates intravascular activity of the thrombocytes and an increase in the reaction of their freeing.

All the functional changes in the thrombocytes, their receptor sensitivity, and tendency of an increased or diminished aggregation are closely related to the cAMP level. The latter in the thrombocytes or IHD patients is lowered to 3.0 pmol/$3 \cdot 10^8$ cells (against 4.1 in healthy persons). Numerous biochemical and biophysical studies revealed that the cAMP is an intracellular mediator of neuromediators and hormones. It determines a definite functional state of a cell by the intracellular redistribution of the calcium ions. A growth in the cAMP level lowers the activity of the thrombocytes. The drop in the cAMP level in thrombocytes we noted in IHD causes a substantial growth in the activity of aggregation.

Thrombocytes have a number of functional properties specific for them:

(1) the ability of changing their shape (a discoid shape may transform into a spherical one);

(2) aggregation (adhesion to one another);

(3) the synthesis of prostaglandin and thromboxan;

(4) secretion of the content from dense granules;

(5) secretion of the content from alpha-granules.

The last two functions are associated directly with the formation and compacting of a blood clot. Of specially great significance is the change in the shape of the thrombocytes (transformation from a discoid to a spherical one), the formation of spurs that attach themselves to various components of the basal membrane (exactly to the collagen of an injured vessel). The adhesion of thrombocytes to the filaments of collagen is a specific property of these cells and distinguishes them from all others. Experimental studies showed that the influence of collagen on thrombocytes and the adhesion process itself have several stages [531, 532]: binding of the collagen filaments to the cells (thrombocytes), adhesion proper, induction of the release reaction, the release reaction itself, and, finally, the formation of thrombocyte aggregates.

The antithrombogenic activity of the endothelium is ensured by numerous factors synthesized by the vessel walls, and also synthesized in other organs, but adsorbed on its surface.

We support entirely the concept of Ross [489, 490, 491, 493, 494] on the pathogenesis of development of atherosclerosis in which he attaches major significance to the proliferation of smooth muscle cells in the intima of vessels and to thrombosis in the development of an atheromatous plaque.

Many investigators experimentally and under clinical conditions are attempting to prove the relation between the change in lipid metabolism and the functional activity of the thrombocytes [57, 97, 428]. However, many links of this relation still remain a puzzle. In particular, it was noted that there is a relation between the level of the membrane cholesterol and the hyperactivity of the thrombocytes [121]. A growth in the concentration of the membrane cholesterol produces hyperactivity of the thrombocytes. A high cholesterol content in the membranes of thrombocytes, in turn, is closely related to hyperlipidemia.

When analyzing the information supplied by various lipid parameters and of glucose in the blood for estimating their influence on the aggregation of thrombocytes, some authors found that the content of total cholesterol in the blood is less significant than an increased level of glucose [65, 84, 111]. In view of the results, investigators concluded that hyperglycemia combined with a high level of insulin boosts the synthesis of cholesterol esters in a vessel wall. It is just these moments that facilitate the development of atherosclerosis. Moreover, the biochemical content of the membrane of a thrombocyte itself is also exceedingly important. In particular, an increased content of esters of cholesterol and the free compound in the thrombocytes is extremely unfavorable. Conversely, a high level of alpha-cholesterol is a favorable metabolic parameter [88, 96, 115].

An increased level of beta-thromboglobulin is still another very important unfavorable factor facilitating the progress of atherosclerosis [34]. It is a specific protein of the thrombocyte granules and is secreted from them in the course of aggregation. The high level of beta-thromboglobulin may point to the participation of the intravascular aggregation of thrombocytes in the pathogenesis of stenocardia in atherosclerosis patients [34].

We attach special significance in the increased aggregation of thrombocytes to spontaneous aggregation, which is absent in healthy persons and appears in IHD.

What are these changes associated with?

There is no doubt that an important role is played by a change in the physicochemical properties of a thrombocyte: first of all, ischemia in the organs and tissues disturbs the energy potential of a thrombocyte, and this was proved in numerous experiments.

A change in the function of a cell (in particular of a thrombocyte) may occur under the effect of hypoxia, ischemia, a progressing hyperfunction, a deficiency of energy, and disturbance of the information relation with the environment [534]. The development of ischemia in a cell may cause depolarization of a membrane with the corresponding loss by the cell of one of the forms of "free energy", i.e. the energy of the transmembrane potential. In this case, homeostasis of the cell is disturbed with a limited possibility of utilizing the energy, substrates, and adequately perceiving the information received. A state appears that in physiology is customarily called metabolic disorders. Here the intracellular regulation of homeostasis suffers. Three main components determine the functional viability of a cell: the exchange of substances, energy, and an exchange of information. These processes under the influence of ischemia, cold, toxicosis, or the action of pharmacological preparations may alter qualitatively and quantitatively.

Disturbance of the activity of cAMP in a cell is the pathophysiological basis of a change in the aggregation of thrombocytes in IHD. Here the level of cAMP is lowered to 3.0 pmol/$3 \cdot 10^8$ cells (in healthy persons it is 4.1).

Disturbance of the intactness of the membranes and a change in their physical state lead to a qualitative disturbance of the processes of the exchange of substances, energy, and information.

On the other hand, if we consider a thrombocyte from the viewpoint of an isolated cell, it is a nonequilibrium system that exchanges substances, energy, and information with its environment (the blood) and ensures the independent maintenance of its nonequilibrium state. The reason is that this cell is a unique nonequilibrium system. A very important role in a thrombocyte reaction (thrombocyte metabolism) is played by cAMP. Depending on its level, there is a whole network of cAMP-dependent reactions. In particular, of special importance is the activation of a whole cascade of proteinkinases responsible for the phosphorylation of the ionic channels of the proteins of the mitochondria, ribosomal apparatus, and membranes of many enzymes. Consequently, cAMP via the system of proteinkinases affects ionic permeability, depolarization of membranes, and activation of the consumption of ATP.

The relation between a cell (thrombocyte) and its environment is determined by a set of nonspecific biochemical reactions appearing in response to disturbance of the exchange of substances, energy, and information. Every cell (including a thrombocyte) has a specific acceptor zone. The latter is physically bound to the cell membrane and it has the ability to perceive external information, modulate it to a level of information capable of functioning, and create a possibility of exchanging nutrient substances and energy with the environment.

The Willebrand factor participates in the adhesion of thrombocytes. Clinicians noted that a deficiency of this factor leads to hemorrhagic diathesis (angiohemophilia, Willebrand's disease) [552]. This factor interacts with thrombocytes by means of two membrane thrombocyte receptors. Thrombocytes are bound to the Willebrand factor when they are stimulated by thrombin and ADP. The thrombin-induced relation of the Willebrand factor with thrombocytes is disturbed in Glanzmann's thrombosthenia that is characterized by disorders of thrombocyte aggregation. In our investigations, we noted an increase in the level of the Willebrand factor to 2.18 un/ml (in healthy persons 1.04). Many clinicians noted that a growth in the level of the Willebrand factor in IHD and DM patients is attended by the rapid progress in atherosclerotic changes. It was noted in experiments that in animals with a deficiency of the Willebrand factor, proliferation of the aorta intima was lower, while fat infiltration was higher than in animals of the control group.

The thrombocytes circulating in the blood channel do not adhere to a normal uninjured endothelium of vessels. From biophysical standpoints, this is explained by the negative electric charge of the thrombocytes and endothelium, which repels them from each other. However, the negative charge is not the only factor inhibiting the adhesion of thrombocytes.

In lesion of the endothelium, the thrombocytes easily adhere to the subendothelium, which is due to the lowering of the content of glycosaminoglycanes having a negative charge and prostacyclin synthesized by the endothelium. The latter prevents adhesion and aggregation of the thrombocytes. It must be stressed that prostacyclin is synthesized in the endothelium not only of the arteries, but also of the veins. The lungs are an especially powerful generator of prostacyclin. Prostacyclin is synthesized and secreted in an organism with stimulation by vasoactive agents (bradikinin, histamine, histamine, angiotensin II, vasopressin, leukotrienes C and D, interferon).

As already noted, prostacyclin forms from arachidonic and eicosapentaenic (series ω-3) acids.

The physiological effect of prostacyclin is very significant: it inhibits the aggregation of thrombocytes, has a vasodilating effect, relaxes the smooth muscles of vessels (including the coronary ones) causing their dilation, lowers the arterial pressure, and has antiarrhythmic, antisclerotic, and antiulcerogenic properties. The pathophysiological mechanism of the antiaggregation effect of prostacyclin is due to its ability of increasing the effect of adenylatecyclase of the thrombocytes. In this connection, the synthesis of cAMP increases, and the ability of the thrombocytes to aggregate lowers. Consequently, the antiaggregation effect of prostacyclin is due to its binding with the specific receptors of thrombocytes and activation of the thrombocyte adenylatecyclase. This is attended by an increase in the level of the cytoplasmic cAMP and lowering of the Ca^{2+} level in the cytoplasm of the thrombocytes and results in diminishing of cell interaction. The antithrombic properties of the vessel wall are especially significant in the development of thrombosis [405]. They are ensured by the synthesis of a number of antithrombotic components, namely, (i) prostacyclin, which is an inhibitor of thrombocyte aggregation and ensures a pronounced vasodilating effect, (ii) anticoagulants (antithrombin III and alpha-2-macroglobulin, (iii) the proactivator of plasminogen produced in the vessels, and (iv) thrombomodulin formed on the surface of the endothelial cells binding thrombin and activating protein C.

It is the functional interaction of all these antithrombotic factors that prevents the deposition of thrombocytes on the intima of vessels and the development of thrombi.

The ratio of the content of thromboxan B_2 and prostacyclin is of major significance. In IHD, the thromboxan level rises to 37.0 pg/ml (against 31.0 in healthy persons), while the prostacyclin content lowers from 65 pg/ml (in healthy persons) to 51.0. A change in the prostacyclin/thromboxan B_2 ratio is very important. If in healthy persons it is 2.1, i.e. with a considerable predomination of PrC, in IHD the ratio sharply shifts toward the predomination of thromboxan,

Clinical and anatomic studies showed that in IHD patients the amount of LDLP in the intima of the coronary arteries grows. These proteins have a multilateral effect on the coagulating system [219, 220, 251, 374, 527]. First of all, the LDLP inhibit fibrinolysis and increase the thromboplastic activity. The decomposition of the LDLP is accompanied by the release of cholesterol, which has a cytotoxic effect on the endothelium. These processes are attended by the synthesis of active thromboplastin. It was noted in immunofluorescence that the sites of deposition of thromboplastin and fibrin on the intima of the coronary arteries coincide in the sections of lipoidosis [228, 273]. This proves the participation of thromboplastin in intramural thrombus formation.

Table 21. Content of Fatty Acids in Thrombocytes in IHD Patients.

Fatty acid, %	Control group	IHD patients
Linolic ($C_{18:2}$, ω-6)	5.70±0.28	4.50±0.31*
Arachidonic ($C_{20:4}$, ω-6)	19.10±0.75	25.47±0.52*
Eicosapentaenic ($C_{20:5}$, ω-3)	0.40±0.09	0.70±0.11
Docosahexaenic ($C_{22:6}$, ω-3)	2.05±0.06	1.80±0.08*
Coefficient K_{ac}	7.80±0.61	10.21±0.78*

* The difference of the parameters is authentic (in comparison with the level in the control group).

The secretion of active thromboplastin into the blood in hypercholesterolemia in IHD patients increases the danger of thrombus formation because the time needed for activation of the tissue thromboplastin is less than for the plasma substance. Immunofluorescence of the intima of IHD patients noted a diminishing of the zones of fibrinolysis [241]. Of a great significance here is the change in the amount of receptors of the intramural nervous system, which is attended by lowering of the information on the threat of thrombus formation.

The glycocalyx layer of the intima surface having an anticoagulation effect, a constituent part of which is profibrin, plays an appreciable role in thrombogenesis [98, 322]. Preservation of the integrity of the glycocalyx ensures the exchange of substances between the blood and the internal membrane of a vessel. In the late stages of atherosclerosis and lesion of the endothelium or de-endothelization of the intima, the continuity of this layer is disturbed, which leads to thrombus formation.

Biochemical investigations of recent years revealed that in the thrombocytes and endothelial cells of the vessels, there is an autonomous system of fatty acid synthesis [30, 31, 128, 311]. Accordingly, the level of fatty acid activity in the plasma and phospholipid of thrombocytes reflects the prognosis of the thrombolytic inclination of a given person and his inclination to atherosclerosis. It is also known that arachidonic acid is the main substrate for the formation of prostaglandins, thromboxan, and prostacyclin. Inhibition of the synthesis of this acid from linolic acid is carried out with the aid of eicosapentaenic and docosahexaenic acids [97, 428].

It is especially important that eicosapentaenic acid has a high affinity for cyclooxygenase and competes with arachidonic acid, inhibiting its transformation into thromboxan A_2 [546].

Apart from the quantitative changes in the components of the coagulating system of the blood, qualitative changes in the thrombocytes occur in IHD. Table 21 shows the content of fatty acids in healthy persons and IHD patients. We can note that the level of linolic acid in the IHD patients dropped to 4.5% (against 5.7% in healthy persons), and that of arachidonic acid increased to 25.4% (against 19.1% in healthy persons).

It is very important to compare the content of eicosapentaenic and docosahexaenic acids. In IHD, the level of the former is increased to 0.70% (against 0.403 in healthy persons), and of the latter is decreased to 1.80% (against 2.05%). The coefficient K_{ac} is increased considerably to 10.2 (against 7.8) This points to the considerable growth in the synthesis of thromboxan with a corresponding growth in the activity of the thrombocytes. In combination with a drop in the prostacyclin level, the growth in that of thromboxan produces a tendency of thrombus formation.

The role of insulin, C-peptide, cortisol, and STH is especially significant in the hormonal regulation of the prostacyclin and thromboxan content. We compared the content of these hormones in healthy patients and IHD patients. A glance at Table 22 shows that in IHD patients the insulin level grows (very significantly – two times) to 13.9 0 μU/ml (against 7.8 in healthy persons). The content of C-peptide grows somewhat from 1.2 ng/ml (in healthy persons) to 1.61 ng/ml (in IHD). The cortisol level grows to 386 nmol/l (against 250 in healthy persons).

Most publications devoted to studying the pathogenesis of atherosclerosis concentrate attention on the functional disconnection of the hormonal systems in IHD attended by the appearance without fail of hyperlipoproteinemia and the frequent development of hyperglycemia. The neurohormonal system of an organism relates factors of the environment (stress, hypodynamia, irrational nutrition) and genetic risk factors of IHD (a change in metabolism, dyslipidemia, hyperglycemia, smoking, obesity, etc.). The hormonal system plays an important role in these relations. It should be noted that the function of every endocrine unit within the limits of one hormonal system alters differently. Functional disintegration of the external environmental, internal medium, and genetic factors appears in the disconnection and polymorphism of functioning of individual units of these systems.

The combination of hypercholesterolemia and hypertriglyceridemia facilitates the aggregation of thrombocytes, especially if these states are accompanied by high values of LDLP. The aggregation ability of thrombocytes lowers under the influence of a high level of the HDLP that have a disintegrating property. The regulating role of the hormones in the aggregation nature of the thrombocyte effect was confirmed in numerous experiments that proved the presence on thrombocyte membranes of specific receptors to hormones [352, 411]. Receptors to insulin are especially important. Their presence confirms the

Table 22. Hormone Content in Blood Plasma in IHD Patients.

Parameter	Control group	IHD patients
IRI, μU/ml	7.8±1.4	13.9±2.0*
C-peptide, ng/ml	1.2±0.17	1.61±0.11*
Cortisol, nmol/l	250±30.0	386±38.0*

* The difference of the parameters is authentic (in comparison with the level in the control group).

regulating role of insulin in aggregation. Moreover, the presence of beta-thromboglobulin in the thrombocytes of DM patients with the presence of microangiopathies was proved [40, 141, 146, 147, 230, 627]. A growth in the amount of arachidonic acid in the thrombocytes ensures a special functional activity for these cells, in particular, a growth in the adhesive and aggregation properties. In these patients, the formation of thromboxan B_2 in the thrombocytes with lowering of the synthesis of prostacyclins increases, which ensures aggregation ability.

Most investigators studying the pathogenesis of atherosclerotic lesions give their main attention to the primary hormonal disorders and the secondary changes in the metabolic processes [453, 519]. Indeed, this is logical: the presence on the one hand of DM, disorders of the estrogenic function, hypothyroidism, hypothalamic disorders, etc., and the development of atherosclerosis on the other.

Much less investigations were devoted to the influence of metabolic disorders (e.g. hyperlipidemia) on the change in the hormonal state in an organism [41].

During many years, we studied the "functional unit of switching over" nerve regulation to humoral. It is in this switching over of the neurogenic influences of negative motions to humoral metabolism that we can reveal various units of the pathogenesis of atherosclerosis.

The role of the endothelium in the development of coronary spasm always attracted the attention of both experimentators and clinicians. Currently, the direct participation of the endothelium in regulating the tonus of the coronary arteries by the secretion of what is called the endothelial relaxing factor, whose nature still remains a puzzle, has been proved [57, 196, 236, 482, 593].

It was noted in investigations that in the atherosclerotic lesion of the coronary arteries, the reactivity of the endothelial cells is disturbed not only of the atheromatous plaque itself and around it, but also at a considerable distance [507]. This gives us grounds to speak of a definite "functional insufficiency of the endothelium" in atherosclerosis that is manifest in the loss of the ability to produce factors relaxing the endothelium. Moreover, it must be remembered that in atherosclerosis an increased secretion (in the adhesion of thrombocytes) of thromboxan having a powerful vasoconstrictor property is noted. Therefore, the presence of an inferior endothelium, spasm of the vessels, and the formation of a thrombus are very important factors underlying the further formation of an inferior endothelium.

An important role in the formation of atheromatous plaques is played by damage to the endothelium, which is noted in hypercholesterolemia, smoking, bacterial and virus infection, or mechanical action [489, 490, 491, 492, 493, 494]. Thrombocytes adhere to the damaged section of the endothelium with the subsequent proliferation of the smooth muscle cells.

Cholesterol, VLDLP, and LDLP have a direct cytotoxic effect on the endothelium. A number of studies showed that an increased deposition of cholesterol in the endothelium correlates with the LDLP level [453, 536, 613]. A leading role in excreting cholesterol from an organism belongs to HDLP. The cholesterol extracted from the endothelial membrane under the influence of LCAT is subjected to esterification and is transported into the liver.

The HDLP inhibits the penetration of cholesterol into the cells of a vessel wall in the composition of the LDLP or VLDLP. There is a biochemical relation (negative correlation) between the content of triglycerides and HDLP due to the change in the lipoproteinlipase [23, 108].

Most publications devoted to atherosclerosis use the following formula of atherogenesis:

$$A = \frac{C_{VLD} + C_{LD}}{C_{HD}}$$

where C_{VLD}, C_{LD}, and C_{HD} are the content of cholesterol in the very low density, low density, and high density lipoproteins, respectively. The formula reflects the probability of incidence of atherosclerosis [234, 351]. Clinicians note the frequent combinations of hyperlipoproteinemia with DM, hypertension, general obesity, and gout [275, 288, 425].

It was shown in experiments that inhibitors of thrombocyte aggregation entering the blood flow are produced in the vessels of a person not injured by atherosclerosis [78, 202, 228]: group 1 includes prostacyclins that form from arachidonic acid and block the aggregation of thrombocytes by raising the cAMP level in the cells; group 2 is the nitrogen oxide forming in the cells from the primary amino groups of proteins, amino acids, or organic nitrocompounds; and group 3 is an unidentified factor of weakening of the smooth

Table 23. Antithrombogenic Activity of Vessel Wall in IHD Patients.

Parameter	Control group	IHD patients
Antiaggregation activity of vessel wall		
Aggregation of thrombocytes induced by ADP 10^{-5} M, %:		
before compression	61.0±2.3	79.1±2.4
after compression	48.2±1.9	90.1±2.0
index	0.8±0.09	1.18±0.07*
Aggregation of thrombocytes induced by ADP 10^{-6} M, %:		
before compression	4.2±0.17	10.3±0.2
after compression	3.4±0.16	13.8±0.21
index	0.81±0.07	1.31±0.09*
Aggregation of thrombocytes induced by collagen, %:		
before compression	57.0±1.4	69.0±2.2
after compression	49.0±1.6	77.0±2.1
index	0.86±0.07	1.12±0.05*
Aggregation of thrombocytes induced by TAF, %:		
before compression	14.5±1	22.3±1.3
after compression	11.2±0.9	27.1±1.3
index	0.78±0.08	1.2±0.06*
Aggregation of thrombocytes induced by adrenaline 10^{-4} M,%:		
before compression	59.3±2.1	79.2±2.4
after compression	52.0±2.0	92.0±2.1
index	0.87±0.07	1.2±0.06*
Thrombocyte factor IV, s:		
before compression	14.9±0.9	18.9±0.9
after compression	11.7±0.8	22.2±0.5
index	0.78±0.07	1.17±0.06*
6-Keto-PGF_1-alpha in plasma, ng/ml:		
before compression	65.0±3.2	51.0±2.9
after compression	109.0±3.0	52.0±3.2
index	1.8±0.08	1.0±0.09*
Anticoagulant activity of vessel wall		
Activity of antithrombin III, s		
before compression	37.6±1.5	27.3±0.9
after compression	49.0±1.6	22.1±0.9
index	1.31±0.07	0.82±0.06*
Fibrinolytic activity of vessel wall		
Fibrinolytic activity of blood, %:		
before compression	20.3±0.9	11.2±0.8
after compression	32.1±1.2	8.4±0.9
index	1.61±0.08	0.76±0.09*

* The difference of the parameters is authentic (in comparison with the level in the control group).

muscles. It was proved that the relaxing factor acts on guanylate cyclase and raises the cGMP level in the cells.

Many investigations were devoted to studying the synthesis and secretion of these factors in the vessel wall [67, 97, 98]. It was proved first of all that the endothelial cells are the main place of synthesizing these cells. Receptors and intracellular mediator signals were discovered that control the synthesis and secretion of these substances. The most important stimulators of the secretion of these inhibitors of thrombocyte aggregation are acetylcholine, bradikinin, vasopressin, and adenine nucleotides. Special significance is attached to the synthesis of prostacyclin in the endothelial cells. It was proved that the regulation of its synthesis is associated with the level of free arachidonic acid. In vessels injured by atherosclerosis, the formation of prostacyclin lowers: the content of cAMP and enzymes in the cells diminishes. The properties of the cells alter: their proliferation and migration grow, the resistance to injury lowers, and processes of anabolism predominate over catabolism. An increased risk of thrombus formation appears.

When animals received a fat load, a growth in the activity of phospholipase A_1 and cathepsin D was proved, while the activity of other lysosome hydrolases did not practically change [262, 272, 500]. It was noted under clinical conditions that in patients with IHD and in obesity the activity of phospholipase A in the adipose tissue grows [519, [64, 250, 312, 343]. When an IHD patient did not have an increased body weight, this relation was not observed, thus confirming the concept that the activation of phospholipases A and cathepsins B and C increase the degradation of HDLP rich in phospholipids, increases the concentration of LDLP, and is an integral and very important link in the pathogenesis of dyslipoproteinemia.

To study the functional state of the endothelium of vessels in IHD patients, we performed a cuff test (Table 23). When analyzing the ADP-induced aggregation of thrombocytes in healthy persons, we found it to lower from 61.0 to 48.2%, whereas in IHD patients it increased from 79.1 to 90.1%. The collagen-induced aggregation of the thrombocytes alters similarly: in healthy persons it lowers from 57.0 to 49.0%, and in IHD patients it increases from 69.0 to 77.0%.

What is the pathophysiological basis of this phenomenon?

Morphofunctional lesion of artery walls (endothelium) involves disturbance of the antithrombogenic activity (antiaggregation, anticoagulant, fibrinolytic). This is exactly the basis on which the initial link of atherosclerosis forms with the relevant development of adhesion and aggregation of thrombocytes, the deposition of fibrin, and the formation of an atheromatous plaque.

LDLP are another important component participating in the formation of a plaque in a vessel wall. They have a considerable procoagulant activity, and their increased concentration in the blood stimulates its coagulation. The thrombocytes affect the metabolism and function of the endothelium quite considerably by stimulating the proliferation of fibroblasts, endothelial, and smooth muscle cells. The integrative relation of thrombin, factor XIIIa, thrombocytes, and LDLP ensures the proliferation of various cells of the vessel walls, and the penetration of lipids into a vessel wall. This penetration may be ensured in a nonregulated way, i.e. without a receptor. Many hematologists relegate an important role in this process to the Willebrand factor whose main function is the binding of thrombocytes to the wall of vessels [34, 57, 197, 441, 442, 552, 581]. In this connection, a growth in the content of the Willebrand factor in the blood of IHD and DM patients stimulates the development of atherosclerotic changes in vessel wells.

Of very great significance is the adrenaline-induced aggregation of thrombocytes. In healthy persons, it lowers under the influence of a cuff test from 59.3 to 52.0%, while in IHD patients it grows from 79.2 to 92.0%.

Adrenaline is one of the very important inhibitors of the antiaggregation ability of a vessel wall [29]. It not only inhibits this ability, but also drastically increases the thrombogenic potential of the blood by accelerating the coagulation and facilitating the aggregation of the thrombocytes. Clinical studies [538] showed that in an emotional stress a large amount of adrenaline is discharged into the blood, and local inhibition of the antiaggregation ability of the vessel walls occurs that disturbs the homeostatic balance between the thrombocytes and vessel walls.

The reaction of the vessels to the production of prostacyclin is very important. In healthy persons, the excretion into the blood of prostacyclin under the influence of a cuff test is very significant: from 65.0 to 109.0 pg/ml. In IHD patients, the prostacyclin level does not practically change (51.0, 52.0 pg/ml).

What is the pathogenesis of this phenomenon?

Antithrombin III ensures the anticoagulant function of the blood. In healthy persons, the reaction of the vessels to an ischemic test is attended by a considerable growth in the activity of AT III. This was also confirmed by our studies. In healthy persons, the activity of AT III after a cuff test increased from 37.6 to 49.0 s; the opposite reaction occurs in IHD patients – the activity of AT III dropped from 27.3 to 22.1 s. A similar reaction was noted for the fibrinolytic activity of the blood: in healthy persons it grows from 20.3 to 32.1%, while in IHD patients it drops from 11.2 to 8.4%.

It is exactly the lowering of the anticoagulation function of hemostasis (AT III and the fibrinolytic activity of the blood) in combination with the inhibition of prostacyclin synthesis that is favorable soil for the appearance of the DIC syndrome in IHD patients.

The DIC syndrome is a clinical nonspecific pathological process of a general biological nature connected with the entry into the blood flow of activators of blood coagulation and thrombocyte aggregation, the formation of thrombin in it, the activation and exhaustion of the coagulating, kallikrein-kinin, and fibrinolytic systems. With this syndrome, laboratory analyses of the blood reveal a multitude of microclots and aggregates of cells blocking microcirculation in organs. This results in thrombohemorrhages, acidosis, and intoxication of an organism. The clinical manifestations of the DIC syndrome are a multitude of hemorrhages, intestine, uterus, and nose bleedings, etc. Lung infarctions are manifest in the form of dyspnea, hemoptysis, cyanosis, and lung edema.

We agree completely with the opinion of Barkagan [34] on the pathogenesis of the DIC syndrome. He writes: "The following mechanisms underlie most forms of the DIC syndrome: (1) activation of the coagulation system of the blood and thrombocyte hemostasis by endogenic factors – tissue thromboplastin, products of the decomposition of tissues and cells of the blood, leukocyte proteases, injured endothelium; (2) the activating effect on the same units of the hemostasis system of exogenic factors – bacterial, viruses, Rickettsias, transfusion and medical drugs; (3) inferior or systemic injury of the vessel endothelium and lowering of its antithrombotic potential; (4) dispersed intravascular coagulation of the blood and aggregation of the thrombocytes and erythrocytes with the formation of a multitude of microclots and blocking by them of blood circulation in the organs; (5) deep dystrophic and destructive disorders in the organs-targets, weakening and dropping out of their functions; (6) deep circulation disorders, hypoxia of tissues, a hemocoagulant shock, acidosis, disturbance of microcirculation, the sludge syndrome; (7) coagulopathy of use with exhaustion of the anticoagulating mechanisms and components of the fibrinolytic and kallikrein-kinin systems, and (8) secondary severe endogenic intoxication by products of proteolysis and destruction of tissues".

2.3. Hemostasis System and Rheological Properties of the Blood in Diabetes Mellitus Patients

2.3.1. Injury of a *V*essel Wall in Diabetes Mellitus

Metabolic disorders in DM, as already indicated, are attended by a number of functional lesions (tissue hypoxia, dyslipoproteinemia, hormonal imbalance, metabolic acidosis, angiopathies, electrolytic disorders, etc.). Special significance is acquired by the relation between disorders of the oxygen-transportation function of the blood, the hemostasis system, and the development of microangiopathies.

Tissue hypoxia in DM patients occurs according to the following scheme: erythrocyte membrane – blood plasma – capillary wall – tissue. An increased lipid content in the blood plasma detracts from the transcapillary diffusion of the oxygen through the lipid-protein ultrafilm to the internal surface of the capillaries.

A genetic defect of a vessel wall triggers off the development of lesions in DM [10, 355, 381, 548, 549]. The manifestation of this defect and the further progress of vessel lesions are due to diabetic disorders of metabolism.

Disorder of the functional state of the system of hemostasis leading to intravascular coagulation of the blood, thrombus formation, or blood flow underlies the entire pathology. There is currently no doubt that disorders in the hemostasis system play a substantial role in the development of atherosclerosis, IHD, and microangiopathies in DM.

In 1956, a hypothesis was advanced according to which lesion of endothelial cells leads to the incidence of atherosclerosis. Later some investigators [428, 467, 469, 489, 490] proposed a hypothesis, quite widespread today, that the development of atherosclerosis includes the following units:

(1) lesion of the endothelial lining of a vessel wall leading to desquamation of the endothelium;

(2) adhesion of thrombocytes to and their aggregation on uncovered connecting-tissue structures and the release of thrombocyte factors (including the mutagenic factor) into the subendothelial layer;

(3) migration of the smooth muscle cells through an internal elastic membrane and their proliferation into the intima;

(4) the formation of a connecting-tissue matrix by the synthesis of collagen, elastic fibers, and glycosaminoglycanes;

(5) the intra- and extracellular accumulation of lipids from the blood plasma.

When studying the various units of atherosclerosis formation in DM, many authors attach great significance to the change in the functional activity of the endothelial cells of vessels, the Willebrand factor, the secretion of prostacyclin and the fibrinolytic activity of the blood [216, 526]. The Willebrand factor is of special significance in disturbing the function of the endothelium in DM. Being a glycoprotein, this factor is present in the plasma as a part of the complex of factor VIII. Since this factor is produced by the endothelium of the vessels, a growth in its concentration in the blood plasma reflects the injury of the endothelial cells. The Willebrand factor is studied most frequently from the viewpoint of the adhesion of the thrombocytes to the subendothelium, and in this connection it is included in the pathogenesis of development of atherosclerosis. In DM, the level of the Willebrand factor in the blood plasma grows.

We can consider as a whole that an increase in the glycoprotein content in the blood (the Willebrand factor) points to the presence of vessel lesions in DM. Prostacyclin synthesized by the vessel walls from arachidonic acid is an important component in the pathogenesis of atherosclerosis in DM. The prostacyclin inhibits the aggregation of the thrombocytes, their adhesion on the vessel wall, and is a powerful vasodilator. Disturbance of the production of prostacyclin in DM patients results in lesion of vessel walls, aggregation and adhesion of the thrombocytes, and the formation of microthrombi.

Many authors find that the crossed binding of plasma proteins with the matrix proteins of vessel walls (with the collagen part) by means of the products of pronounced glycosylation and binding of insoluble complexes of the basal membrane with one another are an important pathogenetic unit of diabetic micro- and macroangiopathies [220, 374, 375, 587, 607]. This relation results in the formation of complexes that are the site of attachment of other proteins of the plasma, which leads to the cascade growth of the process. An important place in the normal homeostasis of a vessel wall is played by macrophages that "purify" the wall from denaturated protein complexes.

In DM patients, a surplus accumulation of the products of pronounced glycosylation appears, and equilibrium of homeostasis shifts in the direction of a lower removal of the denaturated protein complexes. This increases proliferation. Many authors consider the formation of vascular complications in DM to be mainly due to glycosylation and the polyol path of glucose metabolism [181, 182].

In the development of atherosclerosis in DM, great attention is given to the disturbance of lipid metabolism, primarily to an increase in the triglyceride level [39, 235, 376, 381, 468, 578]. A number of authors consider the pathogenesis of this phenomenon to be associated with a lack of insulin and the relevant disturbance of the decomposition of VLDLP. In insulin resistance, a situation also occurs when the decomposition of the LDLP is inhibited and the triglyceride level grows. The atherogenic properties of the LDLP are due not only to the growth in their triglyceride content, but also to a change in glycosylation. It was noted that the LDLP are bound by glycosylized collagen fibers at places of lesion of a vessel endothelium, and it is this process that facilitates the formation of atheromatous plaques.

The products of arachidonic acid metabolism – prostacyclin and thromboxan B_2 – are of great interest when studying the development of atherosclerosis. Of special importance for clinicians is the prostacyclin synthesized by a vessel wall. It is the most powerful endogenic vasodilator and inhibitor of aggregation of the thrombocytes. It is also an antagonist of thromboxan (a vasoconstrictor and aggregation agent), therefore a high prostacyclin level ensures prophylaxis of atherosclerosis. It was proved morphologically that the synthesis of prostacyclin diminishes in arteries effected by atherosclerosis, including the coronary vessels [361]. However, there is a direct correlation between the degree of injury of the arteries by atherosclerosis and the production of prostacyclin.

Synthesis of the latter is affected considerably by a high level of the peroxide oxidation of lipids and a growth in the content of the atherogenic lipids [30, 31, 101, 128, 243]. In atherosclerosis patients in connection with the growth in the free-radical peroxide oxidation of lipids, the synthesis of prostacyclin is suppressed, and the thrombogenic potential of the blood grows. Clinical biochemical studies revealed a correlation between the high level of the LDLP and VLDLP and the drop in the synthesis of prostacyclin in a vessel wall, and, on the other hand, a growth in the HDLP level is an important factor stimulating the formation of prostacyclin [30, 31, 101, 128, 429]. An exceedingly interesting, but insufficiently studied fragment in these investigations is studying of the correlation between the parameters of lipid metabolism, the level of peroxide oxidation of the lipids in the blood plasma and in the membranes of erythrocytes, and the state of the prostacyclin-thromboxan system in DM patients. It was noted under clinical conditions that a very high level of peroxide oxidation of lipids in the plasma and the erythrocytes is attended by lowering of the prostacyclin level [30, 31, 101, 128, 243].

In DM patients with a deficiency of insulin, the biochemical process of glucose transformation is distorted [235]. Glucose is transformed in the insulin-dependent tissues not so much by the glycolytic path as via a polyol shunt by activation of the enzyme aldoreductase. The distorted transformation of glucose results in the accumulation of sorbitol and fructose in the aorta, crystal lens, nerve fibers, and kidneys. These products of polyols that are hydrophilic substances promote the development of hydropic edema of the cells and the disturbance of the active transportation of substances through membranes. These disorders cause a reduction in the supply of ATP and amino acids to the cells.

An important role in the development of angiopathies in DM is played by hyperlipidemia and dysproteinemia [380, 466, 525, 515, 611]. The disorders in fat metabolism in DM can be explained pathophysiologically by lowering of the activity of the pentosophosphate cycle – the main path of glucose transformation. Suppression of the cycle lowers the reduction of NADP. This inhibits lipogenesis with the simultaneous increase in cholesterol synthesis.

Disorders of the functions of other hormones, in particular a growth in the cortisol level, are an important factor in the development of angiopathies [39, 376, 468, 578]. Cortisol at increased concentrations stimulates the synthesis of LDLP having pronounced atherogenic properties. This is done via the activation of the enzymes of the adenylatecyclase system that stimulate the synthesis and decomposition of 3.5-AMP in adipose tissue. Hyperglycemia in DM leads to the increased synthesis of glycoproteins, mucosaccharides, an increase in the content of hemoglobin A_1 in the erythrocytes. and via a polyol shunt to a growth in the transformation of glucose and the accumulation of sorbitol. Hyperlipoproteinemia in addition to lowering the activity of the tissue LPL and increasing the content of peroxide compounds facilitates the deposition of lipoproteins in vessel walls.

Glucose and fructose are an important component of the activation of the pentosophosphate cycle with the formation of a large amount of $NADPH_2$. This is the coenzyme that facilitates the synthesis of lipids. This explains the direct relation between DM and atherosclerosis.

The content of glucose in the blood plasma and erythrocytes was determined in numerous investigations [470, 474, 551, 574]. The ratio of glucose in the blood plasma and in the erythrocytes in healthy persons and in atherosclerosis patients was found to differ. In particular, in atherosclerosis more glucose accumulates in the liquid part of the blood then in the erythrocytes, therefore, the utilization of glucose by tissues is lower, the erythrocytes use less of it, and its amount in the plasma grows. Consequently, in atherosclerosis, hyperglycemia is noted. The pathogenesis of hyperglycemia is associated with the inhibition of glucose utilization by tissues, e.g. by erythrocytes, with damage to their membrane. It is interesting that cholesterol when introduced into an organism for a long time clearly increases the glucose content in the blood plasma and erythrocytes [121, 557]. In DM patients, the glucose level in the plasma grows very often. Disturbance of the glucose tolerance in atherosclerosis patients is associated with this phenomenon.

It was noted simultaneously that the same factors facilitate the production of thromboxan A_2 in thrombocytes [101, 296, 506, 575, 594]. There is an important relation between smoking and the formation of prostacyclin [354, 414, 461, 609]. It was proved that the synthesis of prostacyclin is suppressed in smokers because nicotine inhibits the activity of cyclooxygenase. This is why in smokers, the aggregating ability of the thrombocytes grows substantially, the level of thromboxan B_2 increases, and a coronary spasm develops more rapidly. It is important to note that in DM the tendency of thrombocytes to aggregation increases. This is due to the growth in the formation of thromboxan B_2 in the thrombocytes that mediate the antiaggregation effect of prostacyclin and to the growth in the sensitivity of the thrombocytes to ADP.

In DM, the formation of thromboxan-A_2-synthetase is stimulated, and the reaction of the endothelium of the vessels and thrombocytes is disturbed. Of major significance is the relation between the hormonal and the structural-functional properties of the thrombocytes, especially in the development of IHD. A relation was noted between the growth in the sensitivity of the thrombocytes to catecholamine or to ADP and the increase in the cholesterol content in the membranes of the thrombocytes [30, 31, 128]. A growth in the latter is attended by a sharp growth in the aggregation ability and adhesion of the thrombocytes, which become more sensitive to catecholamine and ADP.

The growth in the aggregation and adhesion properties of the thrombocytes is also related to the level of the lipids and their qualitative composition [57, 243]: an increased level of triglycerides increases thrombocyte aggregation. This is especially significant when hypertriglyceridemia is combined with a high level of insulin in the blood.

Not only thrombocytes, but also erythrocytes play an important role in blood coagulation and fibrinolysis. Experiments showed the presence in erythrocytes of endogenic factors of blood coagulation and fibrinolysis such as the thromboplastic factor, calcium, pro-accelerin, fibrin-stabilizing and antiheparin factors [181, 491, 511]. Under normal conditions, homeostasis exists between the coagulating and fibrinolytic factors of the blood plasma and erythrocytes that is disturbed in IHD.

Clinicians noted, and we agree with them, that erythrocytes in IHD patients have pronounced aggregation and coagulating properties as a result of the presence and increase in them of thromboplastic, pro-thrombin, pro-accelerin, fibrinogen-like, and antiheparin activity. In IHD, these properties of the erythrocytes facilitate the formation of a hypercoagulation syndrome and intravascular thrombus formation.

2.3.2. Rheological Properties of the Blood in Diabetes Mellitus

In recent years, great significance when studying diabetes mellitus is attached to hemorheology, i.e. the properties of the blood flow are studied with a view to the corpuscular characteristics of the formed elements. The viscosity is an integral parameter of the rheological properties of the blood. It depends on the number of formed elements in the blood, their behavior in the flow of the macromolecular components of the plasma, the diameter and state of the vessel walls, and the velocity of the blood flow. The latter is higher at the center of a vessel and gradually lowers nearer to its walls. This retardation is characterized by the magnitude of the velocity gradient and is called the shear rate. The shear stress determines the pressure causing the blood to flow. The viscosity of the blood diminishes with a growth in the shear velocity.

The viscosity of the blood is affected by its formed elements, namely, the leukocytes (very insignificantly) and thrombocytes (they participate in thrombus formation). Of the greatest significance are the erythrocytes that are capable of aggregation and deformation. They determine the viscosity of the blood. Aggregation of the erythrocytes is a result of a drop in the negative potential of their membranes.

In vessels of a moderate diameter, the erythrocytes move at a definite distance and are oriented relative to one another in a definite way, forming radial-annular systems. This structure ensures the minimum resistance to the blood flow and, consequently, the minimum viscosity. When the charge of an erythrocyte drops, the structure of the blood flow is disturbed, and the erythrocytes draw closer and form aggregates. These changes increase the resistance of the blood to motion together with its viscosity [79, 81]. The potential of an erythrocyte membrane is also lowered by changes in the composition of the plasma, especially by hyperfibrinogenemia [160, 180, 181]. The fibrinogen molecules with a low negative charge screen the charge of the erythrocytes, thus facilitating their aggregation and increasing the blood viscosity. The latter is also effected by the shape of the erythrocytes [75, 451]. Normal erythrocytes deform in the direction of the flow, and this creates the minimum resistance to it. A growth in the rigidity of an erythrocyte is attended by a growth in the resistance to the flow with a simultaneous growth in the blood viscosity [60, 394, 590].

The pathogenesis of diabetic angiopathies is of interest. If we proceed from the fact that DM is a combination of two syndromes: hyperglycemia and microangiopathy, the pathophysiology of diabetic angiopathies consists primarily in injury of the endothelium and subendothelium. This results in adhesion of the thrombocytes to the subendothelial structures of a vessel wall. The formed elements are delivered to the injured wall section already prepared for adhesion and aggregation. We must also consider that an important role in these processes is played by humoral factors releasing from the formed elements of the blood or an injured vessel wall and endothelium cells. These factors include serotonin, catecholamines, prostaglandins, kinins, the products of fibrinogen decomposition, and adenine nucleotides. All these factors affect the lumen and permeability of the vessels. The hemorheological disorders in DM are an integrative consequence of functional-biochemical and morphological changes.

A major importance in DM is attached to the state of the erythrocytes in which glycosylized hemoglobin accumulates. Its amount correlates directly with the level of glucosemia [133, 181, 390, 481, 545]. This facilitates the development of hypoxia. A compensating reaction in an organism is the increase in the content of diphosphoglycerate in the erythrocytes, which increases the decomposition of oxyhemoglobin and thus eliminates the lack of oxygen. Special significance in the development of vascular complications in DM is given to lesion of the insulin receptors on the surface of cells of various organs and tissues, including the endothelium [476, 624].

The drop in the number of receptors is due to their blocking by circulating antibodies or immune complexes. Insulin resistance appears.

Most clinicians find the decisive role in the incidence of IHD in DM patients to belong to a disturbance of the rheological properties of the blood: the viscosity grows, hyperfibrinogenemia appears, as well as hyperaggregation of the thrombocytes and erythrocytes [22, 46, 151, 355, 362, 481].

Table 24. Rheological Parameters of Blood in DM Patients.

Parameter	Control group	DM patients
K^2, cP	5.55±0.13	6.96±0.22*
Ht	0.41±0.005	0.46±0.009*
τ_0, dyn/cm^2	0.055±0.002	0.127±0.006*
A, dyn/cm^2	0.75±0.028	1.35±0.068*
T_1, s	9.2±0.28	6.23±0.143*
T_2, s	41.95±0.81	34.11±0.52*
B_1, s^{-1}	34.48±0.85	30.02±0.92*
B_2, s^{-1}	-	57.3±2.0
Er, 10^{12}/l	4.39±0.017	5.15±0.06*
Thr, 10^9/l	245.9±0.96	288.2±9.25*
V_{av}, μm^3	89.9±0.56	89.13±0.55
F, g/l	3.16±0.18	4.96±0.13*

* The difference of the parameters is authentic (in comparison with the level in the control group).

In DM accompanied by microangiopathies, the vessels have a morphological indication: thickening of the glycoprotein basal membrane. The mechanisms triggering off the development of microangiopathies are:

(1) genetic defects of a vessel wall;

(2) diabetic disorders of metabolism;

(3) disorders of hormonal regulation;

(4) disturbance of the rheological properties of the blood;

(5) hypoxia of tissues:

(6) hyperproduction of contra-insular hormones.

In DM, the structure of collagen changes and it becomes glycosylized. This disturbs the function of the erythrocyte membranes – alters their permeability, causes ATP to be discharged from the erythrocytes, and increases the adhesion of the latter [133, 390, 481, 497].

An important role in the pathogenesis of diabetic microangiopathies is played by disturbance of the function of the erythrocytes [22, 46, 151, 362]. This is what sustains tissue hypoxia (owing to the drop in the ability to release bound oxygen), causing an increased thromboplastic activity. The lowering of the elasticity of the erythrocytes in DM and their increased ability of aggregation hinders the blood flow in the zone of microcirculation, boosts the pressure on the walls of capillaries, and leads to the development of diabetic microangiopathy. DM patients show a drop in the content of glycoproteins in the erythrocyte membranes. Accordingly, hemolysis is increased in both types of DM, but it is especially pronounced in the insulin-dependent type I. The antiheparin activity of the erythrocytes is very important. It is absent in healthy persons, while it is quite noticeable in DM patients.

What is the influence of hemolysis on the coagulating system of the blood?

Hemolysis of erythrocytes increases the thromboplastic activity and (because of the excretion of erythrocyte ADP) facilitates an increase in the aggregation of the thrombocytes, hypercoagulation, and an increased thrombus formation.

Moreover, disturbance of lipid metabolism causes a change in the surface charge of the erythrocytes, and this includes the thromboplastic factor of the erythrocytes into hemocoagulation. Hypercoagulation, in turn, increases the hemolysis of the erythrocytes and produces microcirculation disorders.

We consider that the degree of functional-morphological changes in the erythrocytes is greater in type I DM.

In DM patients, the rheological properties of the blood are altered substantially. A glance at Table 24 reveals that the viscosity of the blood in DM patients grows to 6.96 cP (against 5.55 in healthy persona). Very clear distinctions in the fibrinogen content were noted: in healthy persons, the fibrinogen level is 3.16 mg/l, while in DM patients it is 4.96.

The viscosity of the blood is known to depend considerably on the content of various plasma factors such as proteins, lipids, and fibrinogen [133, 307, 390]. The latter is what increases the aggregation abilities of the erythrocytes and creates favorable conditions for blood coagulation.

A growth in the fibrinogen level in DM was noted in a number of investigations, and the rate of aggregation of the erythrocytes was directly proportional to the level of fibrinogen in the blood plasma [151, 307, 359, 390, 403, 448, 504]. Electron microscopy revealed the dropping out of fibrillar proteins of the plasma around erythrocytes in DM patients. Fibrinogen is assumed to be the main aggregating agent of the erythrocytes (at least in DM patients of type II). In DM of type I, this role can be played by the beta-lipoproteins of the plasma that like fibrinogen reacts with the surface of an erythrocyte [526, 557].

In the previous chapter, we noted the considerable changes in the lipoprotein content in DM – this concerned a growth in the level of VLDLP, LDLP, apo-B, etc. It is doubtless that these factors also affect the growth in the blood viscosity. Our investigations proved that the increase in the velocity of the blood in DM patients is affected considerably by a high content of cholesterol, triglycerides, cholesterol of the low-density and very low-density lipoproteins, and fibrinogen. Electron microscopy noted the adsorption of these macromolecules on the surface of erythrocytes with a relevant drop in the negative membrane charge of the blood cells. It also noted adhesion of the erythrocytes to the endothelium of the vessels and erythrocyte aggregation.

An increased glucose level in the blood also appreciably affects its viscosity. A direct correlation was found between the level of the average daily glycemia, the amount of glycosylized hemoglobin with atherogenity of the serum, and the LDLP level. It was also noted that glycosylation accelerates the catabolism of the HDLP and may consequently lower their content in the blood.

It is considered that the greatest changes in the blood viscosity of DM patients is observed in the presence of diabetic angiopathies and in the decompensation of metabolic processes [22, 46, 362, 481].

Considerable changes occur in DM in the aggregation (cohesion) of the erythrocytes. In healthy persons, this parameter is 0.75 dyn/cm^2, while in DM it grows to 1.35 dyn/cm^2.

Aggregation of erythrocytes is an integrative process that is affected by changes in the protein and lipid composition of the blood. Disorder in the lipid composition of the plasma in DM patients (dyslipoproteinemia, hyperfibrinogenemia) alters the electrokinetic charge of the erythrocytes (drastically lowers it), especially in the presence of spherulized forms of the erythrocytes. In healthy persons, the erythrocytes have a considerable surface negative charge that repels the cells from one another. In DM, a growth in the level of fibrinogen and other high-molecular proteins lowers this potential, and this substantially accelerates aggregation (cohesion) of the erythrocytes.

Normally, the beta-lipoproteins have a high molecular weight and negative charge, which prevents adsorption of these particles on the surface of an erythrocyte membrane. In DM deep enzymatic and structural changes of the erythrocyte membranes occur, in particular, the concentration of the glycoproteins drops. The latter include sialic acids responsible for the charge density, which facilitates the adsorption of the beta-lipoproteins on the surface of the blood cell membranes [46, 263, 481, 525, 557]. The negative charge of the erythrocytes in DM also diminishes as a result of the disturbance of protein metabolism in the form of DLP. Here the amount of albumins diminishes the content of globulins in the blood grows [133, 390].

The noticeable growth in the yield point of the blood in comparison with the normal value in DM patients is due to the growth in the forces of cohesion between the erythrocytes that speeds up the formation of both small and large erythrocyte aggregates. In the blood flow of DM patients, small aggregates resisting disaggregation appear that are not encountered in healthy persons. In our opinion, the pathological aggregation of erythrocytes lowers the velocity of the blood flow up to the appearance of an intermittent pendulum-like motion. This disturbs microcirculation in the organs and tissues and is a factor of the formation of metabolic disorders in DM.

It is most likely that the metabolic disorders are secondary with respect to the rheological ones and are factors, in turn, that determine the viscous and elastic properties of the blood.

The increased aggregation ability of the erythrocytes in DM are closely related both to hyperglycemia and to the increased formation and deposition in the basal membrane of glycoproteins (with the subsequent change in the electrical and coagulation potential of the vessel walls). This leads, on the one hand, to the appearance and progress of angiopathies, and on the other to lowering of the rate of the blood flow and the creation of favorable conditions for the aggregation of erythrocytes.

The acceleration of the formation of small and large erythrocyte aggregates is noted in DM. If in healthy persons, the fine aggregates form during 9.2 s, in DM this occurs in 6.23 s. The formation of large aggregates normally takes 41.9 s, and in DM 34.1 s. The pathogenesis of this phenomenon consists in that the regulation of hemostasis in DM is sharply disturbed: both the thromboplastic (heparin) and the fibrinolytic activity decline.

The growth in the aggregation of erythrocytes under conditions of hyperlipidemia and DLP may facilitate disorders of microcirculation and metabolism. This is due to the fact that the serum proteins and lipids can envelope the erythrocyte membranes and protein microfilms on the inner surface of the capillaries. Together with the deposition of glycoproteins in the basal membranes of vessels, this hinders the diffusion of oxygen on the section: erythrocyte membrane – plasma – capillary wall tissue.

In addition, in hyperlipidemia, the permeability of the erythrocyte membranes lowers, the lipid composition alters, and the structure of the lipid bilayer is disturbed [133, 466].

We already noted that in DM the CS/PL ratio and the content of lysophosphatidylcholine grow, and the level of phosphatidylethanolamine in the membranes of the blood cells lowers. This may lead, on the one hand, to a change in the deformability of the blood cells, while on the other. the surplus of lipids in the erythrocyte membranes (similar to fibrinogen and globulins. beta-lipoproteins) adsorbed on the surface of the blood cells change their charge and increase aggregation. In our opinion, the concentration of lipids in the blood plasma and erythrocytes reflects the clinical manifestations of the disease and is related closely to hemorheological pathology.

In DM patients, a growth in hematocrit and the number of erythrocytes is noted. This can be considered as a compensating reaction of the organism aimed at increasing the oxygen capacity of the blood in response to tissue hypoxia. The appearance of the latter is due to an increase in the content of glycosylized hemoglobulin in the erythrocytes with a growth in the affinity for oxygen.

The glycosylizing process grows only under conditions of hyperglycemia and is of a compensating nature, lowering the increased level of free glucose in DM.

It is assumed that the growth in the level of glycosylized hemoglobulin leads to the inadequate compensation of DM during at least two or three months [133, 182].

However, with a growth in the concentration of erythrocytes in the blood, the value of K^2 increased in DM patients ($r=+0.92$, $p<0.001$), which lowered the velocity of the blood flow and led to the formation of the sludge phenomenon.

The hematocrit number was authentically higher in patients with decompensation of the metabolic processes in comparison with this parameter in patients with a compensated course of DM [22, 46]. A similar relation was discovered in patients with diabetic angiopathies, unlike patients without clinical vascular manifestations [390, 391].

Disorders of the rheological properties of the blood appear at the beginning of the development of DM. First of all changes in the erythrocytes are noted [22, 127, 151, 362, 391] that react actively to the metabolism disturbed because of insulin insufficiency. Besides, the metabolic processes of both an intraerythrocyte and membrane nature change.

The nonenzymatic glycosylation of the plasma proteins, erythrocyte membranes, and hemoglobin lowers the deformability of the blood cells, increases their brittleness, and changes the aggregation properties.

Apart from the glycosylation of proteins, in DM there is another specific process that may alter the plasticity (deformability) of the erythrocytes – intensification of the polyolic path of glucose metabolism leading to the accumulation of sorbitol in the cells [181, 182]. The most pronounced drop in the deformability of erythrocytes was noted when DM is combined with IHD [53].

In DM, the microviscosity of the erythrocyte membranes is increased, while the rate of the change in the curvature of a membrane is half that in healthy persons [182]. Possibly, acidosis typical of DM is one of the factors leading to such changes.

The concentration of products of the free-radical peroxide oxidation of lipids (FPOL) was found to be increased in DM patients [181, 557]. A clear dependence of the level of the diene conjugates on the severity of the vascular lesions can be seen.

Insulin facilitates the secretion of the FPOL products from the cell membranes [181, 557]. With insulin insufficiency and weakening of the antioxidant protection, conditions are created for accumulating the FPOL products. This lowers the amount of readily oxidized fractions of the phospholipids (PEA) and increases the content of the ones oxidized with difficulty (phosphatidylcholine and sphingomyelin). This tells on the deforma-

Table 25. Lipid Spectrum of Erythrocyte Membranes in DM Patients

Lipids, %	Control group	Type I DM patients	Type II DM patients
CS_{tot}	44.64±1.08	50.09±2.42*	55.82±3.64*
PL_{tot}	46.23±1.01	30.52±3.06*	29.12±2.08*
TG	1.23±0.39	3.8±0.82*	2.95±2.12
FFA	2.85±0.73	4.34±0.97	4.16±0.97*
ECS	4.87±0.51	15.30±1.81*	9.84±2.02*
CS/PL	1.08±0.04	2.04±0.42*	1.92±0.15*

* The difference of the parameters is authentic (in comparison with the level in the control group).

tion, aggregation ability, and osmotic resistance of the blood cells. Lowering of the latter and lesion of the erythrocyte membranes result in a large amount of free hemoglobin entering the vascular canal. The plasma therefore becomes more osmophilic. Hemoglobin and the iron ions formed in excess in the hemolysis of the erythrocytes are a powerful activator of FPOL whose products act on the membranes and alter their properties.

Consequently, DM is attended by a structural-functional reorganization of the erythrocyte membranes as a result of disturbance of both the lipid fractions directly and their fatty acid residues The accumulation in erythrocyte membranes of the products of metabolism of FPOL, lysolecithin, free fatty acids, etc., and rising of the cholesterol level disturb the rheological properties of the blood. This has the form of an increase in K^2 and the aggregation ability of the erythrocytes τ_0, which may be a factor of the development and progress of the metabolic disorders and diabetic angiopathies.

A role in the complicated mechanism of the incidence and progress of the atherosclerotic process is played by the diverse metabolic disorders, the main ones of which are disorders of lipid metabolism. An increase in the level of cholesterol, triglycerides, FFA, etc. is detected in the blood of patients with atherosclerosis and DM. Such changes also occur in the erythrocyte membranes, which substantially affect microcirculation.

Under conditions of hyperlipidemia and the atherogenic direction of DLP in IHD and DM patients, a considerable growth in the CS/PL ratio of the erythrocyte membranes is observed. This witnesses disturbance of both the microrheological (an increase in the aggregation of erythrocytes, a growth in the viscosity of the blood) and microrheological (a drop in the deformability of the erythrocytes) properties.

The lipid and phospholipid spectra of the erythrocyte membranes in DM patients are presented in Tables 25 and 26. The growth in the content of cholesterol, triglycerides, free fatty acids, and the drop in the fraction of the total phospholipids point to an increase

Table 26. Phospholipid Spectrum of Erythrocyte Membranes in DM Patients.

Phospholipids, %	Control group	Type I DM patients	Type II DM patients
LPC	1.79±0.56	4.24±0.22*	2.28±0.28
SM	19.69±0.86	13.01±2.21*	14.30±2.40*
PC	42.35±1.35	43.89±2.18	44.56±3.83
PEA	34.89±1.33	29.97±0.23*	26.75±1.53*
CL	1.18±0.5	9.79±2.8*	9.31±2.38*
PEA/PC	0.84±0.03	0.92±0.01*	0.83±0.02

* The difference of the parameters is authentic (in comparison with the level in the control group).

in the density of the lipid packing of the membranes and lowering of its deformability, which leads to a defect of the erythrocyte rheological unit.

Such a significant increase in the level of the cholesterol esters in the erythrocyte membranes (almost three times) in type I DM patients. on the one hand, is one way of detoxication or discharge from the metabolic canal of a surplus of FFA, while on the other hand, a growth in the membrane of the fraction of cholesterol esters may be a factor increasing the permeability of the membranes for monovalent ions. The latter will lower the osmotic stability of the erythrocytes and is generally encountered in diabetic angiopathies

In type I DM patients, the proportion of the individual phospholipids in the erythrocyte membranes varies significantly. A drop in the SM fraction possibly reflects the degree of receptor sensitivity of a cell, in particular, to insulin. A drop in the PEA fraction (accordingly, lowering of the PEA/PL ratio) increases the microviscosity of the erythrocyte membranes, lowers their deformability, and alters the functional properties of the erythrocytes.

Similar disorders of the lipid spectrum of the erythrocytes were revealed in type II DM patients. The formation of a lipid microfilm on the surface of an erythrocyte membrane with the following change in the charge of the erythrocyte, the absorption of fibrillar proteins of the blood plasma on the erythrocyte membranes possibly facilitate the accumulation of cholesterol in them and a drop in the fraction of the total phospholipids (mainly at the expense of the PEA and SM). This may also be associated with the presence of hepatosis in DM patients.

A growth in the PEA/PC ratio indicates a disorder in the structure of the erythrocyte membranes and a growth in their density, which affects the physicochemical processes both in the membranes themselves and in their relation to the membranes of the vessel endothelium cells.

2.3.3. Thrombocyte-Vessel Unit of Hemostasis in Diabetes Me llitus

The pathophysiological process of thrombosis in DM patients consists of many components. An increased aggregation ability of the thrombocytes combined with lesion of the vessel walls and a decreased production of prostacyclin are an important initial stage in the development of thrombosis. The disturbed metabolism of lipids in DM patients facilitates the rapid progress of atherosclerotic alterations, especially in type II DM patients. It was proved in recent years that thrombocytes can carry lipoproteins rich in cholesterol into the intimas of vessels [551]. However, the relation between the disturbance of the coagulating system of the blood, the change in the metabolism of the lipoproteins, and the development of atherosclerosis (especially in DM patients) remains to be unstudied.

It was found that LDLP and VLDLP have no appreciable influence on the aggregation ability of thrombocytes, but at a high level of the HDLP its sharp inhibition is noted [57, 97, 428]. Since the level of HDLP in DM is lowered considerably, we can assume this to be one of the factors stimulating the aggregation ability of thrombocytes.

In DM, the biochemical processes in the thrombocytes are distorted. In the thrombocytes of healthy persons, the metabolic process of the transformation of arachidonic acid into thromboxan A_2 takes place. In DM patients, there is an increased synthesis of thromboxan B_2 correlating with the level of glucose in the blood. It is very important that in DM the thrombocytes are less sensitive to the effect of the antiaggregation agents. It was noted that the prostacyclin level is diminished.

Physiological studies revealed that the thrombocytes perform specific functions [67, 98, 405]. First of all, they have their own system of energy supply that is represented by the mitochondria and grains of glycogen. Of very great importance is the ability of the thrombocyte membranes to deform because:

(1) with a change in the morphological properties, the thrombocytes may sharply increase their contact surface; the transition from a disk-shaped form to a spherical one is attended by the formation of numerous pseudopods that (like bristles) create conditions for aggregation;

(2) having a possibility of altering the physicochemical properties under the influence of the aggregating agents (collagen, thrombin, ADP), the thrombocytes carry out reconstruction of the phospholipid enzymes; an important role here is played by cAMP that affects the activity of proteinkinase.

To study the blood coagulation system and antithrombogenic activity of vessel walls, we studied 98 type I DM patients (90 men and 8 women) and 118 type II DM patients (104 men and 14 women) from 35 to 71 years old.

Table 27. Hemostasis System in DM Patients.

Parameter	Control group**	Type I DM patients	Type II DM patients
Fibrinogen, g/l	2.36±0.10	4.2±0.21*	3.9±0.23*
Tolerance of plasma to heparin, s	948±14.0	678.0±18.0*	749±19.0*
Duration of plasma recalcification, s	286±11	235±14	274±11
Antithrombin III, s	37.6±1.5	23±1.3*	26±1.1*
Fibrinolytic activity of blood, %	20.3±0.9	12±1.0*	14±0.9*
Index of intravascular aggregation of thrombocytes, %	1.03±0.03	1.28±0.04*	1.4±0.05*
Spontaneous aggregation of thrombocytes, %	0	5.7±0.5*	4.8±0.4*
Aggregation of thrombocytes, induced by:			
ADP 10^{-5} M, %	61±2.3	72±2.0*	78±2.2*
ADP 10^{-6} M, %	4.2±0.17	9.1±0.2*	10.1±0.2*
collagen, %	57±1.4	67±1.5*	72±1.3*
factor of thrombocyte activation, %	14.5±1	21±1.2*	23±1.3*
adrenaline 10^{-4} M:			
1st wave, %	13.4±1.1	26±1.6*	28±1.7*
2nd wave, %	59.3±2.1	76±2.1*	79±2.3*
aggristine, %	57.2±2.0	74±2.2*	76±2.4*
Beta-thromboglobulin, ng/ml	29.2±1.9	38±2.3*	40±2.2*
Thrombocyte factor IV, s	14.9±0.9	18.2±0.7*	20±0.8*
Thromboxan B_2 in plasma, pg/ml	31±3.1	36±2.7	41±3.6
cAMP in thrombocytes, pmol/$3\cdot10^8$ cells	4.1±0.4	3.2±0.3	2.9±0.25*
6-Keto-PGF_1-alpha in plasma, pg/ml	65±3.2	49±2.8*	47±3.0*
6-Keto-PGF_1-alpha / thromboxan B_2	2.1±0.2	1.3±0.3*	1.15±0.3*
Willebrand factor:			
un/ml	1.04±0.08	2.71±0.17*	2.82±0.1*
%	104±8.9	271±17*	279±14*

* The difference of the parameters is authentic (in comparison with the level in the control group).
** The control group includes persons with functionally active antithrombogenic properties of the vessel wall.

Most of the studied persons smoked (85% of type I and 76% of type II DM patients). Clinical examination showed that among the patients with DM made complicated by proliferative retinopathy, the percentage of smokers was higher than in the groups with preproliferative retinopathy of the retina. We also noted that in 57% of the type I and 42% of the type II patients their parents also suffered from DM.

When analyzing the state of the coagulating system of the blood in DM patients, we noted (Table 27) that the level of fibrinogen is increased considerably in DM patients of both type I (4.2 g/l) and type II (3.9 g/l) in comparison with healthy persons (2.36). The tolerance of the plasma to heparin is increased accordingly: it is 948 s in healthy persons and lowers in type I DM patients to 678 s and in type II to 749 s.

A growth in the fibrinogen level in DM is one of the important links in disturbance of microcirculation that is attended by blocking of the transportation of substances, in particular of the supply to the tissues of oxygen, glucose, lipids, and proteins. Disturbance of the permeability of the membranes is attended by a change in the nature of the oxidizing processes, in particular by a growth in the level of anaerobic oxidation and the development of ischemic and dystrophic changes in the tissues. This is manifest clinically by pronounced dyspnea, edemas, capillaropathic syndromes, polyarthralgias, myopathies, and stenocardiac pains.

Capillaroscopic studies of DM patients revealed an increase in the amount of functional capillaries, development of the network of anastomoses, a growth in the blood viscosity parameters and in the fibrinogen level attended by the progress of atherosclerosis [188, 626]. Of special significance here is the increase in the fibrinogen content and the blood viscosity. This facilitates the aggregation of erythrocytes and thrombocytes and the development of microthromboses, and stases in the system of capillary blood circulation. A result is the poorer diffusion of oxygen, and microcirculation hypoxia appears. These hypoxic changes cause induration of the membranes of vessels and tissues, alter the structure of the connecting tissue, the blood plasma penetrates beyond the limits of the vessels, fibrin and microthromboses form.

The appearance of pericapillary sleeves in the sections of diapedesis of the erythrocytes is noted, and the content of mucopolysaccharides and hyaluronidase in the blood grows. The high level of fibrinogen in the blood is attended by its emergence beyond the limits of the capillary walls. These disorders in the transportation of substances in DM are associated with the growth in the activity of biologically active substances such as histamines, bradikinin, and hyaluronidase. The morphofunctional blocking of the transportation of various biochemical substances changes the tolerance of the tissues to oxygen, glucose, lipids, and proteins, and facilitates the development of hyperlipidemia, especially of hyperproteinemia.

The enzymatic transformation of fibrinogen into fibrin proceeds in several stages. First of all, under the influence of the proteolytic enzyme of thrombin, four peptides detach from the fibrinogen, and fibrin monomers form that combine with one another to form fibrin fibers. Under the influence of factor XIII (with the participation of calcium ions), the transformation of fibrinogen into fibrin is completed. These mechanisms are especially significant in the DIC syndrome.

The combination of a high fibrinogen level and a drop in the activity of antithrombin III is a great danger for the development of thrombosis in DM. In healthy persons, this parameter is 37.6 s, while in DM patients of type I it drops to 23 s, and of type II to 26 s.

Antithrombin III is an important link in the regulation of the relations between coagulation and fibrinolysis, between the coagulating and anticoagulating systems of the blood. Thromboses often appear in patients with a congenital insufficiency of antithrombin III.

A number of disorders of the coagulating system of the blood are noted in DM: lowering of the level of antithrombin III, vascular and nonvascular microcirculation disorders, the appearance in the blood flow of multidimensional aggregates of the formed elements of the blood, stasis thereof, nonuniform caliber of the vessels, their curvature, loop formation, microaneurysms, and microhemorrhages. Disorders of thrombocyte formation are determined very clearly. Changes in the structure of the thrombocytes and erythrocytes point to their hyperaggregation activity, When studying the mucous membrane of the conjunctiva, we noted the diffuse thickening of the capillary walls, dilution of the lumen of capillaries with erythrocyte stasis and edge adhesion of the thrombocytes. Proliferation of the cell elements of the connecting tissue occurred in the perivascular spaces, which confirmed the lesion of the vessel wall and disorder of the tissue metabolism. All these changes in DM lower capillary perfusion, and cause widespread lesion of the vascular system and oxygen deficiency of the tissues.

These phenomena are due to hypercoagulation activity of the blood, including hyperfibrinogenemia. In DM, fibrinogen may be deposited in the volume and on the surface of a vessel wall, be included in microcirculation, be fixed on the surface of the thrombocytes and erythrocytes with a corresponding change in the physicochemical properties of these cells. An increased amount of thrombin acts on the fibrinogen and creates a state of hypercoagulation. An important role of a binding unit between the coagulation and thrombocyte-vascular components of hemostasis is played by the thrombocytes. A growth in their aggregating ability in combination with disturbance of the structure and function of the erythrocytes alter the rheological properties of the blood in DM.

Studying of the functional activity of the thrombocytes in DM patients showed that the amount of thrombocytes diminishes, but their aggregation and adhesive ability grows. In the decompensation of DM, the activity of the prostacyclin-forming system lowers considerably.

In DM, we attached special significance to the aggregation of the thrombocytes. We noted the very important fact of the presence in DM patients of spontaneous aggregation of the thrombocytes. It is absent in healthy persons, whereas in type I DM patients is was 5.7%, and in type II patients 4.8%.

What is the pathophysiological basis of this phenomenon?

Experiments revealed that the inhibition of the formation of prostacyclin in vessel walls is combined with atherosclerosis, especially in DM. This is accompanied by an increase in the production of thrombocytes by thromboxan B_2. It was proved that thrombocytes in DM patients produce a larger amount of thromboxan (in comparison with healthy persons) when they are stimulated by arachidonic acid, ADP, collagen, and thrombin [575].

Morphobiological studies revealed that the thrombocytes of DM patients contain more arachidonic acid in the composition of the phospholipids than of healthy persons [34, 311]. A correlation was noted between the receptor sensitivity of thrombocytes in DM patients and the glucose level in the blood: with a normal glucose level, the sensitivity of the thrombocytes remains the same as in healthy persons. The state of hyperglycemia is attended by a growth in the functional activity of the thrombocytes (the production of thromboxan B_2, its ability of aggregation and adhesion) and their hypersensitivity. Special significance is attached to the penetration of cholesterol into the membranes of the thrombocytes. In DM, the sensitivity of the thrombocytes to prostacyclin declines. Also noted was an increase in the blood of IHD and DM patients of the Willebrand factor of 1.5–2 times. There is a definite correlation here between the frequency of pain seizures in the region of the heart and the increase in the level of the Willebrand factor in the blood. Morphologically, this increase is attended very often by lesion of the endothelium of vessels.

When analyzing the level of thrombocyte aggregation in DM patients and healthy persons, we noted the considerable differences (see Table 27). The aggregation of thrombocytes induced by ATP was 4.2% in healthy persons, in type I DM patients it was 9.1%, and in type II – 10.1%. The same changes were noted in induction by collagen.

Considerable changes in the aggregation of thrombocytes occurred in stimulation by adrenaline. In healthy persons, the first wave of thrombocyte aggregation was 13.4%, in type I DM patients it was 26%, and in type II – 28%.

Although some of the factors explaining the adhesive and aggregation ability of thrombocytes in DM still remain a puzzle, a number of units of the process were deciphered. It was noted, in particular, that in DM stimulators of thrombocyte aggregation such as thrombin, adrenaline, noradrenaline, ADP, and collagen in contacting thrombocytes cause first of all a change in the shape of a cell, and an increased synthesis of endoperoxides, prostaglandins, and thromboxan A_2 [523, 588]. It is these stimulators in DM that activate the thrombocyte phospholipase A_2, and the latter releases arachidonic acid from the phospholipid pool. Under the influence of cyclooxygenase and other prostaglandin-synthetases of the thrombocyte membranes, it transforms into the intermediates of synthesis of prostaglandins – endoperoxides, and also into thromboxan A_2, which are powerful stimulators of the irreversible aggregation of thrombocytes. The aggregating substances react with the relevant receptors on the thrombocyte membranes.

It is considered that calcium ions play the important role of such an inductor, while thromboxan A_2 facilitates their transportation from compact granules into the cytoplasm of the thrombocytes. Thrombocyte aggregation is attended by the appearance of endoperoxides that are used by the microsomes of the endothelium cells for the synthesis of prostacyclin. The latter is what stimulates the adenylatecyclase of the thrombocytes and facilitates the accumulation of cAMP in them. These studies suggested the concept according to which the ability of thrombocytes to aggregation is determined by the relation between the thromboxan-generating system of the thrombocytes and the prostacyclin-generating system of the endothelium. Fibrinogen, and also factors VIII, XIII, and other plasmin factors participate actively in the aggregation and adhesion of thrombocytes.

A number of experiments showed erythrocytes to activate thrombocytes under conditions of hemostasis [181, 605]. The mechanisms of this activation are as follows:

(1) as a result of hemolysis, ADP, an important physiological activator of thrombocytes, is released from the erythrocytes;

(2) the erythrocytes have a physical effect, "pushing" the thrombocytes toward one another and to the vessel wall, thus promoting their adhesion and aggregation.

The following laboratory experiments were run: if enzyme systems utilizing ADP from the plasma (creatinphosphokinase) are added to a thrombocyte plasma, the degree of aggregation of the thrombocytes diminishes. The aggregation ability of the thrombocytes was shown to grow with an increase in the hematocrit. This is associated with the excretion of ADP from the erythrocytes. Definite concentrations of the ADP forming as a result of hemolysis of the erythrocytes were singled out [511, 605]. Consequently, the free ADP in the blood creates conditions for activation of the thrombocytes.

What factors facilitate the release of ADP from the erythrocytes?

First of all, this is the accumulation of cholesterol in the erythrocyte membranes, which increases their rigidity. If a patient suffers from hyperlipidemia, it is often attended by a growth in the content of free hemoglobin, which points to the drop in the stability of the erythrocytes and the appearance of hemolysis. Moreover, the ADP may emerge from the erythrocytes without hemolysis, as a result of deformation of the cells in a turbulent blood flow that appears in atherosclerotic changes in the vessels.

In IHD patients, the amount of spherocytes and mechanically rigid erythrocytes grows. Their fragmentation is accompanied by a large loss of ADP. All these data confirm the chemical mechanism of activation of the thrombocyte aggregation under the effect of the erythrocyte ADP. It was also noted that the erythrocytes act physically on the thrombocytes with a growth in the hematocrit, the rate of diffusion of the thrombocytes to a vessel wall, and the probability of collision of the thrombocytes with one another and the ves-

Table 28. Content of Fatty Acids in Thrombocytes in DM Patients

Fatty acid, %	Control group	Type I DM patients	Type II DM patients
Linolic ($C_{18:2}$, ω-6)	5.70±0.28	4.91±0.44	5.49±0.39
Arachidonic ($C_{20:4}$, ω-6)	19.10±0.75	27.03±1.04*	26.13±1.11*
Eicosapentaenic ($C_{20:5}$, ω-3)	0.40±0.09	0.62±0.21	0.81±0.24
Docosahexaenic ($C_{22:6}$, ω-3)	2.05±0.06	1.45±0.20*	1.76±0.09*
Coefficient K_{ac}	7.80±0.61	13.13±0.94*	10.14±0.71*

* The difference of the parameters is authentic (in comparison with the level in the control group).

sel wall grow. It was found that the adhesion of the thrombocytes is greater when the diameter of the erythrocytes is larger [512].

The role of fibrinogen in processes of adhesion and aggregation of the thrombocytes in DM patients consists in that as a result of the adsorption of this protein on the surface of cells, their negative charge lowers. The adhesive and aggregating ability of fibrinogen in DM grows drastically in the presence of calcium ions which it forms complex compounds with. It is the fibrinogen-calcium bond that creates a strong contact between the thrombocytes. The adhesive-aggregation function of the thrombocytes increases sharply when the thromboxan-generating system of the thrombocytes predominates over the prostacyclin-generating ability of the endothelium.

To check the significance of the metabolic disorders in the thrombocytes in DM, we studied the fat and acid composition if the thrombocytes in type I and II DM patients (Table 28).

The level of linolic acid in the thrombocytes of healthy persons is 5.7%, of type I DM patients is 4.9%, and of type II DM patients in 5.49%. Differences were noted in the level of arachidonic acid: in healthy persons it was 19.1%, while in type I and II DM patients its level grew to 27.03 and 26.13%, respectively. The ratio of the unsaturated fatty acids in the thrombocytes also changed. The fraction of eicosapentaenic acid in healthy persons was 0.40%, and in the type I and II DM patients it was 0.62 and 0.81%. The fraction of docosahexaenic acid in the thrombocytes of healthy persons was 2.05%, and in type I and II DM patients was 1.45 and 1.76%. Of major importance is the determination of the ratio of the fraction of arachidonic acid to that of the unsaturated fatty acids. In healthy persons, the coefficient K_{ac} was 7.8, and in type I and II DM patients it increased to 13.13 and 10.14, respectively.

These disorders of metabolism in the thrombocytes in DM are what cause the incidence in this disease of the DIC syndrome, thromboses, and emboli.

To confirm our data, we compared the results of clinical examination of healthy persons and IHD and DM patients.

Clinical Examples

Example 1

The examined person, Mr. K, 35 years old, made no complaints, plays tennis, considers himself to be a healthy person. During his lifetime he consulted a physician twice, when 5 and 14 years old with respect to angina. He does not smoke. During the last five or six years uses fat mainly in the form of vegetable oils. His parents are healthy. A complex examination was performed on October 24, 1990.

Coagulogram data:

Fibrinogen – 2.9 g/l

Tolerance of plasma to heparin (TPH) – 895 s

Duration of plasma recalcification (DPR) – 253 s

Activity of AT III – 25.8 s

Fibrinolytic activity of blood (FAB) – 16%

Thrombocyte factor IV – 14.5 s

Beta-thromboglobulin – 28.2 ng/ml

Index of intravascular aggregation of thrombocytes – 1.02

Spontaneous aggregation of thrombocytes – absent

Aggregation of thrombocytes induced by ADP 10^{-5} M – 58%, ADP 10^{-6} M – 3.5%, collagen – 52%, thrombocyte activation factor (TAF) – 13%, and by adrenaline – 56%

Thromboxan B_2 in thrombocytes 256 ng/$3 \cdot 10^8$ cells

Thromboxan B_2 in plasma – 29 pg/ml

6-Keto-PGF_1-α/thromboxan B_2 – 2.24

cAMP in thrombocytes – 5.1 pmol/$3 \cdot 10^8$ cells

Willebrand factor – 1.03 un/ml

Index of antiaggregation activity of vessel walls – 0.76

Index of anticoagulation activity of vessel walls – 1.36

Index of fibrinolytic activity of vessel walls – 1.58

Content of polyunsaturated fatty acids in the thrombocytes:

$C_{18:2}$, ω-6 – 5.36%

$C_{20:4}$, ω-6 – 18.11%

$C_{20:5}$, ω-3 – 0.43%

$C_{22:6}$, ω-3 – 2.08%

$$\text{Coefficient } K_{ac} = \frac{C_{20:4}, \omega-6}{C_{20:5}, \omega-3 + C_{22:6}, \omega-3} = 7.22$$

Content of hormones in blood plasma:

IRI – 7.2 μU/ml

C-peptide – 1.2 ng/ml

Cortisole – 241 nmol/l

STH – 1.5 ng/ml

The conducting of veloergometry and electrecardiography revealed no pathology. Consequently, the given healthy person has normal parameters of the coagulating system of the blood that combine with a normal spectrum of the fatty acids in the phospholipids of the thrombocytes.

Example 2.

The examined person, Mr. E, 41 years old, made no complaints. He works as an engineer at a research institute. He did not consult physicians. Pathologies on the part of the cardiovascular system were not revealed. He has been smoking during 25 years (over 20 cigarettes a day). He mainly uses animal fats in his food. His father died as a result of myocardial infarction when 51 years old, his mother suffered from hypertension.

Coagulation data:

Fibrinogen – 3.76 g/l

TPH – 794 s

DPR – 211 s.

AT III – 24.3 s

FAB – 14.1%

Index of intravascular aggregation of thrombocytes – 1.1

Spontaneous aggregation of thrombocytes – 1%

Aggregation of thrombocytes induced by ADP 10^{-5} M – 62%, ADP 10^{-6} M 4.9%, collagen – 62%, TAF – 17%, and by adrenaline – 66%

Thromboxan B_2 in thrombocytes – 300 ng/$3 \cdot 10^8$ cells

Thromboxan B_2 in plasma – 40 pg/ml

Prostacyclin in plasma – 50 pg/ml

Prostacyclin/thromboxan B_2 – 1.25

cAMP in thrombocytes – 3.9 pmol/$3 \cdot 10^8$ cells

Willebrand factor – 1.2 un/ml

Index of antiaggregation activity of vessel walls – 1.16

Index of anticoagulation activity of vessel walls – 0.91

Index of fibrinolytic activity of vessel walls – 1.26

Content of polyunsaturated fatty acids in thrombocytes:

$C_{18:2}$, ω-6 – 5.71%

$C_{20:4}$, ω-6 – 27.11%

$C_{20:5}$, ω-3 – 0.31%

$C_{22:6}$, ω-3 – 1.36%

$$\text{Coefficient } K_{ac} = \frac{C_{20:4}, \omega-6}{C_{20:5}, \omega-3 + C_{22:6}, \omega-3} = 16.23$$

Content of hormones in blood plasma:

IRI – 9.9 μU/ml

C-peptide – 1.5 ng/ml

cortisol – 296 nmol/l

Electrocardiography without exercise and echocardiography revealed no pathology. By the results of veloergometry, the diagnose was IHD (stable stenocardia under tension, functional class 1). Consequently, in the examined person E. disorder of the antiaggregation and anticoagulation activity of the vessel wells was revealed combined with a high content of arachidonic acid, an increased level of cortisol, and a drop in the content of polyunsaturated fatty acids.

The above clinical and laboratory parameters give us grounds to state that the disturbance of the relation between the content of arachidonic acid and unsaturated fatty acids is not only a diagnostic, but also a prognostic feature of the progress of IHD and requires regulation of the metabolic processes.

The thrombocyte aggregates in the system of microcirculation in DM trigger off the formation of thrombi. This is also facilitated by hypercatecholemia and hemolysis of the erythrocytes with the corresponding release of ADP or activation of the process of the intravascular coagulation of the blood. In emotional stress or in a postoperation period, in DM the production of thromboxan A_2 in the thrombocytes sharply grows, the repulsion of thrombocytes from the vessel walls is disturbed, and the DIC syndrome appears.

Many investigators experimentally and under clinical conditions have been attempting to prove the relation between the change in lipid metabolism and the functional activity of the thrombocytes [474, 507, 508, 523]. However, many units of this system have not yet been deciphered. In particular, it was noted that there is a relation between the level of membrane cholesterol and the hyperactivity of the thrombocytes – a growth in the concentration of the membrane cholesterol causes hyperactivity of the thrombocytes. A high cholesterol content in the thrombocyte membranes, in turn, is closely related to hyperlipidemia.

By comparing the information provided by various lipid parameters and glucose in the blood, their influence on the aggregation of thrombocytes, some authors showed that the level of total cholesterol has a smaller significance than an increased content of glucose in the blood. In analyzing the results, the investigators arrived at the conclusion that hyperglycemia causes the relevant hyperinsulinemia, while a high level of insulin, by increasing the synthesis of cholesterol esters in the vessel walls and simultaneously being a growth factor for the smooth muscle cells of the arteries, facilitates the development of atherosclerosis. Moreover, the biochemical composition of a thrombocyte membrane is exceedingly important.

In particular, an increased content of esters and free cholesterol in the thrombocytes is extremely unfavorable. And, conversely, a high level of alpha-cholesterol was found to be a favorable metabolic parameter. An increased level of beta-thromboglobulin is still another very important unfavorable factor facilitating the progress of atherosclerosis [588]. It is a specific protein of the thrombocyte granules and separates from them in aggregation. A high level of beta-globulin may point to the participation of the intravascular aggregation of thrombocytes in the pathogenesis of development of stenocardia in patients with atherosclerosis.

The combined effect of a mechanical decrease in the lumen of the vessels, proliferation of the smooth muscle cells, and the activity of thromboxan A_2 causes a spasm of the coronary arteries. Accordingly, seizures of acute coronary insufficiency with a sharp drop in the tolerance to a physical load appear.

Plasmin and its activators increase their physiological activity under the influence of various factors, among them as a result of ischemia of an extremity and section of plasmin activators into the blood. A great danger is the development of the DIC syndrome, when during a brief time a large amount of activators of blood coagulation and thrombocyte aggregation enters the blood flow, thrombin forms, and activation and exhaustion of the plasma enzyme systems (coagulating, kallikrein-kinin, fibrinolytic, etc.) occur. Under the influence of disintegration of the coagulating system of the blood, a multitude of microclots and aggregates of the cells form that block microcirculation in the organs. Thrombohemorrhages, acidosis, and often death occur. The main factors provoking the development of the DIC syndrome are:

(1) activation of the coagulating system of the blood and thrombocyte hemostasis by endogenic factors (tissue thromboplastin, decomposition of the tissues and cells of the blood, a damaged endothelium);

(2) systemic injury of the vessel endothelium and lowering of their antithrombotic potential;

(3) the appearance of a considerable amount of tissue thromboplastin in the blood canal.

Biochemical studies showed that heparin facilitates and accelerates the neutralization of thrombin with the aid of AT III [564]. The latter is an inhibitor of blood coagulation whose main function consists in the direct inactivation of the factors II, VII, IX, X, XI, XII, and kallikrein. Moreover, being a cofactor of heparin, AT III determines the anticoagulation effect of the introduced heparin. Lowering of the level of AT III is noted in the early stages of the DIC syndrome.

Major significance in DM is attached to the increased formation of thromboxan B_2. It can be noted that the latter (induced by ADP) in healthy persons is 31.0 pg/ml, in DM patients of type I is 36.0 pg/ml and of type II is 41.0 pg/ml.

The metabolic activity of thrombocytes, based among other factors on the disturbed metabolism of arachidonic acid released from the membrane phospholipids of the thrombocytes, ensures accelerated aggregation and adhesion. Prostaglandins form from arachidonic acid under the influence of cyclooxygenase. Thromboxan exists in the blood for a very brief time, but has a very strong influence on the coagulating ability of the blood. As a result of metabolism, thromboxan A_2 transforms into thromboxan B_2. The thrombocyte membranes are activated not only by thromboxan and ADP, but also by hydroperoxides of the polyunsaturated fatty acids in the free radical peroxide oxidation of lipids. These processes are stimulated actively in DM.

The increased formation of thromboxan B_2 is combined in DM with a diminished formation of prostacyclin. In healthy persons, the prostacyclin level is 65.0 pg/ml, in DM patients of type I it is 49.0, and of type II it is 47.0 pg/ml. The predomination of the thromboxan-generating function of the thrombocytes over the prostacyclin-synthesizing ability of the endothelium in DM patients ensures the progress of thrombus formation and the development of the DIC syndrome.

The pronounced lesion of the vessels in DM (diabetic microangiopathies) during recent years attracted attention to studying the pathological processes in a vessel wall (endothelium) of these patients.

Great attention is given to the functional ability of the endothelial cells of vessels of both the muscle and the muscle-elastic type to synthesize and secrete biologically active substances affecting the vessels. Special attention among these substances is given to the consideration of vasodilators (prostaglandins. thrombin, bradikinin, and the substance P_{450}).

Biological studies showed that the receptors of these biologically active substances are located in the endothelium, and participate in the release of the endothelium-dependent relaxation factor [406, 498, 610] that activates guanykatecyclase and thus increases the content of cGMP in a cell. This facilitates the development of vasodilation. The derivative of cyclooxygenase PGI_2 that is produced by endothelial cells and has a vasodilating effect was studied quite well.

An opposite (vasoconstricting) effect appears as a result of the reaction of arachidonic acid and a considerable number of substances called eicosanoids. Moreover, a constrictive effect is produced by angiotensin II, histamine, thromboxan A_2, and catecholamines. In the middle of the 1980s, a new peptide was discovered having vasoconstricting properties. It was called endothelin. Its powerful vasopressor effect exceeds the activity of angiotensin II.

When comparing the level of cAMP in healthy persons and in DM patients, we noted appreciable differences: in healthy persons the cAMP content in the thrombocytes was 4.1 pmol/$3 \cdot 10^8$ cells. while in type I and type II DM patients it dropped to 3.2 and 2.9 pmol/$3 \cdot 10^8$ cells, respectively.

The content of the Willebrand factor in DM patients of types I and II was on an average 2.5 times higher than in healthy persons and ranged from 50 to 450%. In 4% of the patients, the content of the Willebrand factor exceeded 451%. The growth in this parameter in DM patients is a marker of the lesion of the vessel wall endothelial cells and facilitates the adhesion of thrombocytes to the injured endothelium and the subendothelial layer.

We attached special attention in DM patients to the state of the endothelium. In this connection, the conducting of a cuff test is not only of a diagnostic, but also of a prognostic significance.

A relative or absolute insufficiency of insulin in DM patients leads to an increase in the permeability of the cell membranes and capillary wall. and, consequently, exchange occurs between the components of the blood and a tissue. Physicochemical investigations of the mechanism of permeability of a membrane revealed the most important role here of the gialuronic acid in the basic substance of the connecting tissue. This acid (in combination with gialuronidase) is the basic factor of diffusion. Gialuronidase helps detach a protein from gialuronic acid and increase the permeability of a cell membrane. The protein fractions, proteolytic enzymes, acetylcholine, histamine, and hormones play a certain part in regulating the permeability.

When studying the state of the antithrombogenic activity of the vessel walls in DM patients (Table 29), we note considerable changes. The aggregating ability of thrombocytes induced by ADP 10^{-5} M in healthy persons after a cuff test dropped from 61.0 to 48.0%, while in type I DM patients it increased from 72.0 to 84.0% and in type II patients from 78.0 to 94.0%. A similar direction of the reaction is also noted when aggregation of the thrombocytes is induced with the aid of collagen and adrenaline.

The dynamics of the prostacyclin content stands out very well: in healthy persons, its level grows from 65.0 to 109.0 ng/ml, while in type I and II DM patients there is no dynamics (49.0 and 48.0, 47.0 and 48.0 ng/ml).

Lowering of the prostacyclin-forming ability of the endothelium leads to microthrombosis of the vessels in the lower extremities and the fundus of the eye in DM.

It was shown in a number of investigations that an important cause of the incidence of diabetic changes in the fundus of the eye is aggregation of the erythrocytes occurring in the fine capillaries surrounding the macular region [481, 525]. The erythrocyte aggregates clogging the lumen of the vessels cause the appearance of microaneurysms, newly formed vessels, and hemorrhages. Glued erythrocytes and fibrin are detected at the walls of the aneurysms under a microscope. It was suggested that these factors of the process of thrombus formation are what underlies the diapedetic hemorrhages, It was noted that in patients with retinopathy the acceleration of erythrocyte aggregation is sharply accelerated (in comparison with DM patients without changes in the fundus of the eye).

The peroxide theory of the initial stages of atherosclerosis is being discussed widely. Under clinical conditions, the processes of FPOL and biochemical and morphological changes in the vessels are studied in detail. In accordance with O.Voskresenky's peroxide theory of atherosclerosis, the insufficiency of the antioxidant system and the subsequent growth in the FPOL in tissues is considered as the leading unit in the development of IHD. When appraising the criteria of the FPOL, the following parameters are used: the resistance of the membranes of the erythrocytes to spontaneous hemolysis, the content of acylhydroperoxides in atherogenic lipoproteins, and the activity of the antioxidant enzymes (catalase and superoxide dismutase). It was found that in the stimulation of the FPOL, destruction of the elastic fibers, cell proliferation, processes of fibrosis, calcinosis, and lipid infiltration in an arterial wall (the peroxide model of atherosclerosis) were noted morphologically. Hyperlipidemia was discovered biochemically.

Morphological studies show the decomposition of atherogenic lipoproteins on the elastic fibers of the arteries, destruction of the elastic elements, and depolymerization of mucopolysaccharides and calcinosis. A growth in the level of fibrinogen, especially of the products of fibrin decomposition, a growth in the thromboplastic activity of the blood, aggregation of the thrombocytes, and a drop in the products of the antiaggregation factors of a vessel wall are observed.

In patients with a myocardial infarction, the biochemical composition of the thrombocytes changes drastically, especially the energy balance [615]. It was found that in such patients the overall energy balance of the thrombocytes grows, which points to their increased readiness to thrombus formation. In addition to hyperenzymemia in patients with myocardial infarction, substantial changes in the enzymes of the pentosophosphate path are observed. Heterogeneity of the triggering enzyme of glucoso-6-phosphatedehydrogenase appears. Thrombin appreciably lowers the intensity of the oxidative reactions in the thrombocytes.

Diabetic retinopathy is divided into simple and proliferating [380, 476, 488, 515, 551, 557]. The proliferating variety proceeds less favorably. Here new vessels with an increased wall permeability develop in the retina. Progress of simple retinopathy is attended by sclerotic changes in a vessel wall, while in proliferating retinopathy, thromboembolic disorders prevail with extensive zones of retina infarctions and a sharp growth in the permeability of a vessel wall.

The combined disturbance of the prostacyclin-generating activity of the endothelium with lowering of the production of antithrombin III also stimulates thrombus formation in DM. In healthy persons in response to ischemia of a vessel, the activity of an-

Table 29. Antithrombogenic Activity of Vessel Walls in DM Patients.

Parameter	Control group**	Type I DM patients	Type II DM patients
Antiaggregation activity of vessel wall			
Aggregation of thrombocytes induced by ADP 10^{-5} M, %:			
before compression	61.0±2.3	72.0±2.0	78.0±2.2
after compression	48.2±1.9	84±1.6	94±1.9
index	0.8±0.09	1.16±0.07*	1.2±0.06*
Aggregation of thrombocytes induced by ADP 10^{-6} M, %:			
before compression	4.2±0.17	9.1±0.2	10.1±0.2
after compression	3.4±0.16	11.3±0.18	12.9±0.19
index	0.81±0.07	1.21±0.08*	1.27±0.07*
Aggregation of thrombocytes induced by collagen, %:			
before compression	57.0±1.4	67.0±1.5	72.0±1.3
after compression	49.0±1.6	75.0±1.7	78.0±1.4
index	0.86±0.07	1.12±0.05*	1.08±0.04*
Aggregation of thrombocytes induced by TAF, %:			
before compression	14.5±1.0	21.0±1.2	23.0±1.3
after compression	11.2±0.9	25.0±1.4	27.0±1.5
index	0.78±0.08	1.19±0.07*	1.17±0.09*
Aggregation of thrombocytes induced by adrenaline 10^{-4} M, %:			
before compression	59.3±2.1	76.0±2.1	79.0±2.3
after compression	52.0±2.0	89.0±2.2	93.0±2.4
index	0.87±0.07	1.18±0.07*	1.18±0.06*
Thrombocyte factor IV, s:			
before compression	14.9±0.9	18.2±0.7	20.0±0.8
after compression	11.7±0.8	21.0±0.8	24.0±0.9
index	0.78±0.07	1.15±0.08*	1.19±0.06*
6-Keto-PGF_1-alpha in plasma, ng/ml:			
before compression	65.0±3.2	49.0±2.8	47.0±3.0
after compression	109.0±3.0	48.0±2.9	48.0±3.2
index	1.8±0.08	1.0±0.07*	1.0±0.07*
Anticoagulant activity of vessel wall			
Activity of antithrombin III, s:			
before compression	37.0±1.5	23.0±1.3	26.0±1.1
after compression	49.0±1.6	19.0±1.4	23.0±1.2
index	1.31±0.07	0.86±0.06*	0.88±0.09*
Fibrinolytic activity of vessel wall			
Fibrinolytic activity of blood, %:			
before compression	20.3±0.9	12.0±1	14.0±0.9
after compression	32.1±1.2	9.0±1.2	12.5±0.8
index	1.61±0.08	0.73±0.09*	0.9±0.09*

* The difference of the parameters is authentic (in comparison with the level in the control group).

** The control group includes persons with functionally active antithrombogenic properties of the vessel walls.

tithrombin III increases from 37.6 s to 49.0 s, while in DM it lowers from 23.0 to 19.0 s (type I) and from 26.0 to 23.0 s (type II).

Table 30. Hormone content in Blood Plasma in DM Patients.

Parameter	Control group	Type I DM Patients	Type II DM patients
IRI, μU/ml	7.8±1.4	-	15.1±2.3*
C-peptide, ng/ml	1.2±0.17	0.33±0.14*	1.68±0.13*
Cortisol, nmol/l	250±30	490±41*	503±46*

* The difference of the parameters is authentic (in comparison with the level in the control group).

Of major significance in regulating the coagulating activity of the blood and functioning of the endothelium is given to hormonal actions. In this connection, we analyzed the content of IRI, C-peptide, and cortisol in healthy persons and DM patients. Examination of Table 30 shows that the IRI level in type II DM patients considerably (two times) exceeds the normal value and is 15.1 μU/ml (in healthy persons it is 7.8 μU/ml). Hyperinsulinemia affects the development of DLP and provokes the development of microangiopathies.

The main units in the pathogenesis of diabetic angiopathies are (i) changes in the basal membrane of the capillaries, (ii) hemodynamic disorders, and (iii) changes in the parameters of blood viscosity and the function of the thrombocytes.

Changes in the basal membrane and capillaries occur in various organs and tissues, but this is manifest especially in the retina of the eye and the renal glomeruli.

Several factors underlie the pathogenesis of the changes in the basal membrane, including the nonenzymatic glycosylation of proteins, and also a growth in the intracellular content of the osmotic polyol – sorbitol. The latter has a pronounced osmotic effect, causes intracellular edemas and disorders of the function of the cells. A cataract appears as a result of its action on the crystal lens.

A great importance in the appearance of diabetic microangiopathies is attached to free radicals that are powerful oxidizing agents appearing during normal metabolic processes [254, 547]. An excess activity of the free radicals combines with the appearance of microangiopathies in patients. The pathogenetically increased activity of the free radicals injures the cellular endothelium by acting on various biochemical paths, e.g. on the prostaglandin cascade.

Of special significance is the high cortisol level in DM patients. If in healthy persons the cortisol level was 250 nmol/l, in type I DM patients it grew to 490, and in type II to 503 nmol/l.

These data correlate with the changes in the activity of antithrombin III.

The fibrinolytic activity of the vessel walls in healthy persons increased from 20.3 to 32.1%, in type I DM it lowers from 12.0 to 9.0%, and in type II from 14.0 to 12.5%.

The key moment of the relation between IHD and DM is the influence of the excess secretion of insulin on the development of dyslipoproteinemia as the basic pathogenetic unit of atherosclerosis.

A lack of insulin lowers the activity of glycosyltransferase by causing changes in the basal membrane of the vessel walls and increases the aggregation of the thrombocytes by affecting the intracellular distribution of cAMP and thromboxan in DM patients. In hyperinsulinemia, the aggregation of the thrombocytes in healthy persons diminishes after the introduction of insulin [120, 254, 289, 469, 476, 477, 573, 602].

The growth in the cortisol level in DM patients is a clinically very important fact. We noted that the cortisol level correlates with the degree of vascular lesions in DM. Diabetic angiopathies progressed very rapidly in patients with a high cortisol content.

2.4. Correction of Blood Coagulating System in Ischemic Heart Disease and Diabetes Mellitus Patients

In recent years a diversity of methods are proposed for treating patients suffering from DM in combination with atherosclerosis. They are aimed at the large-scale removal of high-molecular compounds in the blood plasma (cholesterol, triglycerides, lipoproteins) from the organism with a corresponding improvement in the quality of blood coagulation. The methods include gravitational plasmapheresis (PPh), which is a nonpharmacological method of treating disorders of lipid metabolism [9, 62, 69, 117, 143, 172, 188, 204, 276, 287, 304, 308, 319, 334, 464, 518].

The mechanism of plasmapheresis in DM is based on the removal from an organism of high-molecular compounds (cholesterol, LDLP, fibrin, immune complexes) and their replacement with vasoactive solutions. Plasmapheresis was performed on a domestic blood fractionator. A heparin solution in a concentration of 10,000 units per 400 ml of physiological solution was used as the anticoagulant. The ratio of the anticoagulant and blood content in the course of extracorporal therapy was 1:5. To conduct the procedure, disposable sterile silicone tube lines were used. The subclavian veins and less often peripheral ones were used to take blood from a patient and introduce erythrocytes and plasma-substituting solutions. After mixing with an anticoagulant, the blood was fed by a pump at a speed of 20–25 ml/min into a rotor, where under the effect of gravitational and centrifugal forces (centrifuging at 1100–1200 rpm) it was separated into two fractions – plasma and erythrocyte matter. After separating the blood, we mixed the erythrocytes with rheopolyglucin and returned the mixture to the patient, while the plasma was collected in a separate reservoir. The course of plasmapheresis consisted of four sessions with an interval of two or three days. During one session, 789.0 ml of plasma was removed. The plasma was replaced with rheopolyglucin in a proportion of 1:1.2.

A labile course of the disease, progressing diabetic retinopathy, insulin resistance, and hyperlipoproteinemia were indications for conducting plasmapheresis in type I DM patients. The preoperation preparation included the following measures: the relative compensation of DM whose criteria were the elimination of ketoacidosis, normalization of hypoglycemic states, a growth in the glucose level in the blood during a day by and not over 12–14 mmol/l, and lowering of the glucose content in the urine to 1.5–2%. Studying of the coagulogram was obligatory. Heparin was prescribed if intravascular coagulation of the blood increased. Immediately after the session of plasmapheresis, the disturbed carbohydrate metabolism was corrected in 6, 12, and 24 h on the basis of determining the glucose level in the blood.

Clinically, the state of the patients when conducting plasmapheresis and after it was satisfactory. During the first session, in three patients there was a certain excitation and stress, in one patient the arterial pressure increased to 190/110 mmHg that become normalized in the second hour of conducting the procedure without using drugs. In six patients after the first session, a sensation of hunger appeared with vertigo, which rapidly disappeared after taking food. Subsequently, these hypoglycemic states were normalized by the abundant taking of food before conducting the procedure.

On completion of the course of treatment, the patients indicated a decrease in torpidity, improvement of sleep, vanishing of headaches, vertigo, and improvement of their mood. After a course of treatment, in all the patients the fluctuations of the glucose level in the blood did not exceed 2–5 mmol/1, while diuresis increased by 30%. The latter was attended by the diminishing or vanishing of edemas: in 10 persons (of 16), the edemas vanished completely by the 10th day of observation. The diuretic effect on the background of treatment is possibly associated both with the removal of high-molecular compounds from the plasma and with the introduction of low-molecular dextran-rheopolyglucin. The latter facilitates movement of the liquid from the tissues into the blood channel, increases the suspension properties of the blood, and restores the blood flow in the small capillaries. In other words, the diuretic effect in plasmapheresis is due to the improvement in the rheological properties of the blood and microcirculation in the tissues.

An important clinical result of plasmapheresis is lowering of an initially high arterial pressure that remained low during a month after a course of plasmapheresis. In our opinion, of significance in the pathogenesis of this phenomenon is improvement of the microcirculation in the tissues, normalization of the rheological parameters, and a growth in glomerulus filtration.

Of very great importance is the diminishing of the features of diabetic polyneuropathy. After a course of plasmapheresis, the pains in the lower extremities of patients lowered considerably, pains at night terminated, the muscular force increased, and paresthesia vanished.

Clinical Example

Patient A. When 66 years old he was admitted with the diagnosis ischemic heart disease, stress stenocardia FK IV. Insufficiency of blood circulation of the 2nd degree; type II DM of moderate severity, compensated. Diabetic angiopathies of the vessels of the lower extremities.

In anamnesis, the patient suffered from IHD during seven years, and from type II diabetes mellitus – 10 years. Regularly takes bukarban (carbutamide) (two pills a day).

Parameter	Norm	Before PPh	After PPh	In 10 days after PPh
Thrombofibrin IV, s	up to 17	18.4	20.2	19.6
Antithrombin III, s	25–35	29.9	36.7	34.1
Fibrinogenic activity of blood, %	12–17	11	13	15
Fibrinogen, mg/l	4.00	4.22	3.33	3.33
Tolerance of plasma to heparin, s	800–1200	640	875	886
Duration of plasma recalcification, s	up to 330	316	314	294
K^2, cP	5.55	7.24	4.9	5.1
A, dyn/cm^2	0.75	1.63	0.9	0.9
τ_0, dyn/cm^2	0.06	0.21	0.09	0.08
Er, 10^{12}/l	4.4	5.16	5.05	5.11
Thr, 10^9/l	250	330	259	270
Ht, %	41	44	43.5	42.0

Was admitted with complaints on retrosternal pains that were relived by taking two or three pills of nitroglycerin (glyceryl trinitrate), on pains and spasms in the lower extremities that appeared together with the pains in the heart after every 50–60 m of slow walking.

His state when admitted was of moderate severity. Cyanosis of the lips. Dyspnea up to 20 a minute. The lower extremities were cold. Trophic changes on the shin and feet. A small amount of moist stagnant rales are heard above the lungs. me heart tones are dull. The pulse rate was 100 bpm, the arterial pressure was 140/90 mmHg.

The patient received five sessions of plasmapheresis with an interval of three days; 2350 ml of plasma were removed and replaced with rheopolyglucin.

After plasmapheresis, the seizures of stenocardia stopped, the pains in the lower extremities diminished substantially. The distance covered without pains in the legs was 2700 m at an ordinary speed.

The laboratory revealed an increase in the level of antithrombin III, lowering of the blood viscosity, aggregating ability of the erythrocytes, and yield point of the blood on the background of constant parameters of hematocrit and the number of erythrocytes.

The duration of clinical remission was seven months.

The results after a course of plasmapheresis indicate improvement of the coronary blood flow and a good effect in the vessels of the lower extremities. There is no doubt that an important role here is played by a drop in the level of fibrinogen, an increase in the content of AT III, and a change in the blood viscosity. Unfortunately, the duration of remission was 6–7 months, after which the course of plasmapheresis must be repeated.

Plasmapheresis considerably improves the clinical state of DM patients in combination with IHD. In 85% of the patients, a drop in the number and intensity of the stenocardia seizures and lowering of the daily demand for nitrates were noted. When patients had the symptom of intermittent limping, the conducting of plasmapheresis was followed by a reduction in the pains in the lower extremities, and a feeling of warmth appeared. Considerable positive results were noted in the state of the vessels of the fundus of the eye.

In most DM patients, we noted the presence of perivascular changes in the form of an edema, focuses of microstasis and hemorrhages. The vascular pathology was determined as the nonuniform diameters of the vessels, an increase in their bending, and diminishing of the arteriole-venule coefficient to 1/4, mainly because of dilatation of the venules. We found substantial changes in the system of capillaries: the presence of zones of their desolation and the absence of clear visibility of the capillaries. The intravascular changes were registered in the form of sludge phenomena with a sharp drop in the space velocity of the blood flow, the appearance of a counterflow, stasis of the blood, and thrombosis of the vessels. After the conducting of plasmapheresis, the phenomenon of the retardation of the blood flow and counterflow vanished, as did the capillary zones of desolation and thrombosis of the vessels. Here a positive dynamics of the perivascular changes was noted.

Clinical Example

Patient M. When 28 years old, she was admitted to a hospital with the diagnosis: type I diabetes mellitus, severe course, diabetic retinopathy, nephropathy, polyneuropathy. Ketoacidosis. Decompensation.

When admitted, she complained of dryness in the mouth, thirst, polyuria, nausea, headaches, great weakness, pains in the legs, numbness of the hands and legs.

She was ill since five. She had a diabetic coma twice, for many years she was given insulin in large doses (up to 80 units). Has frequent hypoglycemic states. The level of the glucose in the blood during 9 day is 16–18 mmol/l, and in the urine – 8–10%. She does not observe a diet strictly. Up to 1985, she was admitted to a hospital annually. The conditions of insulin therapy were 1–2-fold injections. She periodically received additional treatment with angioprotectors, lipolytics, and vitamins.

Examination results: normal constitution (height 156 cm, weight 56 kg), integuments pale, clean, and dry. Puffiness of face, lipodystrophic changes at the places of insulin injections. The skin of the limbs is thin and dry with sections of pigmentations, in palpation is cold, the pulsation on the dorsal arteries of the feet is weakened. On the skin of the limbs, the threshold of sensitivity is lowered on symmetrical sections. Lowering of the Achilles tendon and knee-jerk reflexes.

The insulin dose received when admitted was 60 units a day.

Dullness of the heart tones is noted, the pulse rate is 90 bpm, the arterial pressure is 180/100 mmHg. Vesicular respiration is heard over the lungs. The abdomen is soft, no pains in palpation, the liver protrudes from under the rib edge by 2 cm, the pancreas is not enlarged. An inclination to constipation.

Consultation of an ophthalmologist: diabetic retinopathy, simple exudative form, a plurality of exudates on the fundus of the eye.

Laboratory investigation: clinical analysis of the blood without any special features. In clinical analysis of the urine: glucose – 5%, protein – 4 g/l, leukocytes – 10–12 in V.f., erythrocytes – 3–5 in V.f.

Biochemical investigation of blood: urea – 8.4 mmol/l; creatinine – 133 mmol/l; CS – 6.9 mmol/l; CS of LDLP – 4.66 mmol/l; CS of VLDLP – 0.54 mmol/l; CS of HDLP – 1.7 mmol/l; AI – 3.05.

Erythrocyte membrane, %:

Parameter	CS	TG	PL	FFA	ECS	LPC	SM	PC	PEA	CL
Value	50.5	0	37.7	0	11.7	0	34.0	42.1	23.7	0

The conservative treatment adopted (fractional introduction of insulin, intravenous infusion of rheopolyglucin, vitamin therapy) did not produce the desired effect. After the patient's coagulogram was studied, it was recommended to conduct a course of treatment by plasmapheresis.

Three procedures with an interval of two days were conducted; 30% of the plasma was removed with adequate replacement with rheopolyglucin and albumin.

After the course, the patient felt better: her headaches, edemas, and pains in the legs vanished, sleep became normalized, and diuresis increased. The fluctuations of the level of glucose in the blood during a day did not exceed 2–3 mol/l. It became necessary to lower the insulin dose by the fifth day to 30 units.

Biochemical analysis of the blood on the third and fourth days after the course of plasmapheresis: urea – 6.5 mmol/l, and creatinine – 97 mmol/l.

Parameter, mmol/l	CS	TG	CS of LDLP	CS of VLDLP	CS of HDLP	AI
Value, after						
3 days	6.9	0.97	4.56	0.44	1.9	2.63
10 days	7.3	0.97	4.06	0.44	2.8	1.6
1 month	4.9	1.11	2.6	0.5	1.8	1.7
3 months	6.0	1.2	3.76	0.5	1.7	2.52

Erythrocyte membrane, %:

Parameter	CS	TG	PL	FFA	ECS	LPC	SM	PC	PEA	CL
Value										
	45.1	0	3.8	0	90.9	0	16.1	41.9	36.6	5.02
	50.4	0	36.5	0	13.0	0	22.3	38.8	38.8	0
	47.1	0	37.2	0	15.6	0	14.4	46.1	39.5	0

The clinical example shows that the conducting of plasmapheresis considerably improved the blood flow in the vessels of the fundus of the eye and normalized the patient's vision.

After plasmapheresis, the daily level of glucose in the blood of DM patients lowered from 10.47±0.88 to 6.06±0.66 mmol/l ($p<0.001$) with a constant daily dose of sugar-reducing drugs. In our opinion, the main causal positive factors when conducting plasmapheresis are the improvement of blood supply and metabolism of the pancreas, elimination of tissue hypoxia with subsequent restoration of the sensitivity of the receptor apparatus of the tissues.

Under conditions of disturbed hemodynamics, correction of metabolic disorders becomes useless. Only with the restoration of microcirculation (as a result of a change in the rheological properties of the blood) is adequate transportation of drugs to the target tissues possible. Therefore, solution of the problem of correcting blood circulation (in general) and disturbed microcirculation (in particular) is an important practical matter for lowering the resistance of an organism to the treatment being conducted.

In laboratory analyses after conducting a treatment, moderate thrombocytopenia was noted which vanished by the tenth day. After plasmapheresis, the dynamics of glycemia was bi-directional – in two patients during a day the glucose level grew to 18 mmol/l, while in four persons the daily fluctuations did not exceed 2–3 mmol/l (up to 10 mmol/l). In the remaining patients, the fluctuations of the glucose level during a day were as usual.

The changes in the rheological parameters of the blood after plasmapheresis in IHD patients and with this disease combined with DM were characterized by lowering of the Kesson viscosity of the blood, whose magnitude reaches the parameters of healthy persons (Table 31). This is associated with a drop in the viscosity of the plasma and is due to the

Table 31. Rheological Parameters in Type II DM Patients after Plasmapheresis.

Parameter	Before PPh	After PPh		
		in 3 days	in 1 month	in 3 months
K^2, cP	6.96±0.22	5.32±0.47*	5.06±0.57*	6.53±0.45
τ_0, dyn/cm^2	0.127±0.006	0.07±0.002*	0.055±0.01*	0.076±0.01*
Ht	0.46±0.009	0.45±0.01	0.44±0.03	0.44±0.01
A, dyn/cm^2	1.35±0.068	0.94±0.02*	0.88±0.08*	1.07±0.06*
T_1, s	6.23±0.14	10.13±0.18*	11.5±0.2*	8.7±1.24
T_2, s	34.11±0.52	40.26±0.21*	46.56±1.06*	41.95±1.42*
B_1, s^{-1}	30.02±0.92	38.36±1.65*	36.5±1.05*	35.9±0.85
F, g/l	4.96±0.13	4.41±0.21*	3.84±0.53*	3.80±0.42*

* The difference of the parameters is authentic (in comparison with the level in the control group).

change in its composition.

After plasmapheresis treatment, substantial changes were observed in the cohesion of the erythrocytes. This is due to the change in the composition of the blood plasma, the effect of temporary plasma depletion and artificial washing of the erythrocytes to remove the substances adsorbed on their surface in an extracorporal contour, thus increasing the negative surface charge of the erythrocytes and increasing the forces of repulsion.

The change in the cohesion of the erythrocytes is associated with a decrease in the amount of fibrinogen and other coarsely dispersed proteins. We found that after a course of plasmapheresis, the level of fibrinogen lowers. This corresponds to published information [21, 360, 401], according to which plasmapheresis is followed by normalization of the parameters of the soluble complexes of the fibrin monomers, fibrinogen, soluble fibrin, and the products of fibrin degradation that also affect substantially the viscosity of the blood and the aggregation of the formed elements. Diminishing of the amount of fibrinogen in the blood and the products of its decomposition weakens the "bridge" mechanism of aggregation of the blood cells. A quite important factor in alteration of the cohesion of the erythrocytes in the process of extracorporal circulation is the reduction in the amount of spherulized forms and of discocytes with appendages [60]. This lowers the aggregation ability and increases the electrophoretic mobility of the cells.

Washing of the erythrocytes during centrifuging not only frees the cells of the products adsorbed on their surfaces, but also alters the vital activities of an erythrocyte with restoration of the activity of the enzyme systems and receptors. It is quite possible that the lowering of the resistance to chemotherapy after plasmapheresis is associated with the deblocking of the receptor apparatus.

The cohesion of the erythrocytes after extracorporal therapy may be affected by the redistribution of the formed elements of the blood with excretion from the depots of young cells in response to a brief loss of blood.

The cohesion of the erythrocytes and the segregation ability of the thrombocytes are also affected in plasmapheresis by the plasma-replacing solutions. Rheopolyglucin has a specific action on the formed elements of the blood, disaggregates them by spreading out in a thin layer not only over the surface of the endothelium of vessels, but also over the blood cells. The adsorption of dextran on the erythrocytes is attended by a growth in the electronegativity of their surface, which increases the process of mutual repulsion of the particles from one another and from a vessel wall, this diminishing thrombus formation.

The described mechanisms of the change in the cohesion of the erythrocytes under the influence of the treatment explains the lower rate of formation of the erythrocyte aggregates and their strength. For example, we noted an increase in the duration of forming small and large erythrocyte aggregates.

The growth in the fluidity of the blood as a result of the lowering of its viscosity and the changes in the qualitative composition of the plasma improve microcirculation. Simultaneously, the fine aggregates of erythrocytes resistant to hydrodynamic disaggregation vanish from the blood flow.

After plasmapheresis, we failed to find an appreciable change in the number of erythrocytes in the blood. An insignificant decrease is possibly associated with their being retained in the systems of the extracorporal contour. The drop in the hematocrit level can be explained by the artificial hypervolemia as a result of the treatment connected with the introduction of a large amount of liquid with respect to the volume of the removed plasma. Moreover, the improvement of the microcirculation, and also of the hyperoncotic effect of rheopolyglucin may increase the volume of the plasma in the blood channel at the expense of the intertissue liquid.

In plasmapheresis, the number of thrombocytes in the patients lowered by 24%. The removal of part of the thrombocytes together with the plasma in addition to lowering the number of globulins and fibrinogen, the plasma factors of blood coagulation, improves the thrombocyte hemostasis [21, 290], alters the functional activity of the thrombocytes [60, 401], and increases their ability of disaggregation.

Consequently, plasmapheresis is followed by an improvement in the rheological properties of the blood that is manifest in lowering of the Kesson viscosity, the forces of cohesion interaction between the erythrocytes, and the rate of their aggregation. The forming aggregates of the erythrocytes can easily break up. When the number of thrombocytes diminishes, their functional activity grows. The effect of plasmapheresis is associated with improvement of the composition of the plasma as a result of the elimination of pathological components, increasing of the metabolism of the lipids and proteins in the cells and tissues, and the disaggregating effect of the plasma substituents.

In a month after plasmapheresis, the parameters of the Kesson viscosity of the blood and the yield point remained within the level of the physiological norm. This may indicate the secretion into the blood flow of young erythrocytes with a preserved deformability.

Individual analysis of the rheological parameters of the blood revealed that in one month after plasmapheresis the rate of formation of the erythrocyte aggregates diminishes. This is attended by a further drop in the strength of the aggregates and the concentration of fibrinogen.

Table 32. Lipid Spectrum of Blood Plasma in Type I DM Patients after Plasmapheresis.

Time of observation	Lipids, %					AI
	Total CS	TG	CS LDLP	CS VLDLP	CS HDLP	
Background	5.71±0.60*	2.50±0.46*	3.46±0.21*	1.14±0.21*	1.11±0.15*	4.48±0.66*
After PPh:						
in 3 days	5.18±0.37	2.50±0.44	2.86±0.46	1.14±0.20	1.18±0.09	3.59±0.33
in 10 days	4.54±0.48*	1.98±0.35*	2.22±0.48*	0.90±0.16	1.49±0.12*	2.03±0.23
in 1 month	5.45±0.45	1.94±0.39*	3.34±0.47	0.88±0.18	1.23±0.14	3.60±0.53
in 3 months	5.96±0.50	2.18±0.42	3.57±0.48	0.99±0.19	1.39±0.1	3.34±0.20

* The difference of the parameters is authentic (in comparison with the background level).

On the background of the drop in the level of fibrinogen with a simultaneous growth in the volume of the cell pool, the amount of fibrinogen per erythrocyte lowers with preservation of the zeta potential of the cells. This is also indicated by the absence of an authentic change in the magnitude of erythrocyte cohesion.

In three months after plasmapheresis, the changes in the rheological parameters of the blood consisted in a growth in the cohesion properties of the erythrocytes with an increase in their ability of forming small aggregates (with the absence of changes in the rate of formation of large ones). The strength of the aggregates remained within the limits of the normal level.

Plasmapheresis is especially efficient in the normalization of lipid metabolism (Table 32). We can note that in type I DM patients (16 persons), the initially high values of cholesterol, triglycerides, and lipoproteins lower. By the tenth day of observation, the fraction of cholesterol lowered from 5.71% to 4.54%, of triglycerides from 2.50 to 1.98%, and of the cholesterol of LDLP from 3.46 to 2.22%. The fraction of cholesterol of the HDLP increased from 1.11 to 1.49%. It is very important to give attention to the drop in the atherogenic index to 2.03.

The phospholipid spectrum of the plasma also changed by the tenth day – the fraction of LPC lowered from 4.46 to 3.58%, of SM increased from 8.93 to 11.73%, and of PC lowered from 75.4 to 68.5% (Table 33).

Of major significance is normalization of the composition of the erythrocyte membranes under the influence of plasmapheresis. In 10 days after conducting it, the fraction of cholesterol in the membranes dropped from 50.09 to 40.20%, of PL increased from 30.5 to 37.5%. The CS/PL ratio changed accordingly. After treatment it was 1.53 against its initial value of 2.04.

Table 33. Phospholipid Spectrum of Blood Plasma in Type I DM Patients after Plasmapheresis.

Time of observation	Phospholipids, %			
	LPC	SM	PC	PEA
Background	4.46±0.96	8.91±1.35*	75.45±2.41*	11.00±1.07
After PPh:				
in 3 days	3.16±0.65	11.33±1.59	73.69±2.38	11.67±1.16
in 10 days	3.58±0.45	13.73±4.15*	68.58±3.53	13.49±2.48
in 1 month	5.73±1.39	7.88±1.46	71.56±2.96	14.77±1.66
in 3 months	6.19±0.48	6.59±2.02	71.03±3.44	16.09±1.98

* The difference of the parameters is authentic (in comparison with the background level).

We attach special significance to the growth in the level of the cholesterol of the HDLP under the influence of plasmapheresis. In DM patients, the level of free CS in the tunic of the HDLP grows, the activity of LCAT lowers, as does the ability of the HDLP to accept cholesterol from the surface of a cell membrane. Plasmapheresis caused these phenomena to normalize: the receptor sensitivity of the erythrocyte membranes increased, the rigidity of a membrane diminishes and microcirculation improved.

Clinical Example

Patient Sh. When 53 years old, he was admitted with the clinical diagnosis: IHD, stress stenocardia FC VI. Postinfarction cardiosclerosis. NC II. Type II Diabetes mellitus, moderate severity, subcompensated. Diabetic macro- and microangiopathies (macroangiopathies of the leg vessels and microangiopathies of the eye ones).

When admitted, he complained of frequent retrosternal pains under a small physical load and at rest attended by dyspnea lasting 5–7 min, poorly relieved by nitroglycerin (glyceryl trinitrate), of pains and spasms in the lower extremities when walking and at rest.

In anamnesis, the patient during three years suffered from IHD, in 1988 and 1989 had myocardium infarctions with the subsequent development of chronic aneurysms of the left ventricle and insufficiency of blood circulation. From 1987 suffered from insulin-independent diabetes mellitus, from 1989 – intermittent dysbasia.

On the background of the treatment conducted including prolonged nitrates (sustak-forte – 12 pills a day), calcium antagonists (korinfar – 9 pills a day), cardiac glycoses (digoxin – 1.5 pills a day) and diuretics, a large number of stenocardia seizures remained requiring the additional taking of up to 35–40 pills of nitroglycerin a day.

His state when admitted was of moderate severity. Cyanosis of the lips. Dyspnea when resting up to 24 a minute. The pulsation of the muscle arteries of the foot is reduced, the extremities are cold. Flaccid shins and legs. Harsh breathing is heard over the lungs, and moist small bubbling rales in the posterior and bottom sections. The boundaries of the relative cardiac dullness are expanded to the left by 2 cm. Tachycardia up to 90 bpm, at the top a systolic noise is heard. The arterial pressure is 140/90 mmHg. In palpation of the abdomen, pains are noted in the right hypochondrium. The liver protrudes from under the rib arc edge by 2 cm, slightly morbid.

Echocardiography revealed an increase in the cavity of the left atrium up to 4.5 cm, of the left ventricle (EDD – 5.3 cm), lowering of the expulsion fraction to 47.9%, hypokinesis of the myocardium, inferoposterior and inferoseptal segments, dyskinesia of the myocardium, apical segment of the left ventricle with development of an aneurysms.

Rheovasography of the vessels of the lower extremities revealed the peripheral type of blood circulation with lowering of the value of the pulse blood flow on the shins and feet (RI=0.3–0.4 units) as a result of lowering of the elastic properties of the vessels. After a nitroglycerin test, there was no authentic increase in the pulse blood flow.

Conjunctival biomicroscopy: perivascular disorders revealed in the form of an edema, focuses of microstasis and hemorrhages; the vascular pathology is determined in the form of nonuniformity of the diameter of the vessels as a result of dilatation of the venules, the presence of desolation and the absence of clear visibility of the capillaries. Intravascular changes in the space velocity of the blood flow. stasis of the blood, and thrombosis of the vessels were revealed.

In laboratory examination in addition to an increase in the concentration of fibrinogen and disorders of the lipid composition of the blood plasma, a growth in the viscosity of the blood, the aggregation ability of the erythrocytes with lowering of the duration of formation of the erythrocyte aggregates was determined.

The patient received three sessions of plasmapheresis with an interval of two days. A total of 2200 ml of plasma was removed with its following replacement with rheopolyglucin.

The course of plasmapheresis was followed by vanishing of dyspnea, pains and spasms in the lower extremities, the number of seizures of stenocardia lowered to 2–5 a day, which made it possible to expand the motor activity and lower the daily dose of drugs.

Control echocardiography: reduction of the size of the left ventricle cavity (EDD – 6.6 cm; ESD – 5.2 cm), of the left atrium to 3.7 cm, an increase in the expulsion fraction to 51%. Rheovasography of the leg vessels: a growth in the pulse blood flow on the vessels of the shins and feet (RI=0.5–0.7 units) with an increase in the blood flow after a nitroglycerin test to 0.6–0.8 units.

Conjunctival biomicroscopy: vanishing of the focuses of stasis, edema and hemorrhages, capillary

Parameter	Norm	Before PPh	1st course of PPh				2nd course of PPh			
			3 days	10 days	1 mon	3 mon	3 days	10 days	1 mon	3 mon
Hb, g/l	15.7	14.2	14.0	14.2	14.0	14.5	13.8	14.0	14.0	14.2
Ht, %	41.0	41.7	40.2	41.0	42.5	41.8	40.8	39.4	41.0	41.2
Thrombocytes, 10^9/l	250	359	200	240	300	320	210	200	230	260
Erythrocytes, 10^{12}/l	4.4	4.89	4.7	4.5	4.8	5.2	4.9	5.0	5.1	5.0
Fibrinogen, g/l	400	660	300	440	480	600	340	300	390	440
K^2, cP	5.55	6.67	3.7	4.0	5.2	6.4	2.8	4.4	4.6	5.0
A, dyn/cm^2	0.75	1.3	0.7	0.9	1.0	1.2	0.9	0.9	1.0	0.9
T, s	9.2	6.5	9.2	8.7	7.7	6.9	9.6	9.4	8.8	8.8
Total CS, %	4.78	5.7	4.4	5.2	5.8	6.0	4.4	0.5	5.2	5.6
TG, %	0.87	1.37	1.08	1.2	1.54	1.3	0.9	0.5	0.8	1.2
LDLP, %	3.6	7.8	4.5	8.0	8.4	9.4	6.0	5.8	5.6	5.6
VLDLP, %	0.55	3.7	1.7	3.9	3.6	3.6	2.9	1.8	1.7	1.7
HDLP, %	1.13	1.14	1.4	1.9	1.73	1.13	1.5	1.7	2.0	2.0

zones of desolation and thrombosis of small vessels.

After plasmapheresis, the level of fibrinogen lowered, the disorders of the lipid composition of the blood plasma diminished, and the rheological potential of the blood became normalized.

In three months after the course of plasmapheresis, the clinical picture of the disease returned to the initial level, which was an indication for repeating the course of plasmapheresis: 2500 ml of plasma was removed with the subsequent replacement with rheopolyglucin.

After the second course of plasmapheresis, the clinical remission was five months (after the first one it was three months).

After plasmapheresis, we noted a change in the level of the total protein in the plasma, mainly as a result of lowering of the concentration of fibrinogen and globulins. The positive shifts in lipid metabolism were characterized by the restoration of the normal ratios of the lipoprotein fractions as a result of lowering of the total cholesterol, triglycerides, and a growth in the level of the HDLP cholesterol. We noted the dependence of the effect on the initial level of the plasma components. The change in the level of lipids, as of the protein composition of the plasma, is associated not only with their elimination from the blood, but is also due to their redistribution between the tissues and plasma, and to the action of plasmapheresis on their metabolism in an organism. As a result of the change in the protein and lipid composition of the plasma, its viscosity lowers and improves the rheological properties of the blood.

An important matter in correcting the coagulating system of the blood is the use of chemotherapy. Having in view that in type I and II DM patients the intravascular aggregation of the thrombocytes is disturbed, the level of beta-thromboglobulin, thrombocyte factor IV, and thromboxan B_2 grows, and the content of prostacyclin and antithrombin III lowers, it was necessary to find a combination of drugs for restoring the coagulation activity of the blood. It was noted that dipiridomol, fitin, glutamic acid, and also korinfar and vitamin E restore the disturbed relations [117, 618].

Pharmacological studies revealed the following.

Curantyl *(*dipiridomol*)* inhibits phosphodiesterase and thereby increases the content of cAMP in the thrombocytes. The level of the cytoplasmic Ca^{2+} lowers, while the thrombocytes retain a disk shape at which their ability of aggregation is minimum. The synthesis of prostacyclin grows, the activity of the vessel walls is restored and stimulates the secretion into the blood flow of prostacyclin and the tissue activator plasminogen.

*K*orinfar *(*nifedipin*)* raises the level of cAMP, and lowers that of Ca^{2+} in the cytoplasm of the thrombocytes, which inhibits their aggregation, increases the antithrombogenic activity of the vessel walls as a result of the greater synthesis of prostacyclin and its secretion into the blood; the calcium blocking agents have a vasoprotecting property that inhibits the development of atherosclerosis.

Table 34. Hemostasis System in Type I DM Patients after Treatment.

Parameter	Control group	Type I DM patients before treatment	Type I DM patients after treatment
Fibrinogen, g/l	2.36±0.10	4.24±0.24*	4.0±0.21
Tolerance of plasma to heparin, s	948±14	670±20*	796±19*
Duration of plasma recalcification, s	286±11	230±15	263±22
Antithrombin III, s	37.6±1.5	21±1.4*	27.0±1.6*
Fibrinolytic activity of blood, %	20.3±0.9	11.5±1.1*	16.1±1.3*
Index of intravascular aggregation of thrombocytes, %	1.03±0.03	1.29±0.04*	1.11±0.05*
Spontaneous aggregation of thrombocytes, %	0	6.1±0.6*	2.4±0.5*
Aggregation of thrombocytes induced by:			
ADP 10^{-5} M, %	61.0±2.3	74.0±1.9*	64.0±2.1*
ADP 10^{-6} M, %	4.2±0.17	9.3±0.2*	6.5±0.4*
collagen, %	57.0±1.4	68.0±2.2*	62.0±2.4
factor of thrombocyte activation, %	14.5±1.0	22.0±1.4*	17.0±2.5
adrenaline 10^{-4} M:			
1st wave, %	13.4±1.1	28.0±1.8*	20.0±2.0
2nd wave, %	59.3±2.1	79.0±2.9*	68.0±3.0*
aggristine, %	57.2±2.0	75.0±2.4*	66.0±2.4*
Beta-thromboglobulin, ng/ml	29.2±1.9	39.0±2.5*	30.8±2.4*
Thrombocyte factor IV, s	14.9±0.9	19.0±0.6*	16.7±0.7*
Thromboxan B_2 in plasma, pg/ml	31.0±3.1	37.0±2.9	33.0±3.0
cAMP in thrombocytes, pmol/$3 \cdot 10^8$ cells	4.1±0.4	3.0±0.3	3.9±0.24*
6-Keto-PGF_1-alpha in plasma, pg/ml	65.0±3.2	48.0±2.2*	60.0±2.3*
6-Keto-PGF_1-alpha/thromboxan B_2	2.1±0.2	1.2±0.2*	1.8±0.2*

* The difference of the parameters is authentic (in comparison with the level in the control group).

Fitin: activates the Ca-dependent ATPase, increases the accumulation of ATP binds the free ions of calcium.

Glutamic acid: activates the synthesis of ATP and increases its accumulation in a cell.

Vitamin E: is a strong antioxidizing agent; increases the synthesis of prostacyclin and prevents lowering of the antiaggregation activity of the vessel walls.

In accordance with these studies, we compiled the following variant of chemotherapy consisting of two complexes.

The first complex includes the following drugs:

(1) curantyl, one pill (0.025 g) three times a day;

(2) fitin, one pill (0.25 g) three times a day:

(3) glutamic acid, one pill (0.25 g) three times a day.

The second complex Includes the same drugs with the addition of 100 mg of vitamin E once a day and korinfar, one pill (0.01 g) three times a day.

Table 35. Hemostasis System in Type II DM Patients after Treatment.

Parameter	Control group	Type II DM patients	
		before treatment	after treatment
Fibrinogen, g/l	2.36±0.10	4.0±0.3*	3.5±0.3*
Tolerance of plasma to heparin, s	948±14	740±21*	850±19*
Duration of plasma recalcification, s	286±11	269±12	278±11
Antithrombin III, s	37.6±1.5	25.0±1.3*	31.0±1.4*
Fibrinolytic activity of blood, %	20.3±0.9	13.5±0.9*	15.1±1.0*
Index of intravascular aggregation of thrombocytes, %	1.03±0.03	1.36±0.05*	1.13±0.05*
Spontaneous aggregation of thrombocytes, %	0	5.1±0.5*	2.0±0.3*
Aggregation of thrombocytes induced by:			
ADP 10^{-5} M, %	61.0±2.3	77.0±2.4*	66.0±2.5*
ADP 10^{-6} M, %	4.2±0.17	9.9±0.2*	6.3±0.3*
collagen, %	57.0±1.4	70.0±2.4*	63.0±2.0
factor of thrombocyte activation, %	14.4±1	23.0±1.4*	16.5±1.8*
adrenaline 10^{-4} M:			
1st wave, %	13.4±1.1	29.0±1.8*	18.0±2.0*
2nd wave, %	59.3±2.1	77.0±3.0*	64.0±2.8*
aggristine, %	57.2±2.0	78.0±3.9*	67.0±4.8
Beta-thromboglobulin, ng/ml	29.2±1.9	40.0±2.4*	31.0±2.6*
Thrombocyte factor IV, s	14.9±0.9	19.5±0.7*	17.1±0.6*
Thromboxan B_2 in plasma, pg/ml	31.0±3.1	42.0±3.0	34.0±3.1
cAMP in thrombocytes, pmol/$3\cdot10^8$ cells	4.1±0.4	2.8±0.3*	3.8±0.22*
6-Keto-PGF_1-alpha in plasma, pg/ml	65.0±3.2	45.0±3.0*	58.0±2.9*
6-Keto-PGF_1-alpha/thromboxan B_2	2.1±0.2	1.18±0.25*	1.85±0.2*

* The difference of the parameters is authentic (in comparison with the level in the control group).

The duration of each complex was 30 days. Treatment was conducted for 77 type I and 47 type II DM patients.

Tables 34 and 35 show the positive dynamics of some parameters of the hemostasis system in type I and II DM patients. The amount of fibrinogen lowered moderately, more noticeably in type II patients (from 4.0 to 3.5 g/l). The growth in the activity of AT III was registered clearly in type I DM patients from 21 to 27 s, and in type II from 25 to 31 s. The intravascular aggregation of thrombocytes diminished noticeably under the effect of treatment: in type I patients from 1.29 to 1.11%, and in type II from 1.36 to 1.13%. Positive dynamics was also noted in the spontaneous aggregation of the thrombocytes.

Of great importance is the lowering of the level of thromboxan B_2 in type I DM patients from 37.0 to 33.0 pg/ml and of type II from 42.0 to 34.0 pg/ml. Correspondingly, the cAMP level in the thrombocytes increased, and the amount of prostacyclin increased considerably – in type I patients from 48.0 to 60.0 pg/ml, and in type II from 45.0 to 58.0.

From our viewpoint, the influence of the pharmacological drugs on the functional activity of the vessel walls is exceedingly significant. Table 36 reveals that in a cuff test in

Table 36. Antithrombogenic Activity of Vessel Walls in Type II DM Patients after Treatment with Drugs of First Complex.

Parameter	Before treatment			After treatment		
	before compres-sion	after compres-sion	index	before compres-sion	after compres-sion	index
Antiaggregation activity of vessel wall: aggregation of thrombocytes induced by:						
ADP 10^{-5} M, %	78.0±2.2	90.0±2.3	1.17±0.07	66.0±3.1	57.0±2.8	0.86±0.07*
collagen, %	72.0±1.3	77.0±1.5	1.09±0.05	61.0±2.0	54.0±1.7	0.9±0.04*
Thrombocyte factor IV, s	20.0±0.8	23.0±0.6	1.17±0.07	17.1±0.7	15.1±0.6	0.9±0.06*
6-Keto-PGF$_1$-alpha in plasma, pg/ml	47.0±3.0	46.6±3.2	1.0±0.07	58.0±3.6	76.0±3.1	1.31±0.08*
Anticoagulant activity of vessel wall: Activity of antithrombin III in blood, s	26.0±1.1	21.9±1.3	0.85±0.09	32.8±1.3	42.1±1.7	1.29±0.08*
Fibrinolytic activity of vessel wall: Fibrinolytic activity of blood, %	14.0±0.9	12.1±0.8	0.86±0.07	15.7±0.93	22.2±1	1.41±0.1*

* The difference of the parameters is authentic (in comparison with the level before treatment).

type II DM patients, a positive dynamics was found. Prior to treatment in a cuff test, the level of aggregation increased from 78.0 to 90.0%, while after chemotherapy the parameter lowered from 66.0 to 57.0%. The level of prostacyclin in type II DM patients did not change in a cuff test (47.0, 46.6 pg/ml), while after the treatment the content of prostacyclin increased from 58.0 to 76.0 pg/ml. The activity of AT III changed in a similar way. If before treatment, the activity of AT III in response to a cuff test lowered from 26.0 to 21.9 s, after chemotherapy the picture changed: in response to a cuff test type II DM patients reacted by a growth in the AT III activity from 32.8 to 42.1 s. The fibrinolytic activity of the vessel walls changed similarly.

Clinical Examples

Example 1

Patient K., 24 years old, visited a clinic with complaints on poorer vision in both eyes, pain in the gastrocnemius muscle at a moderate physical load (when walking at a moderate speed 100 m), and a feeling of cold in the feet in cold weather. Has been suffering from type I DM during eight years. He had had the measles,

Parameter	Before treatment	After treatment	
		1st course	4th course
Fibrinogen, g/l	3.9	3.7	3.2
Tolerance of plasma to heparin, s	703	791	813
Duration of plasma recalcification, s	241	259	291
Antithrombin III, s	23	29	32
Fibrinolytic activity of blood, %	13	19	21
Index of intravascular aggregation of thrombocytes, %	1.36	1.08	1.07
Spontaneous aggregation of thrombocytes, %	5.9	2.1	1.9
Aggregation of thrombocytes induced by:			
ADP 10^{-5} M, %	76	62	62
ADP 10^{-6} M, %	10	6.6	6.1
collagen, %	70	61	59
thrombocyte activation factor, %	24	18	19
adrenaline 10^{-4} M:			
1st wave, %	29	20	21
2nd wave, %	80	66	66
aggristine, %	77	65	60
Beta-thromboglobulin, ng/ml	40	30	29
Thrombocyte factor IV, s	20	16	15.6
Thromboxan B_2 in plasma, pg/ml	38	33	30
cAMP in thrombocytes, pmol/$3 \cdot 10^8$ cells	3.0	4.1	4.3
6-Keto-PGF_1-alpha in plasma, pg/ml	46	59	63
6-Keto-PGF_1-alpha/thromboxan B_2	1.21	1.79	2.1
Antithrombogenic activity of vessel walls, index:			
antiaggregation	1.18	0.89	0.81
anticoagulant	0.89	1.16	1.21
fibrinolytic	0.77	1.58	1.48

rubeola, frequent respiratory diseases, and influenza. His father is healthy, his mother has high blood pressure. The family uses more animal fats in their food.

Examination of K. provided information on the state of the system of hemostasis before and after treatment with drugs of complex 1.

Content of polyunsaturated fatty acids in the phospholipids of the thrombocyte membranes: $C_{18:2}$, ω-6 – 4.5%; $C_{20:4}$, ω-6 – 30.11%, $C_{20:5}$, ω-3 – 0.5%, and $C_{22:6}$, ω-3 – 1.61%. The coefficient

$$K_{ac}=\frac{C_{20:4},\omega-6}{C_{20:5},\omega-3+C_{22:6},\omega-3}=14.27$$

Content of hormones in the plasma:

IRI – not determined;

C-peptide – 0.35 ng/ml;

cortisol – 480 nmol/l.

The patient's thrombogenic potential of the blood is increased: the procoagulant activity of the blood is increased, the anticoagulant and fibrinolytic activity is lowered, intravascular aggregation of the thrombocytes was discovered, the antithrombogenic (antiaggregation, anticoagulant, and fibrinolytic) activity of the vessel walls is reduced. Disorder of the state of the hemostasis system is combined with a high content of polyunsaturated fatty acids in the thrombocytes, the absence of IRI in the blood, the extremely low level of the C-peptide, and a high cortisol content in the blood plasma.

Consequently, in patient K. apart from restoration of the antithrombogenic activity of the vessel walls, functional improvement of the clinical state of hemostasis, improvement of the clinical state after the fourth course of treatment with drugs of the 1st complex, positive dynamics of the microcirculatory canal of the fundus of the eye was also noted.

Example 2

Patient S. 36 years old, visited a clinic with complaints on poorer vision (more in the right eye), a feeling of cold in his arms and legs, pain in the gastrocnemius muscles, more at the right side, periodically appearing under a physical load. Has been suffering from DM during 12 years. He had had measles, sinistral infralobal pneumonia (in 1985). He has no information on the health of his parents.

Examination of S. provided information on the functional state of the hemostasis system before and after treatment with drugs of complexes 1 and 2.

Content of polyunsaturated fatty acids in the phospholipids of the thrombocyte membranes:

$C_{18:2}$, ω-6 – 4.3%, $C_{20:4}$, ω-6 – 31.16%, $C_{20:5}$, ω-3 – 0.56%, and $C_{22:6}$, ω-3 – 1.48%. The coefficient

$$K_{ac}=\frac{C_{20:4},\omega-6}{C_{20:5},\omega-3+C_{22:6},\omega-3}=15.27.$$

Content of hormones in the plasma:

IRI – not determined;

C-peptide – 0.31 ng/ml;

cortisol – 465 nmol/l.

The patient's procoagulant activity of the blood is increased, the anticoagulant and fibrinolytic activity is reduced, the functional activity of the thrombocytes is increased. Intravascular activity of the thrombocytes is revealed, the antithrombogenic (antiaggregation, anticoagulant, and fibrinolytic) activity of the vessel walls is reduced. Pronounced unbalance with respect to the polyunsaturated fatty acids is noted in the phospholipid spectrum of the thrombocytes. The hormonal spectrum is noted by the absence of IRI and an extremely low level of the C-peptide, which confirms the diagnosis of type I DM. The cortisol level in the blood is increased.

Diagnosis: type I diabetes mellitus, moderate degree of severity in the compensation phase. Preproliferative diabetic retinopathy, exudative-hemorrhagic form, central edema of eye retina. The first complex of drugs was prescribed, and in 30 days clinical and laboratory studies were performed. After the course of treatment with the drugs of the first complex, no restoration of the antithrombogenic activity of the vessels and correction of the state of the hemostasis system occurred. Positive dynamics in the clinical state of the patient was also not noted.

In this connection, a second 30-day course of treatment was prescribed with drugs of the second complex (curantyl, 0.025 g three times a day, glutamic acid, 0.25 g three times a day, korinfar, 0.01 g three times a day, and vitamin E, 100 mg once a day). After the course of treatment with drugs of the 2nd complex, the antithrombogenic activity of the vessel walls was restored, the procoagulant activity lowered, the anticoagulant activity and fibrinolytic activity of the blood increased, and the functional activity of the thrombocytes diminished. The cortisol level in the blood dropped to 363 nmol/l.

Subsequently, similar courses of treatment with drugs of the second complex were conducted three times (every two months after completion of the preceding course). After the fourth course, clinical laboratory analyses were repeated. The antithrombogenic activity of the vessel walls and the functional state of the patient's hemostasis system were within the limits of the normal values.

Parameter	Before treatment	After treatment	
		1st complex	2nd complex
Fibrinogen, g/l	3.7	3.8	3.5
Tolerance of plasma to heparin, s	709	725	805
Duration of plasma recalcification, s	250	249	263
Antithrombin III, s	25	26	30
Fibrinolytic activity of blood, %	12	13	19
Index of intravascular aggregation of thrombocytes, %	1.42	1.36	1.15
Spontaneous aggregation of thrombocytes, %	5.1	4.6	1.1
Aggregation of thrombocytes induced by:			
ADP 10^{-5} M, %	73	72	62
ADP 10^{-6} M, %	9	8.1	5.3
collagen, %	68	67	64
thrombocyte activation factor, %	26	26	20
adrenaline 10^{-4} M:			
1st wave, %	27	25	20
2nd wave, %	79	70	71
aggristine, %	79	80	72
Beta-thromboglobulin, ng/ml	40	38	31
Thrombocyte factor IV, s	21	20	14.5
Thromboxan B_2 in plasma, pg/ml	39	38	31
cAMP in thrombocytes, pmol/$3 \cdot 10^8$ cells	3.3	3.5	4.3
6-Keto-PGF_1-alpha in plasma, pg/ml	44	46	60
6-Keto-PGF_1-alpha/thromboxan B_2	1.13	1.21	1.94
Antithrombogenic activity of vessel walls, index:			
antiaggregation	1.21	1.18	0.79
anticoagulant	0.82	0.89	1.23
fibrinolytic	0.79	0.83	1.51

Positive dynamics in the clinical state of the patient was noted after treatment with drugs of the second complex. It consisted in improvement of the visual acuity, a considerable reduction of the pains in the gastrocnemius muscles and of the feeling of cold in the arms and legs.

Simultaneously with the growth in the thrombocyte-vascular activity in DM patients, we noted the positive dynamics of the clinical state in the form of diminishing or vanishing of the pains in the feet, gastrocnemius muscles, and a drop in the feeling of numbness. The visual acuity improved.

All the patients were subjected to fluorescent angiography. After treatment, we noted a positive dynamics in the state of the vessels of the fundus of the eye in the form of a reduction in the twisting of the veins, lowering of the amount of microaneurysms, hemorrhages, and extravasal secretion of fluorescein. In many patients, partial resorption of the solid exudate occurred.

2.5. Conclusion

We must stress in concluding that the incidence of DM is attended by the development of two syndromes – hyperglycemia and microangiopathies. In DM, lesions of the endothelium and subendothelium are the pathophysiological basis of injury of the vessels. Our functional-biochemical analysis of the blood of DM patients showed an important unit in the development of microangiopathies – glycosylation. We found that a deficiency of insulin in DM distorts the biochemical transformation of glucose, which occurs not so much glycolytically as via a polyol shunt, by activation of the enzyme aldoreductase. The result is the accumulation of an excess amount of products of pronounced glycosylation and binding of the insoluble complexes of the basal membrane with one another. Here the structure of the endothelium of the vessel wall changes, which causes disturbance of the production of prostacyclin. The increased formation of a peroxide of lipids suppresses the synthesis of prostacyclin, and accordingly the thrombogenic potential of the blood grows together with aggregation and adhesion of the thrombocytes, and microthromboses develop.

We find that the formation of vascular lesions in DM occurs in two ways: by glycosylation and by the polyol way of glucose metabolism.

An important role in disturbing the rheological properties of the blood is played by a change in its viscosity, hyperfibrinogenemia, hyperaggregation of the erythrocytes and thrombocytes. In DM, glycosylized hemoglobin accumulates in the erythrocytes, which causes the development of hypoxia. Of major significance in DM is lesion of the insulin receptors of the endothelium that lead to the development of insulin resistance. It is these changes in the morphofunctional activity of the erythrocytes in DM that are attended by a change in their surface charge with the development of hypercoagulation and microcirculatory disorders.

An important moment in the formation of microangiopathies in DM is the disorder of lipid metabolism – a growth in the content of cholesterol, triglycerides, LDLP, VLDLP, and apo-B. This is attended not only by quantitative changes in the individual components of lipid metabolism, but also by a growth in their biochemical activity. For example, in the LDLP, not only the content of triglycerides grows, but also the activity of glycosylation increases. These biochemical disorders in DM patients result in binding of the glycosylized collagen fibers at the places of injury to the endothelium of a vessel, and the atheromatous plaques form very rapidly.

In DM, the thrombocytes participate in activation of the thrombogenic potential of the blood. It was proved that in DM the metabolic process of transformation of arachidonic acid into thromboxan B_2 activates considerably, which correlates clearly with the glucose level in the blood. These disorders cause an increase in the aggregation and adhesive ability of the thrombocytes in DM.

Our determination of the content of linolic acid and unsaturated fatty acids in thrombocytes of DM patients showed that the level of unsaturated fatty acids (eicosapentaenic and docosahexaenic) in the thrombocytes drops sharply, which causes the development of the DIC syndrome, thromboses, and emboli.

A growth in the Willebrand factor is an important unit in the process of development of disorders in the coagulating system of the blood. This factor is a marker of the lesion of the endothelial cells of the vessel walls and facilitates the adhesion of the thrombocytes.

CHAPTER 3.
DIABETES MELLITUS AND ISCHEMIC HEART DISEASE

A very important complication in DM is the development of IHD. Most authors consider that atherosclerosis of the coronary vessels in DM patients is encountered in 70% of the cases [291, 351, 431, 453], and the coronary vessels and not the aorta are what is injured. Very often "mute" forms of IHD and myocardial infarction are noted [5, 50, 190,

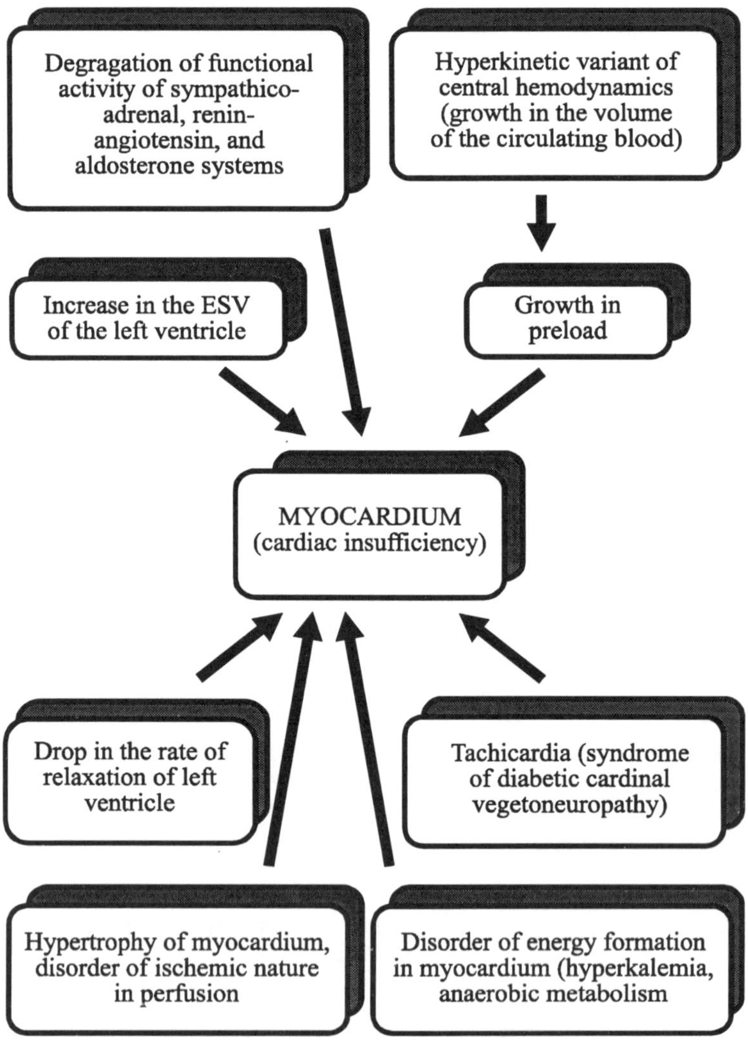

Fig. 33. Development of Cardiac Insufficiency in DM.

380, 466]. Pathoanatomical studies revealed that the extent of the atherosclerotic process is not associated with the duration of the course of DM [198, 201, 381, 431, 453]. Generally, in type II DM patients hyperinsulinemia was noted, while IHD developed twice as frequently as in persons with a normal insulin level [110, 145, 260, 280, 537, 548].

Clinicians attach special importance in the progress of atherosclerotic changes in the coronary vessels to the combined influence of DM and obesity [26, 42, 205, 465].

Many authors note that the level of the total cholesterol is less important for the development of IHD in DM than an increased content of glucose in the blood [27, 112, 132, 198, 201, 537]. On analyzing the results, scientists arrived at the conclusion that hyperglycemia causes the relevant hyperinsulinemia, while a high insulin level by increasing the synthesis of cholesterol esters in the vessel walls facilitates the development of atherosclerosis.

3.1. Pathophysiological Fundamentals of Development of Cardiac Insufficiency in Diabetes Mellitus Patients

A glance at Fig. 33 reveals that the contractility of the left ventricle in DM patients (especially of type II) is sharply disturbed. First of all, the formation of energy in the myocardium drops: ischemia of the myocardium and metabolic disorders entail the predomination of anaerobic metabolism in combination with hypokalemia of the myocardium.

Table 37. Parameters of Hemodynamics in IHD Patients (Data of Electrocardiology).

Parameter	Control group	IHD patients
End systolic volume, ml	32.64±1.93	72.5±7.43
End diastolic volume, ml	104.8±4.25	168.4±11.2
Minute volume, l/min	4.85±0.42	6.36±0.32
Cardiac index, $l/min \cdot m^2$	2.78±0.12	3.48±0.2
Stroke volume, ml	72.65±4.28	103.06±4.5
Stroke index, ml/m^2	40.57±2.78	55.8±3.6
Unit power, $kgm/min \cdot m^2$	3.34±0.26	4.87±0.33
Peripheral resistance, $dyn \cdot s \cdot cm^{-5}$	1508.4±97.6	1338.8±95.4
Maximum mass, g	115.2±4.63	167.6± 9.85
Index of maximum mass	66.42±2.1	82.24±6.05
End diastolic vol./max. mass	1.07±0.075	1.14±0.07
Ejection fraction, %	70.42±1.94	62.26±2.4
%ΔS	40.07±1.62	32.84±2.63
V_{cf}, s^{-1}	1.21±0.07	1.19±0.09
Contraction rate, cm/s	5.25±0.37	3.85±0.31
Relaxation rate, cm/s	13.2±0.68	6.82±0.815

Disorder in energy formation causes (Table 37) a growth in the end diastolic volume (EDV) of the left ventricle and a drop in the relaxation rate. In type II DM patients, a sharply pronounced hypertrophy of the myocardium is noted that leads to violation of the coronary blood flow with the presence of a perfusion defect of an ischemic nature. Progress of hypertrophy and dilatation of the left ventricle lowers its functional ability. The coronary arteries fail to deliver the required amount of oxygen to the hypertrophied myocardium, dilatation of the cavity of the left ventricle results in tension of its walls during the systole, and, consequently, in the myocardium needing oxygen. The development of cardiofibrosis is attended by disorder in filling of the left ventricle during diastole.

Diabetic cardiac vegetoneuropathy is attended by tachycardia, and ineffective contractions of the myocardium exhaust the energy resources. The hyperkinetic variant of central hemodynamics with a growth in the volume of circulating blood is an important pathological factor in DM (especially combined with obesity). These changes are attended by a growth in the preliminary load and lead to the incidence of blood circulation insufficiency.

The disintegration of the functional sympathico-adrenal, renin-angiotensin, and aldosterone activity is significant [14, 54, 85, 262, 368, 539]. Disturbance of the water-electrolyte balance in an organism with the retention of sodium in it causes the development of an edematous syndrome. Physicians are especially concerned by a combination of type II DM with obesity [205, 289, 365, 465]. Here the hyperkinetic syndrome is accompanied by disturbances of the precapillary permeability with the relevant growth in the load on the contracting myocardium.

Like any cell of a living organism, a cardiac myocyte performs the main physiological functions: the specific function of contrition, the conversion of energy to the form of phosphate bonds of nucleotides (glycolysis and oxidative phosphorylation), and also the synthesis of protein structures [12, 187, 387, 388, 434]. These functions in thyrotoxicosis experience considerable changes [338, 535]. The deformation of the structural and biochemical organization of this cell touches on the energy aspect (the energy devaluation of tissue respiration), electrolytic metabolism, the electrophysiology of a cell (shortening of the potential of action), the contractive function, and plastic supply of the myocardium (inhibition of protein synthesis).

Disturbance of the oxygen homeostasis in an organism causes a change in various physiological and metabolic constants, the most important of which is disorder in energy metabolism. Organic phosphates and erythrocytes are of major significance in the oxygen-transportation function of the blood. They regulate oxygenation [180, 257, 277, 317, 364, 386].

Table 38. Parameters of Hemodynamics in DM Patients (Data of Electrocardiology).

Parameter	Type I DM patients	Type II DM patients
End systolic volume, ml	39.37±4.68	39.45±4.62
End diastolic volume, ml	130.21±9.32	136.25±7.55
Minute volume, l/min	6.66±0.38	6.97±0.42
Cardiac index, $l/min \cdot m^2$	3.67±0.2	3.46±0.23
Stroke volume, ml	90.84±5.1	96.8±6.34
Stroke index, ml/m^2	49.76±2.22	48.13±3.63
Unit power, $kgm/min \cdot m^2$	4.85±0.3	4.65±0.36
Peripheral resistance, $dyn \cdot s \cdot cm^{-5}$	1246.66±86.3	1217.0±77.5
Maximum mass, g	116.82±6.46	167.84±6.21
Index of maximum mass	69.13±2.5	84.18±3.69
End diastolic vol./max. mass	1.06±0.06	0.82±0.05
Ejection fraction, %	70.5±2.27	65.17±1.94
%ΔS	40.99±1.83	41.12±1.69
V_{cf}, s^{-1}	1.47±0.08	1.38±0.07
Contraction rate, cm/s	5.04±0.46	5.25±0.3
Relaxation rate, cm/s	9.81±0.57	9.53±0.85

The glycogen of the liver is one of the main reserves of energy sources in an organism. A human organism contains about 70 g of glycogen. Under conditions of fasting, it rapidly vanishes. Glycogen is formed in the liver and muscles with the aid of glucose-6-phosphate. Biochemical studies showed that the energy of glycogen is used in two stages. In the first, glycogen is decomposed to glucose (glycogenolysis), and in the second it is transformed into CO_2 and water (glycolysis) [96, 386, 387, 388, 389, 431].

Numerous investigations were devoted to studying the ability of DM patients to perform physical work (especially type II patients). The disease activates the development of atherosclerotic changes in the myocardium [8, 190, 198, 201, 298, 544]. Disturbance of the contracting ability of the myocardium is considered to be the pathophysiological basis of the drop in the physical workability in DM. In the opinion of a number of authors, diffusion lesions of the myocardium, its hypertrophy, and disturbance of its contractibility are noted morphologically in DM [4, 190, 227, 535]. Other authors consider the development of cardiopathy with a drop in the rate of diastolic relaxation of the myocardium to be the main pathophysiological indication of lesion of the myocardium [102, 179, 190, 198, 201, 256]. To reveal the pathophysiological causes of lesion of the myocardium in DM, we compared the physical workability of healthy persons and type II DM patients in the compensation stage (from 39 to 63 years old). For the same persons, we also studied the central hemodynamics by echocardiography.

In DM patients, the power of the threshold load (M_{th}) is lowered considerably: in healthy persons it is 13.78 W/min, while in DM patients it is 7.62 (accordingly, the total power of the performed load is 7423 and 4000 kgm). The chronotropic heart reserve (CHR) lowers correspondingly, in the DM patients it is 58.5 units, and in the healthy persons it is 66.1. The difference in the energy expenditure index (EEI) is also significant – it in 4.2 units in healthy persons and 14.08 in DM patients.

In DM patients of type II (Table 38), a hyperkinetic syndrome is noted. The high level of the cardiac index – 3.46 l/m^2 (against 2.78 in healthy persons) is mainly ensured by a growth in the stroke index. In type II DM patients, the total peripheral resistance is lowered to 1217.0 dyn·s·cm^{-5} (in healthy persons it is 1508.4). It is quite logical to assume that a growth in the ventricular ejection and drop in the peripheral resistance are aimed at increasing the supply of oxygen under conditions of lowering its partial stress in the tissues. The growth in the ventricular ejection in type II DM patients is doubtlessly due to the increase in the mass of the myocardium of the left ventricle. Hypertrophy of the myocardium in DM is explained by disturbance of metabolism and dystrophic changes, especially by diabetic microangiopathies. When comparing other parameters of echocardiography (the ejection fraction, extent of shortening of the front-rear dimension of the left ventricle during systole), we failed to note considerable distinctions. Most likely, the growth in the mess of the myocardium of the left ventricle in type II DM compensates the drop in the contractility of the cardiac myocytes.

We agree with the opinion of Starkova *et al.* [548, 549] that in hyperinsulinism there is a tendency in the vessel walls of developing atherosclerosis under the effect of insulin that increases lipogenesis by stimulating the proliferation of the smooth muscle cells and hyperlipoproteinemia. The lipoproteins are subjected to nonenzymatic glycosylation in the same way as the other proteins of an organism. These processes stand out especially in DM [157, 182]. Glycosylized LDLP have a higher negative charge than normal ones, which disturbs their reaction with the specific receptors of fibroblasts, lowers their catabolism, and increases the nonspecific capture by endothelial cells and macrophages. Insulin correction of carbohydrate metabolism in DM normalizes the reaction of the LDLP with the specific receptors of the fibroblasts [49, 176, 247, 358, 516].

Glycosylation also affects the main components of the coagulating system of the blood [22, 67, 78]. In particular, the glycosylation of antithrombin III blocks the residue of the aminogroups of lysine by means of which antithrombin III reacts with heparin. Of special significance is the glycosylation of fibrinogen and the intrathrombocyte proteins. Morphologists noted that the degree of glycosylation of fibrinogen increases as it is deposited in a vessel wall [63, 176]. This is attended by a drop in the fibrinolytic activity of plasmin and a growth in the aggregation of thrombocytes, parietal thrombus formation, and proliferation of the smooth muscle walls.

Nonenzymatic glycosylation is associated primarily with hyperglycemia. The latter is typical of DM and is a biochemical process synthesizing (integrating) carbohydrate, protein metabolism, and the consumption of oxygen [112, 182].

Most authors consider thickening of the glycoprotein basal membrane as a pathomorphological indication of microangiopathies in DM [355, 466, 549]. Genetic defects of a vessel wall and diabetic disorders of metabolism trigger off the development of microangiopathies. Thus, a change in the morphobiochemical structure of collagen that is the main component of the basal membranes of capillaries is one of the important morphological features of microangiopathies [10, 46, 355]. It has the longest lifetime and is subjected the most to the unfavorable effect of glucose in DM patients with the presence of hyperglycemia. An increased content of nonenzymatic modified collagen is attended by lowering of its solubility and elasticity, and its functional sensitivity to proteolytic enzymes alters. This is why the formation of glycosylized collagen is an exceedingly significant additional unit of pathogenesis of microangiopathies. Since collagen is a structural component of many organs and tissues, its glycosylation causes a premature disturbance of their functions and aging.

Glycosylation is what produces, as it were, "glucosotoxicity" in DM patients. Glycosylation of collagen in the basal membranes disturbs the function of the capillaries and is attended by their thickening. The accumulation of collagen in the tissues alters their functional properties. In DM patients, glycosylation of the collagen in the skin sometimes leads to the appearance of a sclerodermo-like syndrome [96].

The glycosylation of collagen disturbs the function of the erythrocyte membranes, changes their permeability, and this is attended by a growth in their adhesion [22, 46].

Glycosylation is especially undesirable in the renal glomeruli because it is attended by thickening of the main membrane of the glomeruli, a growth in their permeability, and development of what is called diabetic glomerulosclerosis [285, 355, 466].

Fibrin is also glycosylized. The attachment of glucose to fibrin inhibits its degradation and facilitates its accumulation in various tissues in DM patients [63, 182].

Considerable amounts of glycosylized fibrin are found in the basal membranes of kidney capillaries in DM patients.

The proteins of erythrocyte membranes play an important role in the development of diabetic microangiopathies. It was noted that as a result of glycosylation, the permeability of the membranes is disturbed, the consumption of ATP grows, the deformability of the erythrocytes lowers, the microviscosity and formation of "coin stacks" grow, and the adhesion of erythrocytes on the endothelial membranes becomes stronger [22, 182, 285]. An important role in the development of microangiopathies is played by the glycosylation of lipoproteins [80, 166, 176, 247, 249].

It was proved that the glycosylized lipoproteins circulating in the blood plasma are bound to the glycosylized collagen via active centers and accumulate in the vessel walls in connection with the development of hypercholesterolemia and lipid infiltration. Atherosclerotic lesion appears most often on the injured sections.

Morphobiochemical studying of glycosylation in DM patients revealed that the development of microvascular lesions and atherosclerosis is based on glycosylation [46, 63, 466, 549, 550]. Nonenzymatic glycosylation is the leading unit in the pathogenetic chain of processes on the way to the development of microangiopathies.

We agree with Efimov's opinion [134] that diabetic microangiopathies develop as a result of genetic, immunological, hormonal, metabolic, hemorheological, and other disorders. But nevertheless, the glycosylation of collagen and lipoproteins is an important additional link in the pathogenesis of microangiopathies.

In hyperglycemia, the glycosylation of proteins alters their structure and functions. These, on the one hand, cause disorders of metabolism in a vessel wall and the adjoining tissues, and on the other, alter the structure of the membranes themselves – they become thicker, and their elasticity diminishes [15, 56, 80, 157, 176, 182, 355]. These disorders are attended by a change in the tonus and structure of the vessels of the microcirculation channel, disturbing of the rheological properties of the blood, lowering of the oxygen balance of the tissues, and disorders of the vessel permeability and resistance.

3.2. Energy Metabolism in Diabetes Mellitus

Energy metabolism is an involved thermodynamic system integrating the interaction of physiological and biochemical components, the metabolic characteristics of metabolism (especially of carbohydrates and fats), and also regulatory hormonal relations.

Clinicians noted long ago that in DM patients myocardial infarction develops without a pain syndrome or it is insignificant [28, 201, 380, 466, 535, 615]. Among the reasons of the painless course of myocardial infarction, most authors indicate generalized capillaropathy, lesion of the small intramural vessels with disturbance of the function of the neuroreceptor apparatus [190, 355, 548]. Hindrances in the development of collateral blood circulation in the myocardium in DM are also associated with the same pathogenetic factors. This creates favorable conditions for repeated and extensive infarctions and aneurysms [201, 355, 380, 466, 535].

The development of atherosclerotic changes in the myocardium in DM patients can be considered from the following viewpoints:

(1) genetic determination of metabolic disorders in an organism facilitating the development of atherosclerosis (disturbance of carbohydrate and fat metabolism);

(2) development of dyslipidemias with a change in the content of cholesterol, triglycerides, LDLP, VLDLP, HDLP, and also apo-B and apo-A-1.

(3) disorder of the morphological structure and functional activity of a vessel wall;

(4) change in the sensitivity of the receptor apparatus of the cells of the myocardium and vessels.

A lack of insulin in DM sharply accelerates the disturbance of metabolism and the development of atherosclerosis. The presence of hyperglycemia in DM patients facilitates the formation of simple and complex polysaccharides in a vessel wall. The presence of hyperlipidemia (an increased amount of cholesterol and triglycerides) increases the formation of lipids in the intimas of vessels [32, 45, 284, 365, 380, 381, 558].

Disorder of lipoprotein metabolism in DM is closely related to nonenzymatic glycosylation [10, 18, 166, 189, 247]. The attachment of glucose to the HDLP with the formation of a glycosylized form of HDLP (GHDLP) prevents the transportation of cholesterol from the tissues into the liver and the participation of HDLP in the synthesis of lipoproteinlipase that carries out catabolism of the atherogenic lipoproteins. Owing to glycosylation in DM, the hyperproduction of glycosylized LDLP and VLDLP occurs. Here their interaction with the specific receptors is suppressed, the capture by fibroblasts increases, hypercholesterolemia and lipid infiltration of a vessel wall appear.

In DM (a lack of insulin), lipolysis grows together with the excretion of free fatty acids into the plasma. The acids (especially palmitic, oleic, and linolic) are deposited in the adipose tissue in the form of triglycerides and phospholipids [15, 42, 45, 120, 205, 432, 542, 558].

The development of IHD proceeds especially actively in type II DM patients [190, 201, 291]. The explanation is that in this type of diabetes hyperinsulinemia, insulin resistance, and simultaneously a pronounced lowering of the HDLP are generally noted in patients. The pathogenesis of the drop in the HDLP is due to the fact that in type II DM insulin is applied not to the peripheral tissues, but to the liver. The activity of the liver lipase grows and stimulates catabolism of the HDLP.

The pathomorphological studies in DM are attended by numerous functional features: the increased accumulation in a vessel wall of phospholipids, cholesterol esters, and mucopolysaccharides [10, 189, 257, 381]. The metabolic disorders in DM are very diverse. Among them are tissue hypoxia, DLP, hormonal unbalance, metabolic acidosis, angiopathy, and electrolytic disorders [63, 284, 355, 381]. Special significance is acquired by the relation between disorders of the oxygen-transportation function of the blood, the coagulating system and, as a consequence, by the development of microangiopathies [22, 46, 355, 380].

The incidence of IHD in DM consists in the breaking of the metabolic and hormonal relations with an accent to the disturbance of carbohydrate and fat metabolism.

A physical load sharply alters the reaction of the cardiovascular system and increases metabolism: the activity of the heart grows, the blood flow is redistributed among the various vascular zones, perfusion changes in favor of functionally loaded muscles, gas metabolism in the lungs increases, and the oxidative processes in the tissues grow. The extent of all these changes depends both on the intensity of the load and on the functional state of all the organs and systems.

The physiological systems of an organism are genetically determined for the timely replacement of periods of intense functioning with ones of relative physiological rest.

Physiological studies of cardiac activities (both under conditions of rest and under a physical load) revealed that the increase in the amplitude and force of heart contraction at any frequency thereof occurs on the background of the complete diastolic relaxation of the myocardium [12, 109]. Under conditions of a physical load, when the amplitude, force, and frequency of heart contractions increase simultaneously, this result can be achieved only by increasing the rate of the relaxation process. Experiments with animals did show this relation [388, 434]. It was proved that the rate of relaxation increases not only with a growth in the frequency, but also with a growth in the amplitude or force of contraction within the physiological range. It is this clear-cut physiological relation between the contraction and relaxation of the cardiac myocytes that preserve the diastolic pause and the stability of the contractive function of the heart under any physiological loads.

Physiologists proved that the relation between contraction and relaxation occurring at the level of the cardiac myocytes ensures the most important law of functioning of the heart – its adaptation to the changing conditions of the environment [388, 389, 434]. We proved that in DM patients, realization of Starling's mechanism is disturbed, the contracting function is preserved, but the rate of relaxation lowers on the background of a high force and rate of contraction.

It was shown in clinical and experimental studies that adaptation to a physical load is combined with an increase in the activity of the adenylatecyclase system and in the amount of adrenergic fibers per unit of myocardium mass. This increases the adrenoreactivity of the myocardium [109, 129, 386, 387]. A growth in the number of H-chains that are carriers of ATPase activity is observed in the heads of myosin. Simultaneously, the power of the calcium pump of the sarcoplasmic network grows, and this increases the rate and depth of diastolic relaxation of the myocardium. A growth in the power of the system of energy supply lawfully boosts the resistance of the heart to hypoxia during a physical load.

Disorder of the metabolic processes in DM sharply alters the adaptation of the organism to the performing of a physical load.

Clinical studies of the tolerance to a physical load in DM patients revealed numerous pathological changes [118, 298, 544, 619, 623]. First of all, the systolic pressure during a physical load grows to a greater extent than in healthy persons. The growth in the frequency of heart contractions underlies the increase in the minute volume of blood discharged by the heart and is an important physiological mechanism of the adaptation of the heart in DM patients to a physical load.

The changes revealed in an electrocardiogram in a veloergometric test are quite diverse – extension of the electrical systole and disturbance of the relation between T_{V_1} and T_{V_6}. Disorders appear especially frequently during the period of repolarization of the tooth T, namely, from its lowering to the appearance of a negative tooth T. It was noted that a myocardial infarction proceeds very severely in DM patients – lethal outcomes in these patients are noted three times more frequently. Sometimes aneurysm of the left ventricle is uncovered [8, 27, 28, 198, 201, 615]. Coronarographic studies showed that in DM patients, complete occlusion of the left and right coronary arteries is encountered with the same frequency [625].

To study the relation between the clinical picture of IHD and the metabolic changes in an organism, we performed coronarographic examination of 15 type II DM patients, in most of them, two large arteries were injured simultaneously. Six patients had changes in three coronary arteries, and the right coronary artery was often involved in the process. Generally, when three vessels were injured, one of them was narrowed by 75%. These patients most often suffered from stenocardia of rest, and in a number of cases the fits appeared up to 10 times a day.

A relation was noted between the frequency of the fits and the development of decompensation of DM. When comparing the coronarographic and electrocardiographic data for patients with IHD and DM, we found that in lesion of the right coronary artery and the anterior interventricular branch of the left coronary artery, changes were revealed on the anterior or anteroseptum walls of the left ventricle.

Clinical Example

Patient K., 54 years old, during 15 years has been suffering from type II DM with periodic phenomena of decompensation. During eight years, fits of sharp retrosternal pains were noted that irradiate into the left shoulder and scapula. The fits occur several times a day and are relieved by taking nitroglycerin.

When admitted, a state of moderate severity was noted. Because of the frequent fits of pains, he was forced to spend most of the time in bed. In objective examination, the heart tones were dull, the pulse was rhythmic – 84 bpm, with a satisfactory filling, the arterial pressure was 190/100 mmHg. The electrocardiogram showed indications of insufficiency of blood circulation in the region of the anterior wall of the left ventricle and interventricular partition. To clarify the features of injury of the coronary arteries and solve the question of the possibility of operative interference, selective coronarography was performed by Jadkins's procedure. When contrasting the coronary arteries, deepening of the myocardial ischemia was noted that was manifest by pains and changes in the ECG.

The fit was relived with nitroglycerin and the intravenous introduction of analgesics. The coronarogram showed severe injury of the coronary arteries, especially of the right one. Its complete occlusion was noted.

Numerous clinical observations showed that disorders in the activity of the cardiovascular system in IHD and type II DM patients have very many similar characteristics [3, 8, 16, 28, 392, 619]. These comparisons suggested the pathogenetic similarity of various changes in the myocardium and in the vessels in these diseases. However, very few clinical and experimental investigations were devoted to studying the pathology of the myocardium and vessels in DM patients [198, 201, 234, 298].

To clarify these matters, we conducted veloergometric studies of patients with IHD, type II DM, and of healthy persons. We observed 27 IHD patients from 38 to 60 years old (who had had a myocardial infarction from one to six years ago), 22 DM patients of type II (the compensation stage) from 39 to 63 years old, and 18 healthy persons from 22 to 49 years old.

When running the veloergometric test, we guided ourselves by the generally adopted criteria – stepwise application of the load in three-minute steps: step 1 – 75W, step 2 – 100W, step 3 – 150W, and step 4 – 175W. The following were the generally adopted criteria of the ischemic reaction to a physical load: lowering of the segment S-T (horizontally or slanting downward) by at least 1 mm, or rising of this segment, typical anginal retrosternal pains during the test or during the first five minutes of rest, serious disturbances of the rhythm, lowering of the arterial pressure during a load, an increasing dyspnea, and sharp muscular fatigue.

We also determined the power of the threshold load per kg of body mass, the ratio of the achieved power of the threshold load to its required value (the energy coefficient of tolerance), the total power of the applied load, the maximum consumption of oxygen per kg of body weight, the ratio of the maximum consumption of oxygen to its required value, the pulse coefficient of tolerance,, the increment of the beats per minute at the threshold load (the chronotropic heart reserve), the increment of the systolic arterial pressure at the threshold load (the inotropic heart reserve), and the index of the energy expenditures.

In analyzing the results, we noted first of all lowering in the DM patients of the total power of the performed load. In healthy persons, this value is 7423.0 kgm, in type I DM patients – 4515.1, and in type II – 4223.3 kgm. The power of the threshold load is lowered accordingly: in healthy persons it is 167.2 W/min, in type I DM patients – 125.8, and in type II – 113.3 W/min.

What are these changes associated with?

Physiological studies in recent years showed that there is a definite metabolic level in a human organism below which the content of lactic acid in the blood does not grow, but it does grow considerably above this level [333, 388, 389, 434]. This level is designated as the anaerobic threshold, and it appears at such a consumption of oxygen beginning from which anaerobic metabolism is supplemented with aerobic metabolism. Physiologists consider that the anaerobic threshold is the level of oxygen consumption under a physical load whose exceeding is accompanied by insufficiency of the aerobic mechanisms of energy formation and the inclusion into the process of energy supply without fail of an anaerobic mechanism with the development of metabolic acidosis.

It was shown in physiological studies that the following metabolic disorders are observed under a physical load [109, 129, 356, 388]:

(1) the incidence of metabolic acidosis;

(2) a change in the ratio between the production of carbon dioxide and the consumption of oxygen;

(3) a nonlinear change in ventilation in response to an increase in metabolism;

(4) a change in the kinetics of capture of oxygen;

(5) acceleration of glycolysis and glyconeogenesis;

(6) a pronounced growth in the level of catecholamines in the blood;

(7) activation of renin in the blood plasma.

Under a physical load, the utilization of glucose by muscle tissue increases from 10 to 15 times, while lactic acid is the main factor limiting the ability of muscles to perform work. The reason is that lactate blocks the key enzymes of glycolysis, nerve-muscle transmission, and binding of Ca^{2+} with the contracting proteins of the myofibrills. A change in the pH is of major significance (especially in DM patients}: in the development of acidosis and lowering of the pH to 6.4, complete inactivation of 6-phosphofructokinase occurs together with blocking of glucose utilization [63, 333, 355].

When continuing to analyze the parameters of the veloergometric test in healthy persons and DM patients, we noted that the maximum oxygen consumption in healthy persons (MOC) was 3.26 l/min, while in type I and II DM patients it lowered to 2.24.

How can these changes be explained?

The relation between carbohydrate and fat metabolism was described in detail by Randle et al. [473]. The antagonism between them exists in a clear biochemical proportion: hyperglycemia inhibits the release of the free fatty acids in the fat cells, and in hyperlipacidemia there is observed a delay in the utilization of glucose in the muscle cells spending energy. In DM, this equilibrium is disturbed: if normally 50% of the glucose is oxidized to carbon dioxide and water, and 40% is accumulated in the form of triglycerides, in DM the content and utilization of fatty acids in the blood always grow, and their oxidation increases in accordance with Randle's glucose–fatty acid cycle, the utilization of glucose is inhibited, and glyconeogenesis in the liver increases.

A physical load in DM sharply increases glyconeogenesis in the liver, while the utilization and oxidation of glucose is lowered. Of special significance is the decompensation of DM. With compensated DM, the utilization of glucose and fatty acids under a physical load occurs normally, while in decompensation of DM (especially of type II), a physical load does not increase the utilization of glucose and produces hyperglycemia and ketonemia as a result of the increase in glyconeogenesis and ketogenesis in the liver.

Considerable changes in the chronotropic heart reserve were also noted. In healthy persons it was 85.7 bpm, in type I DM patients – 60.7, and in type II – 62.4 bpm. Definite distinctions were also noted in the level of the systolic and diastolic arterial pressure (both of the initial and threshold values). The inotropic reserve also changed appreciably, being 66.1 mmHg in healthy persons, 59.0 mmHg in IHD patients, and 58.5 in DM patients.

The energy expenditure index is a very outstanding parameter of disturbance of the energy balance of an organism in DM. In healthy persons it was 4.2 units, in IHD patients it increased to 6.97, and in DM patients to 14.08 units. This parameter shows the high significance of the metabolic disorders in DM.

Considering the substantial differences in the clinical picture of the development of IHD in type I and II DM patients, we compared the tolerance to a physical load of the two types. We observed 20 type I and 27 type II DM patients. They were all in the compensation state.

Veloergometry revealed both the common features and distinctions in the hemodynamic parameters of the type I and II patients. In comparison with the parameters in the control group, in the DM patients, the total and threshold powers of the performed load dropped authentically. The drop in the reserve possibilities of the myocardium in DM was indicated by diminishing of the chronotropic heart reserve (CHR) and of the rate of growth of the pulse – the pulse coefficient (pct). DM patients spent more energy on performing the threshold load (EEI) then healthy ones. The drop in energy formation in DM is associated with disturbance of oxygen consumption.

A growth in the oxygen exchange at rest, which was indicated indirectly by a higher double product (DP_{in}) resulted in rapid exhaustion of the myocardial reserve under a physical load and, consequently, in diminishing of the DP index (DPI) and DP increment (ΔDP) for the threshold load. The maximum oxygen consumption (MOC) and its ratio to the normal value (MOC/NMOC) are also sharply lowered in DM patients, which properly reflects the disturbance of the redox processes and oxygen-transport function of the blood.

In addition to the tendency of impairment of hemodynamics common for both types of DM, we also discovered appreciable differences between them. For example, the energy and pulse coefficients of tolerance, and also their ratio (ECT/PCT) in type I DM patients correspond to the parameters for healthy persons, whereas in type II patients they were considerably lower than the normal values. By individualizing the values of the veloergometry parameters via the ratios of M_{th} and MOC to the body weight, we also uncovered a more pronounced drop in the physical ability to work in type II DM patients.

In both subgroups of patients, we noted a low chronotropic heart reserve as a result of the insufficient growth in the beats per minute under a load. The cause of the drop in the chronotropic heart reserve in type I DM was tachycardia at rest, which underlies the rapid exhaustion of the chronotropic heart reserve in the type I patients, and it can be considered as a manifestation of the diabetic cardiac vegetoneuropathy syndrome. It is caused by weakening of the parasympathico-tonic influences on the function of the sinus-precardiac ganglion at the beginning of the disease, and later by the setting in of sympathetic functional denervation of the heart. In type II DM, the drop in the chronotropic heart reserve depended less on the quicker initial rhythm, but chiefly on the drop in the tolerance of the pulse to a physical load. A feature of type II DM was the considerable growth (in comparison with type I) of the initial systolic, diastolic, average, and average threshold arterial pressure due to the more pronounced changes in the vessels of these patients.

Arterial hypertension is one of the factors making the course of DM more complicated [18, 156, 259]. It is of special significance in IIDM, when diabetic retinopathy progresses rapidly, and in IHD. In the pathogenesis of development of arterial hypertension in DM, major significance is attached to insulin resistance and hyperinsulinemia. It was proved in experiments that an excess of insulin increases the reabsorption of sodium in the proximal canaliculi, stimulates the production of catecholamines and the sympathico-adrenal system [345, 355, 589, 592]. All these factors increase the arterial pressure. The relation between IIDM and essential hypertension was noted [18, 151, 156, 259]. DM often appears in this form of hypertension. It was shown under clinical conditions that arterial hypertension in IIDM has a number of features [18, 151, 156, 259]:

(1) the presence of hyperinsulinemia and insulin resistance;

(2) the presence of hypertriglyceridemia, an increase in the level of VLDLP, hypercholesterolemia;

(3) lowering of the level of HDLP;

(4) the hypokinetic type of hemodynamics with a growth in the specific peripheral resistance and lowering of the cardiac discharge;

(5) lowering of the renin level in the blood plasma.

The unusualness of the ratios of the pulse frequency and arterial pressure level in the groups of DM patients was expressed in the authentically higher figures of the pressure/pulse index for both the initial and threshold pulse in type DM patients. In type I DM, an increased functional excitability of the myocardium at rest prevailed. A high initial heart rate was not compensated by a certain drop in the initial systolic arterial pressure. This resulted in one of the types of degradation of the cardiovascular system (a drop in the pressure/initial pulse index). In type II DM patients, the hypertensive vessel reaction to a physical load manifested itself in degradation of a different kind – a growth in the pressure/threshold pulse index (in comparison with the figures for healthy persons). The features of the reactions of the cardiovascular system to a dosed physical load that we found in type I and II DM patients are related to a certain extent to the different sensitivity of the insulin receptors. The higher tolerance to a physical load in type I DM patients is due to the preserved affinity of the receptors for insulin.

It is noteworthy that the veloergometry parameters for IIDM are identical to those in IHD patients (the postinfarction period) in whom the ability of the heart to increase the rate of contractions in accordance with the physical load is also limited. Here arterial hypertension was noted in all the stages of conducting veloergometry.

The similarity of the changes in this case is explained by the common nature of the units in the pathogenetic chain. For instance, the primary atherosclerotic changes in the pancreas arteries impair its perfusion and lower the production of insulin or its quality. On the other hand, the insufficiency of insulin disturbs carbohydrate metabolism that changes the metabolism of the vessel walls and thus speeds up the development of atherosclerosis.

In view of the major significance of decompensation in DM patients, especially under a physical load, we analyzed the level of glycemia in type II DM patients in the stage of compensation (15 persons) and in that of decompensation (5 persons).

The background level of glucose in the blood differs in the patients. In patients with compensation of DM, the glucose level was 9.6 mmol/l, while with decompensation it was 16.8. Under the influence of a physical load in patients with compensated DM the concentration of glucose in the blood lowered insignificantly in 30 min after a load to 8.8 mmol/l, and then returned to the initial level.

The reverse reaction occurs in DM patients with decompensation. In 30 minutes after a load, the glucose level grows to 18.3 mmol/l and remains at this level after 60 min.

What are these disorders in carbohydrate metabolism under a physical load due to in type II DM patients?

Physiologists showed that oxidation in the Krebs cycle cannot supply the required amount of energy under an intense physical load, therefore glycolysis is activated in an organism at the level of 6-phosphofructokinase, which although is less economical as regards energy production, owing to the high rate of the reactions has a greater energy-producing capacity than oxidative phosphorylation [109, 129, 387, 388, 389]. If under moderate loads an organism uses glucose and fatty acids to produce energy (oxidative phosphorylation), under the maximum load carbohydrates are chiefly oxidized.

In DM, energy metabolism is distorted considerably: in connection with lowering of the insulin level in the blood (type I) or its inactivation (type II), the activity of hexokinase and glucokinase is suppressed in a number of organs and tissues (muscles, myocardium, adipose tissue [134, 333]. In this connection, a growth in the glucose concentration in the blood three times increases its utilization only 1.5 times, while the oxidation of glucose in the Krebs cycle to carbon dioxide is halved. The decrease in the peripheral utilization of glucose in DM sharply activates glyconeogenesis in the liver, and an important role here is played by fatty acids.

It was noted that in DM (especially of type II), glyconeogenesis in the liver is carried out with the use of all the precursors of glucose, namely, amino acids, lactate, pyruvate, glycerol, and fructose [56, 96, 284, 550]. In the liver, all the enzymes of glycolysis, especially glucokinase, are activated. Type I DM is characterized by lowering of the activity of the glycolysis processes, the Krebs cycle, the synthesis of glycogen, proteins, lipids, and also an increase in glyconeogenesis and ketogenesis – this is all a consequence of a lack of insulin.

The pathophysiologically described processes are realized by suppression of the transportation of glucose. The metabolic changes in type II DM patients are absolutely different: in addition to hyperglycemia, glycosuria, and ketonemia, there are noted lowering of the tolerance to glucose and insulin, and also obesity and hyperinsulinemia. With this type of diabetes, the secretion of insulin grows, and its content in the blood increases sharply (in combination with hyperglycemia, glycosuria, and ketonemia). The utilization of glucose grows, as does the body weight as a result of the increased lipogenesis from glucose. This picture of hyperglycemic obesity is typical of type II DM.

Experiments with animals (in which diabetes was produced) revealed that glyconeogenesis increases in the liver from all the basic precursors of glucose (amino acids, lactate, pyruvate, and fructose) [96, 284, 289, 333, 549]. Ketosis and acidosis (the most serious metabolic shifts in the decompensation of DM) may be a result of glyconeogenesis. Here a disproportion appears between the lack of oxaloacetate directed along the path of glyconeogenesis, and the excess of acetyl-CoA formed as a result of the increased oxidation of fatty acids as the energy substrates of glyconeogenesis.

Glucose is excreted into the blood in hypermetabolism (especially in the first minutes) at the expense of the activity of the liver glycogenolysis. It was shown in experiments that insulin mainly affects the penetration of glucose into the muscles and does not affect its penetration into the hepatocytes [56, 280, 289, 566, 595]. However, this does not correspond completely to what actually occurs. Insulin affects the processes determining the rate of excretion and absorption of glucose by the liver. It activates the liver glucogenase and glucosyltransferases and inhibits glucoso-6-phosphatase. Consequently, insulin may cause changes opposite to the processes that appear in hypermetabolism.

It was noted in clinical observations of sportsmen that a growth in the glucose level in the blood (hyperglycemia) appears very sharply even under a severe physical load [109, 120, 129, 387, 388]. Under a brief load, the production of glucose increases from three to five times and, accordingly, the utilization of glucose by the muscle tissue grows. Consequently, there is a correlation between the consumption of glucose by the muscle tissues and its production by the liver. In DM patients with a substantial lack of insulin, homeostasis of glucose in the blood is disturbed under a physical load, and this is especially pronounced in the decompensation of diabetes.

These clinical observations suggest practical conclusions on the expediency of physical activity in decompensated diabetes.

Many clinicians and experimentators showed that insulin is the key hormone in energy metabolism [14, 49, 96, 205, 280, 465, 567]. Insulin reacts functionally with a cell at the level of the specific receptors of the cytoplasmic membranes. It was indicated that insulin receptors are a complicated group of proteins. In the membranes of the adipose tissue, the system of receptors and the transportation system of glucose have different physicochemical characteristics. Both the glucose carriers and insulin receptors are polypeptides having a different chemical structure. The transportation and metabolic effects of insulin are a result of binding of the hormone to the membrane receptors. In hyperinsu-

Table 39. Biochemical Parameters of Blood in Patients with IHD and Type II DM.

Parameter	Control group	Type II DM patients	IHD patients
Glucose, mmol/l	4.82	10.33	5.13
IRI, μU/ml	13.92	25.13	12.9
Pyruvic acid, μmol/l	59.03	105.4	70.8
Lactic acid, μmol/l	0.8	0.55	1.025

linemia of any etiology, the amount of receptors in the adipose tissue, liver, and muscles drops substantially, while in normalization of the insulin level, the number of receptors is restored [109, 120, 153, 171, 200, 289, 358, 465, 595].

To confirm the relation between the level of insulin in the blood and the metabolism of carbohydrates, we compared the concentration of insulin in the blood with the level of glucose, pyruvic (PRA) and lactic (LA) acids in healthy persons and in IHD and type II DM patients. We observed 52 healthy persons, 31 IHD patients, and 14 type II DM patients in the compensation stage.

A glance at Table 39 reveals that the level of glucose in the blood in the DM patients is naturally higher than in the other two groups (10.33 mmol/l). We can also note that the concentration of insulin in the blood in the DM patients is 25.13 μU/ml, while in the IHD patients (12.9) and healthy persona (13.92 μU/ml) it is considerably lower. This points to the sharply reduced sensitivity of the receptor apparatus of DM patients to insulin, In the DM patients, the content of pyruvic acid is very sharply increased (in comparison with the other two groups) with a lower level of lactic acid. The quantitative ratio of these acids in the blood is an indication of the intensity of the glycolytic and oxidizing transformations of carbohydrates.

In type I and II DM patients, metabolic distinctions are noted in carbohydrate metabolism.

Glyconeogenesis in type I DM patients is a result of the insulin deficiency and the subsequent diminishing of the phosphorylation of glucose in the liver. In this type, glyconeogenesis is easily corrected with insulin.

In type II DM, glyconeogenesis develops on the background of hyperinsulinemia and, consequently, depends only slightly on the level of insulin in the blood.

In type I DM, a paradoxical reaction appears in the patients, when the utilization of glucose in tissues depending on insulin is sharply lowered, while its production in the liver is sharply increased at the expense of glyconeogenesis. This leads to hyperglycemia, lowering of the synthesis of lipids and their mobilization, and also to plastic and energy imbalance.

In type I DM, ketosis and acidosis are a result of increased glyconeogenesis. It follows that hyperglycemia in DM of type I is a result of the combined influence of glyconeogenesis and inhibition of the peripheral utilization of glucose.

In type II DM patients, the cause of hyperglycemia is insulin resistance and disturbance of the transportation of glucose into the tissues. The pronounced insulin resistance of muscle tissue in type II DM patients is due not to disturbance of the hormonal-receptor interactions, but to disorganization of the mechanisms of glucose transportation and metabolism.

The reduced tolerance to glucose in type I and II DM patients is due to different reasons. In type I, it is caused by a lack of insulin, while in type II by lowering of the sensitivity of the tissues to insulin. In the pathogenesis of insulin resistance in type II DM, an important role is played by hypertriglyceridemia, especially in combination with obesity. An increased amount of VLDLP in the blood in type II DM causes increased secretion of insulin, while in hyperinsulinemia the synthesis of the triglycerides contained in the VLDLP increases, and the vicious circle closes.

With a deficiency of insulin (type I DM), glucose cannot be supplied in a sufficient amount into fat cells. The free fatty acids are transported into the liver, where owing to their increased supply, lipoproteins are synthesized and discharged. The deficiency of insulin is due to the following biochemical processes: inhibition of lipogenesis and an increase in lipolysis.

A growth in the pyruvic acid/lactic acid ratio in type II DM patients indicates the predominating use of anaerobic ways of utilizing the main energy sources. Under anaerobic conditions, the formation of ATP is less efficient than under aerobic ones, and this fact is an indicator of a drop in the ability to do physical work in DM patients in the process of veloergometry.

Physiological studies showed that a high level of lactic acid in the blood inhibits the oxidation of fatty acids in the mitochondria by reducing lipolysis and stimulating reesterification of the fatty acids in the adipose tissue [15, 333, 386, 389]. In working muscles, the lactate is removed by increasing glyconeogenesis. Biochemical studies of the blood in DM showed that a change in the activity of insulin in the blood (insulin resistance) inactivates hexokinase and glucokinase in the muscles, myocardium, and adipose tissue [56, 63, 96, 176, 280].

Under a physical load, the level of the triglycerides in the blood lowers in connection with the fact that the activity of the LPL in the muscles and adipose tissue grows, and, accordingly, the rate of excreting of triglycerides from the organism increases. Moreover, physical activity increases the level of HDLP, which is very important for DM patients. This is why DM patients must include physical exercises in their dally rounds.

In recent years, much attention is being given to studying the relation between the activity of the endocrine glands and the tolerance to a physical load [544]. It was assumed that persons with an increased activity of the nervous system (with predomination of activating reticulo-hypothalamic influences over inhibiting ones) have a high activity of the sympathico-adrenal, hypophyseal-adrenal, and insular systems on the background of a drop in the functions of the thyroid and hypophysis-gonad systems. Such an organization of the endocrine system is a reflection of the high activation of the hypothalamo-hypophyseal complex.

To study the hormonal conditions in DM patients when performing physical work, we studied the level of catecholamines in the blood.

In DM patients, the background level of adrenaline in the blood is increased considerably to 12.65 ng/min (against 6.82 in healthy persons). The content of noradrenaline is approximately the same: 7.69 ng/min in healthy persona and 6.15 ng/min in DM patients. The dopamine level is increased somewhat, but the content of dioxyphenylalanine (DOPA) is especially high – up to 26.5 ng/min (against 14.8 in healthy persons). When healthy persons performed physical work, the amount of adrenaline increased to 14.05 ng/min, of noradrenaline to 15.8. After completing the work, these parameters gradually became normalized. In DM patients, such a physiological reaction failed to occur: the adrenaline level dropped somewhat under a physical load (to 10.55 ng/min), while that of noradrenaline did not change (6.04 ng/min). In healthy persons, the dopamine level grew sharply (to 487.8 ng/min), while the DOPA content lowered moderately. In DM patients, the increase in the dopamine level was very moderate (up to 316.5 ng/min), while the DOPA content increased very insignificantly.

What do these results show?

Catecholamines (adrenaline and noradrenaline) have a contrainsular property [205, 301, 589, 592]. In DM patients, especially of type II, the hyperglycemic effect of the catecholamines increases from five to six times owing to the sharp drop in insulin activity. The contrainsular effect of the catecholamines is especially clear when the level of NEFA in the blood rises (the lipolytic effect of catecholamines).

Under a physical load, the reactions of DM patients differ sharply from those of healthy persons in the nature of the metabolic disorders. Under a physical load, in DM patients, the utilization of glucose (its metabolism) depends on the degree of compensation of diabetes: in compensation, a physical load causes utilization of glucose and fatty acids (as in healthy persons); on the other hand, with a state of decompensation, a physical load causes hyperglycemia and ketonemia as a result of a growth in glyconeogenesis and ketogenesis in the liver. In connection with the presence of hyperglycemia in diabetes, a physical load increases the oxidation of the fatty acids, the formation and utilization of ketone bodies.

Many authors note the drop in the contracting ability of a cardiac muscle in DM [4, 28, 179, 298, 392, 619]. An important role in the development of energy insufficiency of the myocardium is played by its ischemia. These changes are especially pronounced with a combination of DM and IHD [118, 190, 198, 201, 619]. Most often, registered myocardiodystrophy is manifest morphologically by the disturbance of metabolism in the cardiac myocytes, by diabetic nephropathy and microangiopathy [119, 152, 190, 198, 248, 256]. Ischemic lesion of the myocardium in DM appears with a combination of coronarogenic and metabolic disorders. This relation results in energy insufficiency of the myocardium and development of cardiac insufficiency.

Recent years saw the widespread introduction in cardiological clinics of the method of myocardium perfusion – scintigraphy with ^{201}Tl. The features of the pharmacodynamics of this radionuclide preparation make it possible under conditions of a physical load to reveal the concealed disorders of perfusion of the myocardium, determine its reserve possibilities and the state of the vascular basin at the microcirculation level. In DM patients, it is especially important to reveal the causes and mechanisms of the lack of a balance between the demand of the myocardium for oxygen, the presence of both anatomical and functional disorders, the nature of the myocardial blood flow, and the structure of the coronary arteries.

The introduction into clinical practice of the noninvasive radionuclide method of investigation (scintigraphy of the myocardium) made it possible to visualize the distribution of the preparation in the heart and reveal the contingent of patients with a high risk of disturbance of the main coronary blood flow.

Scintigraphy with thallium is an exceedingly important method in connection with the fact that ^{201}Tl is an analogue (physicochemical) of a potassium ion. When it gets into an organism it is included in the potassium-sodium ATPase system, replaces a potassium ion without harm for the ionic pump, and thus enters into undamaged muscle cells.

Perfusion of the myocardium makes it possible to uncover the zones of transient ischemia. Analysis of the changes found in a scintigram under a physical load allows one to determine the state of the expanding reserve of the diastolic part of the coronary channel. The scintigrams can be used to determine sections with disorders of the myocardial blood flow.

The biokinetics of ^{201}Tl is as follows. After intravenous introduction, it is included in the potassium-sodium ATPase cell pump, where it replaces a potassium ion. Consequently, the distribution of ^{201}Tl in the myocardium, in principle, reflects the degree of preservation of the enzymatic activity of the sodium pump, and this with a sufficient approximation characterizes the functional intactness of the cardiac myocytes. With intravenous introduction, ^{201}Tl in 80% of the cases is taken up by the myocardium and accordingly reflects its perfusion.

There are two viewpoints as regards the interpretation of the scintigrams. Some authors consider that the preparation is retained by the "live" myocardium and when there is more of the latter, it retains the preparation longer. Others consider that the thallium should be washed out of the myocardium, and its accumulation in a section indicates its poor removal, i.e. that perfusion is disturbed. It is considered that the processes of distribution and redistribution proceed simultaneously, but the state of perfusion is reflected more on the initial scintigrams, whereas a delayed scintigram reflects the state of metabolism. This suggests that when more time elapses from the moment of introduction of the thallium to the preparation of a scintigram, it reflects metabolism to a greater extent and perfusion to a smaller one.

The blood flow in the myocardium is 80 times greater than in the skeletal muscles, and therefore the thallium concentration will be the highest in the myocardium of the left ventricle. The blood flow in separate sections of the myocardium varies (from 1.5 to 2 times) because perfusion of the top and basal parts of the myocardium is normally always smaller than that of the remaining regions. The right ventricle cannot be seen in the scintigrams because of the small thickness of its walls. Thallium chloride is accumulated actively by the healthy tissue of the myocardium and is transported through the membranes of the mitochondria by means of a potassium-sodium ATPase pump. It is distributed in the myocardium depending on the blood supply, i.e. on the perfusion of the myocardium. The highest intensity of thallium accumulation is noted in the region of the partition, which is due to the greatest mass of the myocardium in this part. Under a physical load, the following changes are noted in IHD patients in scintigraphy: (i) expansion of the focuses existing at rest and the appearance of new "cold" focuses, i.e. zones of disturbance of myocardium perfusion, (ii) the absence of changes in the scintigram (in comparison with the results obtained at rest), and (iii) diminishing of the cold focuses existing at rest [66, 306, 329, 356, 619].

The first variant is noted the most frequently, i.e. disorder in the supply of blood to a cardiac muscle.

Table 40. Parameters of Perfusion Defects of Myocardium in Type I and II DM Patients in Veloergometry (Data of Scintigraphy)

Parameter, %	Type I DM patients	Type II DM patients
Id	8.8	4.2
r	0.46	3.04
Cl	25	34.2
Id+r	1.4	2.3
Id+Cl	19.4	20.5
r+Cl	2.8	2.3
Id+r+Cl	6	11.4

To study the perfusion of the myocardium under conditions of a physical load, we investigated 20 patients suffering from DM combined with IHD from 30 to 64 years old (8 persons – type I and 12 type II DM). The duration of suffering from diabetes in both groups ranged from four to eight years. The disease was of moderate severity, in the compensation stage, without clinical indications of insufficiency of blood circulation.

Scintigraphy of the myocardium with ^{201}Tl was performed for all the patients combined with veloergometry. We started the load from 25 W and increased it by the same amount every three minutes. The test was stopped when the submaximal heart rate was reached with consideration of the age of each patient or when a seizure of stenocardia appeared. When the threshold load was reached, we introduced intravenously 2 ml of ^{201}Tl with an activity of 55–74 mBq, after which the load on a patient was continued for one minute. Scintigraphy of the myocardium was performed in 5–7 min and in 4 h after the intravenous introduction of ^{201}Tl. We determined the following parameters: the defect of adsorption of ^{201}Tl in the initial scintigrams (Id), the defect of absorption of ^{201}Tl in the initial scintigrams with disturbance of the clearance (Id+Cl), the initial defect plus the disorder in redistribution (Id+r), the perfusion defect in redistribution (r), the perfusion defect in the initial scintigrams plus the defect of redistribution plus disorder of the clearance (Id+r+Cl), the perfusion defect in redistribution plus disturbance of the clearance (r+Cl), and only disturbance of the clearance (Cl). In addition, the total capture of ^{201}Tl by a ventricle was appraised.

In the scintigrams registered directly after a veloergometric test (the initial loading scintigrams), in all the patients the myocardium of the left ventricle and in 11 of both the left and right ventricles was visualized. In two type I and three type II DM patients, the distribution of the thallium was visually uniform, while in the remaining patients zones of

Table 41. Perfusion Defects in Accumulation and Removal of ^{201}Tl in Type I and II DM Patients and Type II DM + IHD Patients (Data of Scintigraphy)

Patients	Perfusion defects, %	
	in accumulation of ^{201}Tl	in removal of ^{201}Tl
Type I DM	45.83	71.52
Type II DM	34.02	69.44
Type II DM + IHD	52.0	84.37

reduced perfusion of various localization were determined, but mainly in the region of the apex, septum, and side wall of the left ventricle.

Inspection of Table 40 shows that in DM patients of type I defects of absorption of ^{201}Tl on the initial scintigrams were registered more frequently – 0.8% (in type II DM – 4.2%). In the type II patients, a greater percentage of perfusion defects in redistribution was noted – 3.04% (in type I 0.46%). Defects of perfusion in the initial scintigrams plus defects in redistribution plus disorder of clearance were found in type II patients in 11.4% of the cases, and in type I – 6%. The defects in the initial scintigrams plus disturbance in redistribution amounted to 2.3% in type II DM and to 1.4% in type I. Disturbance of clearance in type II was 34.2%, and in type I was 25%.

Considering that the normal accumulation of ^{201}Tl exceeds 75%, while its washing out normally exceeds 30%, we counted the number of myocardium segments with perfusion defects in the stages of accumulation and removal of ^{201}Tl in the anterior, apical, lower, and posterior regions of the heart in three standard projections (first – front, second – left front oblique 45°, and third – left side 90°) in type I and II DM patients and in six type II DM patients combined with IHD (Table 41).

Of 144 studied segments of the myocardium, the greatest number of injured regions was found in patients with a combined pathology (52% of accumulation defects and 84.37% of defects of elimination of the thallium). When comparing the groups of DM patients without additional IHD, the initial scintigrams registered a drop in the accumulation of thallium in 45.83% of the myocardium segments in type I patients and 34% of the segments in type II. Analysis of the delayed scintigrams revealed an approximately identical percentage of disorders in washing out of the preparation (about 70%).

A comparison of the initial and delayed scintigrams provides information on the degree of the drop in the perfusion of the myocardium, and also shows whether the disorders of blood supply are stable (when there are zones with lower perfusion on both the

Table 42. General Capture of ^{201}Tl by Left Ventricle in DM Patients in Veloergometry (Data of Scintigraphy).

Type of scintigramm	Heart position	Total capture of ^{201}Tl in patients, %	
		Type I DM	Type II DM
Initial	1	2.13±0.57	1.62±0.06
	2	1.35±0.05	1.16±0.02*
	3	2.62±0.12	1.83±0.07*
Delayed	1	1.71±0.07	1.27±0.1*
	2	1.95±0.14	1.38±0.1*
	3	1.48±0.07	1.04±0.09*

* The difference of the parameters is authentic (in comparison with the level in type I DM patients).

initial and delayed scintigrams) or transient (when there are zones of lowered perfusion in the initial scintigrams and none in the delayed ones).

The stable zones of lower perfusion (perfusion defects) may be a result of cardiosclerosis of myocardiodystrophy of any etiology. The transient perfusion defects are due to stenosed changes in the coronary vessels, and also disturbance of their ability of vasodilation at a microcirculation level. In computer processing of the initial and delayed scintigrams, in all the patients we studied both stable and transient perfusion defects were registered. However, in type I DM patients, stable zones of hyperperfusion were registered two times more frequently in type I DM patients than in type II ones (owing to metabolic disorders in the vessel walls), whereas transient defects of the "ischemic" type were found in type II patients.

The maximum capture of ^{201}Tl was normally about 3%, and it lowered in all three positions being studied in DM of both types (7 patients of type I and 11 of type II). The total capture of ^{201}Tl by the left ventricle in the group of type II patients was authentically lower than in type I (Table 42). It depended linearly on the myocardial blood flow. There is no doubt that the standing out of the transient perfusion defects in the scintigrams of type II patients is related closely to the disorders of the central hemodynamics noted in veloergometry and is the cause of the more appreciable than in type I DM lowering of the tolerance to a physical load.

In patients with a combined pathology (IHD and type II DM), perfusion defects were registered more often than in DM patients, and accordingly the parameters of the physical workability dropped more.

Correlation analysis also showed the presence of a close relation between the frequency of the perfusion defects and the tolerance to a physical load.

Analysis of the correlation coefficients of the capture of ^{201}Tl on the initial ($^{201}Tl_{in}$) and delayed ($^{201}Tl_{del}$) scintigrams with the pulse and pressure parameters (heart rate, systolic, diastolic, and average arterial pressures, and double product index) in veloergometry showed the following. In type I DM patients at rest, the correlation being studied was absent (the results are authentic): in the group of type II DM patients, a close correlation was noted between the degree of capture of ^{201}Tl in the initial and delayed scintigrams with the heart rate and diastolic arterial pressure, and of $^{201}Tl_{del}$ with the average arterial pressure and the double product index.

At a minimum load (25 W), in type I DM patients only three authentic correlations appeared: $^{201}Tl_{in}$ with the double product index, and $^{201}Tl_{in}$ and $^{201}Tl_{del}$ with the diastolic arterial pressure. At the same time, in type II DM patients the number of close correlations increased sharply and became four times greater than in the group of type I DM patients. From this stage, the authenticity of the correlations was observed only on the delayed scintigrams. This indicated a more involved mechanism of metabolism regulation in type II DM patients, in particular slower adaptation of the cardiovascular system to a physical load.

At a load of 50W, the number of authentic correlations for type I patients remained the same and was noted only in the initial scintigrams, which pointed to the adequacy of adaptation of the cardiovascular system to a dosed physical load at the given step. On the other hand, the number of authentic correlation coefficients for type II patients dropped sharply to two. At a threshold load, in type I patients the number of authentic correlations of $^{201}Tl_{del}$ with the heart rate and diastolic arterial pressure increased to four, whereas for type II patients only two correlations remained – with the systolic and average arterial pressure.

Thus, for type I DM, we noted the absence of close correlations of the myocardium perfusion parameter (the total capture of ^{201}Tl) with the parameters of the pulse and arterial pressure at rest, and an increase in the strength of the relations with a growth in the physical load. This corresponds to more adequate adaptation of coronary blood circulation to a physical load than in type II DM. In the latter, we noted high coefficients of the correlation between the capture of ^{201}Tl by the myocardium and the parameters of hemodynamics at rest (background) being studied. We also noted a sharp growth in the number of authentic correlation coefficients for the initial stage of the load, and a just as sharp loss of strength of the correlations at the threshold load level.

Our previous investigations (groups of healthy persons with a high and a moderate tolerance to a physical load) revealed that more active functioning of the system of blood circulation at rest leads to more rapid exhaustion of the myocardium reserves under load tests and to a lower tolerance to a physical load. Mathematical analysis confirmed the clinical results on the lower parameters of the physical workability in type II DM patients in comparison with type I. Our studies showed a close relation to exist between DM and IHD. The diminishing of the contracting ability of the myocardium in DM and IHD is due to cardiac and noncardiac reasons.

Diabetic myocardiodystrophy is based on disturbance of metabolism of a cardiac myocyte, diabetic vegetoneuropathy, and microangiopathy that may be encountered separately, but more often combine with one another. IHD develops as a result of coronary atherosclerosis and metabolic changes, and complications are introduced by cicatricial lesions of the myocardium during the postinfarction period. Coronarogenic and metabolic disorders mutually enhance the energy insufficiency of the myocardium and have an unfavorable effect on its functional state.

A very important parameter of the myocardium activity is the value of the coronary reserve (the ability of the coronary vascular channel to increase the coronary blood flow with an increased load on the myocardium). Clinical experiments showed that the myocardium at rest consumes 10 ml of oxygen a minute per 100 g of myocardium [434]. Under stress situations, this value can be increased to 65 ml/min.

Coronary insufficiency appears as a result of the following disorders in the main physiological relations:

(1) a drop in the coronary blood flow because of a spasm or occlusion of the coronary arteries attended by normal consumption of oxygen by the myocardium;

(2) an extremal increase in the consumption of oxygen by the myocardium exceeding the magnitude of the coronary reserve;

(3) a limited coronary reserve with a physiological increase in the level of oxygen consumption by the myocardium.

Opposite opinions exist on the pathogenesis of a coronary spasm in IHD patients, namely, (i) a spasm is not related to atherosclerosis and may appear in arteries not injured by atherosclerosis, and (ii) a spasm is a result of coronarosclerosis [59, 202, 386, 406, 434, 505]. Coronarographic analysis does not yield a clear-cut result. It was noted that ischemia of the myocardium may develop because of a spasm of the small vessels in the basin of one coronary artery. Moreover, a spasm may be caused by an increased discharge into the blood flow of histamine or serotonin [206, 388, 505], an increased content of which in the coronary arteries was proved for IHD patients.

The mechanism of regulating the tone of the coronary arteries was not studied in detail to date, especially in type II DM patients. It is known that various neuromediators, neuropeptides, and metabolites of the factors circulating in the blood are included into this regulation. All of them influence directly the smooth muscles of the coronary vessels, part of the factors being produced by the endothelium. These influences may be directed oppositely in connection with the postsynaptic and presynaptic influence of various agents, directly or via the endothelium. A spasm of the coronary arteries is based on contraction of the smooth muscles [59, 232, 410, 610].

What is the degree of their contraction?

It was considered in recent years that the endothelium is a direct modulator of the tonus and reactivity of the coronary vessels, and it was found that apart from the production of prostacyclin, an endothelium relaxation factor is excreted in the endothelium [35, 202]. It was shown in experiments that many neuromediators and pharmacological substances have a double effect on a vessel, namely, they cause its dilatation by secreting an endothelial relaxation factor and its contraction in the secretion of other factors [187].

The development of an important trend in studying a coronary spasm was based on the idea of the relation between thromboxan and prostacyclin [35, 40, 57, 322]. Thromboxan is a powerful vasoconstrictor excreted from activated thrombocytes. However, the idea of a lack of balance between the content of thromboxan and prostacyclin is not confirmed. A greater significance in recent years is attached to the production by the endothelium of a special substance the relaxation factor [59, 65, 202]. It was proved morphobiochemically that the endothelium of the coronary vessels is a very important regulator of their tonus: it modifies its reaction in response to the effect of many vasoactive substances [59, 65, 202, 206].

What occurs in the coronary vessels of DM patients also suffering from IHD?

To answer this question, we performed coronarographic investigations in healthy persons and type II DM patients also suffering from IHD. We found that in the healthy persons, the coronary reserve of the myocardium ensures the accurate supply of blood in both the systole and diastole periods. In type II DM patients, pronounced occlusion of the right coronary artery is noted that sharply limits the inflow of blood to the myocardium during both the systole and the diastole.

In DM (especially of type II) on the background of hyperlipidemia, disorders of the lipid metabolism in the myocardium occur, in particular the content of free fatty acids in the myocardium increases [102, 152, 249, 291, 453], which leads to a pronounced cardiotoxic effect. The pathogenesis of these disorders is associated with the ability of the free fatty acids to alter the permeability of the intracellular membranes with subsequent disturbance of the function of the ionic pumps. This is just what creates the prerequisites for the appearance of electrical instability of the myocardium and lowering of its contractile ability. The permeability of biological membranes depends to a greater extent on the functioning of their lipid component, in particular on the content of phospholipids. Even an insignificant increase in the content of the atherogenic fractions of lipoproteins in the blood alters the lipid composition of the membranes in the subcellular fractions of the cardiac myocytes.

A number of experimentators proved that under conditions of atherogenesis, a disturbance in the composition of the membranes of the mitochondria in the cardiac myocytes appears already in the early stages of atherosclerosis and creates a mechanism of disturbance of the contracting ability of the myocardium [179, 227, 231, 249, 544].

3.3. Lesion of Myocardium in Diabetes Mellitus

In the matters of regulation of the metabolic processes in an organism, major significance during recent years is attached to diseases such as DM and atherosclerosis [157, 176, 249, 284]. Using the example of these diseases, it is very important to trace the pathogenetic relations between the nervous and hormonal regulation of the cardiovascular system, lipid and hydrocarbon metabolism. The interactive nature of the pathological changes in DM, its relation to IHD, hypertension, and obesity create considerable difficulties in conducting therapeutic measures in such a category of patients.

In DM patients with obesity, as already indicated, myocardial infarction generally develops without a pain syndrome, or it is very insignificant [66, 190, 198, 201, 248, 416]. Among the reasons of a painless syndrome of myocardial infarction, most authors indicate generalized capillaropathy, lesion of the fine intramural vessels with disturbance of the function of the nervous-receptor apparatus. Difficulties in the development of collateral blood circulation in the myocardium are associated with these pathogenetic factors, which creates a foundation for the repeated extensive infarctions and the appearance of aneurysms

DM is generally attended by lesion of the cardiovascular system whose symptoms may be noted at the incidence of the disease (these lesions take first place among the causes of mortality of DM patients [14, 118, 179, 190, 198, 201, 256, 259]. In DM patients of both types, diabetic microangiopathies, atherosclerosis of various localization, and arterial hypertension are noted – these are the factors that in combination with primary metabolic changes lead to the formation of "diabetic myocardiodystrophy". The difficulty of diagnosis of it consists in the absence of specific changes in hemodynamics.

The bioenergetic processes in the myocardium include the following phases: (i) oxidizing phosphorylation and the cumulation of energy, (ii) the deposition and transportation of the energy-producing substances, and (iii) utilization of the energy [109, 129, 187, 386, 388, 389, 434].

The first phase of energy formation includes the release of ionized hydrogen from the carbohydrates, fatty acids, and amino acids being oxidized. The formation of energy mainly occurs in the Krebs cycle. The second phase or cumulation and transportation of energy is associated with its deposition in the form of the energy of ATP, the formation of creatine phosphate, and the transfer of the macroergic phosphate bond to ADP. The third phase of utilizing energy includes its transformation, shortening of actomyosin, and work of the myofibrills. The energy produced by glycolysis is a sufficiently important factor of the contracting ability of the myocardium and ensures the following functions of a myocardial cell: (i) functioning of the calcium pump, (ii) transportation of the macroergic phosphates to the contractile proteins at the expense of activation of the cytoplasmic creatinphosphokinase, (iii) preparation of amino acids for inclusion in a Krebs cycle, and (iv) maintaining of the physiological positive potential of the action of a cardiac myocyte.

It is very important to maintain the potential of a myocardial cell because the lack of ATP drastically increases the danger of the appearance of fibrillation of the ventricles. Clinical studies revealed that ischemia of the myocardium is accompanied not only by disturbance of the synthesis of ATP, but also by hindrances in the transportation of the macroergic compounds of ATP to the places of its utilization [129, 187, 206, 386, 389, 434]. Under conditions of ischemia of the myocardium, anaerobic metabolism appears, oxidizing phosphorylation is sharply inhibited, and the following biochemical changes occur: a sharp growth in the lactate content, an increase in the lactate/pyruvate ratio, and a growth in the consumption of glucose by the ischemized myocardium.

Under normal conditions, an important feature of the myocardium energetics is the assimilation of fatty acids. The ability of the myocardium to extract and consume fatty acids is ensured owing to the high activity of the beta-hydroxybutyrate-dehydrogenase. In the incidence of ischemia of the myocardium, free fatty acids are practically not oxidized, and a change in the system of using the fatty acids occurs. In ischemia, a growth in the synthesis of triglycerides in the myocardium and fat infiltration into it are noted. The utilization of NEFA in ischemia of the myocardium is drastically inhibited. Disturbance of the contracting function of the myocardium is the pathophysiological basis of lowering of the physical workability in type II DM (in which atherosclerotic changes in the myocardium develop especially actively).

A number of authors consider that insulin-dependent DM (juvenile) and insulin-independent DM (late) differ in the features of their combination with the pathology of the cardiovascular system: insulin-dependent DM is attended mainly by the development of specific microangiopathies, while patients with insulin-independent DM usually suffer from cardiovascular diseases of an atherosclerotic genesis [120, 134, 166, 259, 537]. At the same time, many authors do not find a direct relation between the incidence of a vascular pathology of atherosclerotic genesis and the duration and severity of DM [120, 134, 380, 548]. As a result of a number of observations, some investigators arrived at the conclusion that both diseases – DM and atherosclerosis – should be considered as parallel and independent pathological states [256, 381].

Diseases of the heart and vessels in DM patients are encountered two times more frequently, and IHD six times more frequently than in persons of the same age having no disorders of the carbohydrate metabolism [166, 380, 537].

The formation of pathological changes in the heart and vessels both in atherosclerosis and in DM is based on metabolic disorders [190, 198, 201, 256, 262]. However, to date the mechanisms of these disorders and the sequence of their inclusion into the pathological process are still being discussed widely. The adaptation reorganization of the endocrine and metabolic status in DM patients is manifest in the activation of carbohydrate and lipid metabolisms accompanied by a growth in the glucocorticoid, sympathico-adrenal activity [120, 134, 166, 249, 548].

In DM, the myocardium is injured regardless of the presence of atherosclerosis of the coronary vessels. This lesion was called diabetic cardiopathy (or cardiomyopathy) [179, 190, 198, 201, 256, 312, 416]. Pathophysiological diabetic cardiopathy is expressed in diabetic changes in the vessels of the heart in the form of microangiopathies. Here microcirculation is disturbed with morphological and functional changes in the myocardium. Special significance is attached to disorders of the bioelectrical activity of the myocardium, lowering of its contractility, and development of decompensation of the heart. In addition to lesion of vessels, the neurogenous regulation of the heart changes in connection with diabetic vegetoneuropathy. Here an adaptation reflector reaction of the cardiovascular system to the action of various factors is noted. Among them are a physical load and an orthoposition.

Microscopic studies revealed a drop in the number and diameter of the capillaries with a growth in the severity and duration of the disease, while the diameter of the cardiac myocytes and the radius of diffusion increased [227, 320, 232, 350, 481]. These factors indicate the presence of hypoxia of the myocardium.

With a prolonged duration of DM (over 10 years), especially of type II, atherosclerotic changes are revealed not only in the main arteries, but also in ones of a medium and small size [179, 222, 227, 256]. Studying of the tissue basophils in type II DM patients revealed an increased accumulation of sulfogroups of heparin, which points to the protective reaction of the myocardium in response to hypoxia. Morphologically, two types of changes were revealed in the myocardium of DM patients [185, 232, 256, 332, 395, 457, 458], namely, (i) nonuniform hypertrophy of the muscle fibers with atrophied cardiac myocytes, features of vascular and granular dystrophy of the myocardium, necrobiosis and necrosis, and (ii) disseminated perivascular and interstitial fibrosis, microreticular cardiosclerosis with compression and atrophy of the cardiac myocytes, and thickening of the perineurium. The parietal endocardium in DM is considerably thickened and sclerosed. A growth in the mass of the myocardium was noted.

The progress of atherosclerotic changes in the myocardium in DM developed with an increase in the age of the patients [5, 50, 119, 190, 380, 466]: up to an age of 40, atherosclerosis developed insignificantly, single lipid spots and strips were noted in the aorta, mainly in the abdominal part, in the myocardium the descending branch of the left coronary artery and the right coronary artery were injured most often, and fibrosis and calcinosis were mainly found in them. In a prolonged duration of DM, the presence of fibrosis and calcinosis was attended by stenosis of the lumen of the coronary vessels. These changes were especially pronounced in the descending branch of the left coronary artery.

Microscopic studying of the internal elastic membrane revealed its cleavage and atrophy, and frequently the breaking through of atheromatous matter into a vessel was discovered [190, 338, 381, 548]. The intima had quite a large thickness. Considerable trophic processes were noted in the sections of localization of the large atheromatous plaques in the middle membrane of a vessel. They resulted in its becoming thinner and vanishing completely. When the intima was more sclerous, the atrophic processes in the middle membrane were more pronounced. Along the small veins, moniliform inflations were often noted that were replaced by interceptions reminding one of a goffered pipe. In the lumens of the small venules, parietal thrombi were often found, especially in the decompensated forms of DM. We often noted clinical manifestations of the DIC syndrome in these cases.

The number of capillaries and muscle fibers in the myocardium (especially in type II DM) diminished as DM proceeded, while the diameter of the cardiac myocytes increased. A growth in the radius of diffusion was noted with an increase in the diameter of the cardiac myocytes, which resulted in hypoxia of the myocardium. The capillaries were oriented along the muscle fibers, and in some places fitted loosely to them. Cardiac myocytes were encountered with phenomena of focal myocytolysis (dissolution of the myofibrills) over a limited length. Vanishing of the lateral marking and the presence of coagulation necrosis were noted [225, 227].

The myocardium in the form or a disseminated (chiefly perivascular and interstitial) fibrosis with the accumulation in the grown out connecting tissue of alternating larger and smaller amounts of mucopolysaccharides was injured the most [4, 190, 227, 409, 535].

The development of pronounced tachycardia is an important disorder in the cardiovascular system in DM. This is why most of the DM patients complain of tachycardia and arrhythmia. The pathogenesis of these phenomena is associated, on the one hand, with an increased number of circulating catecholamines, and on the other with the presence of neuropathy [102, 179, 190, 198, 201, 256]. Lesion of the parasympathetic section of the vegetative nervous system facilitates the appearance of all these changes. Seizures of paroxysmal tachycardia appear especially frequently in DM patients. Their frequency grows in the period of the development of hypoglycemia or in orthostatic hypertension produced by neuropathy. Ketoacidosis and disturbance of electrolytic metabolism are an important cause of the disorders. These factors facilitate the development of disorders such as ventricular extrasystole, ciliary arrhythmia, intraventricular and atrioventricular blockades. All these disorders of rhythm and conduction are typical of DM.

To study the pathological changes in the activity of the myocardium in DM, we studied 67 patients with DM of moderate severity in the compensation stage with a duration of the disease from two to six years, without clinical manifestations of insufficiency of blood circulation, namely, 28 type I patients (12 women and 16 men whose age averaged 36) and 39 type II patients (17 women and 22 men whose average age was 49). The control group included 30 healthy persons (16 women and 14 men) whose average age was 29). Echocardiographic analysis was performed for all three groups. Analysis of the results showed that there is a significant difference in the echocardiographic parameters of the hemodynamics in healthy persons and DM patients (see Tables 37 and 38). In healthy persons, the end systolic volume was 32.64 ml, in type I patients it was 39.3, and in type II – 39.4 ml. This increase is noted especially clearly when analyzing the end diastolic volume – in healthy persons it was 104.8 ml, in type I patients – 130.2, and in type II DM patients it was 136.2 ml.

The pump function of the heart depends primarily on the pre- and postload, on the contractility of the myocardium and the heart rate. It is very important to determine the contractility of the myocardium by echocardiography. And here such parameters are very important as the end systolic volume, end diastolic volume, stroke volume, ejection fraction rate of circular contraction of the myocardium fibers, degree of shortening of the anteroposterior size of the left ventricle, and the volume rate of ejection.

It was noted in clinical studies that the stroke volume, rate of circular contraction of the myocardium fibers, volume rate of ejection are very significant, while the end diastolic volume reflecting the preload state of the myocardium is less significant. An attempt to compare the heart rate and contractility of the myocardium did not yield positive results, and we did not succeed in establishing a law here.

What is the pathophysiology of these changes in the DM patients?

It is quite difficult to determine the contractility of the myocardium under conditions of the real activity of a healthy and ailing organism because of the complex influence of the load on the contraction, relaxation, and stretching of the myocardium. Determination of this parameter from the ratio of the volume and pressure in each heart cycle is the most popular method. It provides an integral idea of the pump function of the heart because it characterizes not only the conditions of a load, but also to a definite extent the state of contraction. This determination is based on the fact that at different levels of resistance, one can determine the relation between the systole volume and pressure at the end of systole.

An important characteristic of the contractility of the myocardium (especially in DM patients) is the relaxation phase, which is determined by analyzing the pressure in the cavity in the phase of isovolumic relaxation

Physiological studies showed that there is a correlation between the energy parameter of myocardium contraction and the relaxation energy [388]. Cardiologists proved experimentally that the energy of isometric relaxation combines into a single entity the end diastolic pressure, the change in the pressure in the left ventricle, and the duration of relaxation.

Relaxation of the myocardium is an important component of its activity. A major significance is attached here to the sucking action of the left ventricle when it is filled from the atrium during diastole. From these facts, cardiologists conclude logically that the diastole is the active phase in realization of myocardium contraction. It was shown in clinical studies that the process of weakening of the myocardium of the left ventricle is determined by its inotropic state. Consequently, the energy of isometric relaxation combining the end diastolic volume, the change in the pressure in the left ventricle, and also the duration of relaxation are the most informative parameters of this process. The nature of the cellular metabolism of calcium is the pathophysiological basis of the relaxation process.

The parameters of the contraction and, especially, relaxation of the myocardium are very sensitive to the level of the calcium and its removal from the myofibrills into the sarcoplasmic network. In the initial period of disturbance of energy formation, retardation of relaxation occurs that may be associated not only with the improper absorption of calcium from the myofibrills, but also with lowering of the concentration of Ca^{2+} functioning in the myoplasma and activating the Ca-dependent ATPase of the sarcoplasmic network.

The rate of relaxation of the myocardium is more sensitive to the concentration of the Ca^{2+} in the myocardium than the rate of contraction. This is associated with the fact that activation of the membrane-bound proteins (calmodulin) with a growth in the Ca^{2+} concentration in the myoplasma may be the cause of an increased sensitivity in the rate of relaxation of the myocardium. Lowering of the rate of relaxation in comparison with that of contraction may occur in hypoxia (ischemia) of the myocardium [386, 434]. The factors participating in the development of ischemia of the myocardium (inhibition of oxidative phosphorylation, lesion of the myocardium, disturbance of adaptation) also lower the rate of relaxation. Ischemia lowers the content of ATP and the contracting function of the myocardium.

When analyzing the echocardiographic parameters of the hemodynamics of healthy persons and type I and II DM patients (see Tables 37 and 38), we noted significant distinctions: apart from an increase in the end diastolic volume of the left ventricle in DM, the rate of relaxation diminished. In healthy persons it was 13.2 cm/s, while in type I DM patients it was 9.81 and in type II – 9.53 cm/s.

The dependence of the pump function of the heart on the level of the end diastolic pressure in the left ventricle is of a nonlinear nature, while the increase in the end diastolic pressure to above the optimal value does not practically increase the level of the stroke work. Unlike this, the dependence of the stroke work on the end diastolic volume is linear under the same conditions.

Further analysis of the echocardiographic parameters in healthy persons and DM patients revealed a difference in the stroke and minute volume of blood in these groups. In healthy persons, the minute volume of the blood was 4.85 l/min, in type I patients it was 6.66 and in type II it was 6.97 l/min. Accordingly, the stroke volume in healthy persons was 72.65 ml, in type I patients it was 90.8, and in type II was 96.8 ml.

What factors determine this hyperkinetic type of myocardial contraction in DM patients?

When comparing the parameters of an echocardiogram in healthy persons and DM patients, we noted first of all that the mass of the myocardium increased sharply in type II patients. In healthy persons it was 115.7 g, while in type II patients it increased to 167.8 g. The compensating hypertrophy of the myocardium in DM patients is an important factor ensuring a growth in the minute volume. The hyperfunction of the myocardium, exhibiting itself in a growth in the stress, involves a marked growth in the consumption of oxygen, resynthesis of ATP, and no less marked activation of the synthesis of nucleic acids and proteins. In the process of the continuous hyperfunction of the myocardium in DM, there occurs a consecutive mobilization of the energy and structural reserves of the myocardial cells with subsequent wear (early) of the myocardium and disturbance of its functions.

An important factor in the development of the hyperfunction of the myocardium in DM is the drop in the overall peripheral resistance in type I and II DM patients. In healthy persons, its level was 1508.4 $dyn{\cdot}s{\cdot}cm^{-5}$, in type I patients it was 1264.6, and in type II – 1217.0.

The adaptation of the pump function of the heart to various resistance is one of the most complicated problems of the physiology of the heart. The reaction of the myocardium to an increased load resisting expulsion differs considerably from that of an isolated cardiac muscle. The inverse relation between the load and shortening of an isolated cardiac muscle enables us to predict that with a growth in the resistance, the stroke volume or the rate of expulsion from the left ventricle should decrease accordingly.

It is known from physiology that the conditions of functioning of the myocardium are determined by the magnitude of the resistance to the blood flow. With a very great resistance, when the power of the left ventricle cannot surmount the resistance in the aorta, all the energy of contraction is realized in the form of development of the pressure, while the conditions of contraction are designated as isovolumic. If we imagine the opposite phenomenon, when there is no resistance to the expulsion of blood, in this case the conditions of functioning of the myocardium are said to be isotonic because the ventricle must develop a certain force to impart a pulse to the blood contained in its cavity.

In a real organism, contraction occurs against a gradually growing resistance because the blood expelled into the aorta does not have time to instantaneously pass over into the arterial channel. This is why the given conditions of contraction are called auxotonic. Unlike an appraisal of the functioning of an isolated heart, analysis of myocardium functioning in an organism involves some difficulties. To characterize the stretching of a ventricle, it is necessary to determine the relation between its volume and the pressure appearing in its cavity.

An important matter in the pathology of the cardiovascular system in DM is the studying of the relation of the changes in the myocardium under conditions of compensation and decompensation of diabetes. It was noted that the greater the degree of decompensation of DM, the more the rate of diastolic relaxation of the myocardium lowers [118, 152, 179, 232, 619].

Conversely, after the treatment and compensation of diabetes, the rate of diastolic relaxation grows. The end diastolic volume of the left ventricle reacts very accurately to the decompensation of diabetes. Under compensation conditions, there also occur (on the background of a growth in the stretching ability of the myocardium) an increase in the filling of the coronary channel and stretching of the muscle fibers connected to the arterioles. The increased stretching of the myocardium (with compensation of DM) under the influence of the increased force of contraction ensures the adaptation of the pump function of the heart to the process of adaptation of the blood circulation.

Diastolic filling of the left ventricle depends on the influence of the extra- and intracardiac factors, among which the following are distinguished:

(1) age;
(2) heart rate;
(3) ejection fraction;
(4) geometry of the left ventricle cavity;
(5) activity of sympathico-adrenal system;
(6) preliminary load;
(7) interaction between heart cavities;
(8) influence of pericardium.

The causes of disturbance of the diastolic function of the left ventricle are as follows:

(1) slowing down of myocardium relaxation;
(2) growth in the rigidly of the cavity of the left ventricle;
(3) increase in the rigidity of the left ventricle wall;
(4) weakening of the elastic rebound of the myocardium.

In DM, the ability of the myocardium to relax completely and be filled with blood during the diastole is disturbed. From the pathophysiological viewpoint, this feature is explained by the growth in the rigidity of the myocardium and lowering of its elastic properties. The pathogenesis of the increase in rigidity is considered by most authors to be associated with disturbance of the transportation of calcium, electromechanical unbalance attended by the asynchronous nature of relaxation, and with mechanical factors [4, 225, 227]. Special significance is attached to the intramyocardial factors (disorganization of the muscle fibers of the hypertrophied myocardium).

It is exactly hypertrophy of the myocardium that causes disturbance of the diastolic relaxation. Hypertrophy of the myocardium in type II DM prevents a deficiency of energy in the myocardium and acute cardiac insufficiency. Physiologists proved that the maximum strength and rate, and also the energy efficiency of contraction in a hypertrophied myocardium lower, and its functional reserve diminishes in pronounced hypertrophy [388, 434]. A complex of the increased wear of a hypertrophied myocardium develops.

It was shown in morphofunctional studies that the hypertrophy of the myocardium is attended by disturbance of the density of the sympathetic innervation of the myocardium, disorder in the consumption of noradrenaline that exceeds its resynthesis, and the concentration of noradrenaline in the hypertrophied myocardium drops from 3 to 5 times [388, 434]. This entails lowering of the positive inotropic influence of the sympathetic part of the nervous system on the hypertrophied myocardium, i.e. the inotropic function of the heart lowers.

Clinical observations of various authors confirm the presence of hypertrophy of the myocardium in DM patients, in particular, a growth in the thickness of its rear wall and interventricular septum [102, 118]. It was also suggested that the mass of the myocardium lowers in the decompensation of diabetes and is restored under the influence of normalization of metabolism [179, 392].

Of significance in the pathogenesis of the increase in the mass of the myocardium in DM are not only the specific (dystrophic) changes in a cardiac myocyte under the influence of disorder of metabolism. A known role in the genesis of cardiomegalia is played by the diffusion injury of the vessels in the form of diabetic microangiopathy, disturbance of the neurohumoral regulation of cardiac activity, and the early development of coronarosclerosis. These factors facilitate the development of hypoxia of the myocardium and its compensating hyperfunction. The increased work of the heart leads to its hypertrophy, which causes a still greater lack of oxygen.

In the hypertrophy of the myocardium (in type II DM patients), an unbalanced growth of the myofibrills is attended by lagging behind of the growth in the number of arterioles and capillaries [152, 623]. This lowers the number of coronary capillaries per unit mass of the myocardium. In hypertrophy of the myocardium, together with a drop in the total number of coronary capillaries, the diffusion distance of oxygen grows, and hypoxia of the myocardium appears accordingly. As a result, already at rest most of the reserve capillaries are open, and the coronary reserve lowers substantially. In a hypertrophied myocardium, the increased consumption of oxygen is not ensured by the timely growth in its transportation into the intercapillary spaces, and zones of ischemia appear.

When DM is combined with obesity, hypertrophy of the adapted heart occurs, and the adaptation reaction transforms into a pathological one [190, 289, 365, 465]. With this combination, the functioning of the mechanisms responsible for the synchronization of excitation, contraction, and relaxation is disturbed. The lowering of the transmitting capacity of the coronary arteries in DM diminishes the capacity of aerobic resynthesis of ATP.

The state of microcirculation of the myocardium is an exceedingly important component in the development of IHD and atherosclerotic lesions. An appreciable role here is played by the functional interaction between the contracting function of the myocardium, the microcirculation channel, and the morphological substrate of the cardiac muscle. In the pathogenesis of IHD in DM patients, of significance is the morphofunctional state of the contracting myocardium, the hemotransportation possibilities of the capillaries and their permeability, and the functioning of the smooth muscle component in the walls of the lymphatic vessels of the myocardium. Disorder of the transportation and trophic provision of the myocardium tissues and disorder of the homeostasis of the intratissue medium are important links in the process of the formation of IHD and cardiosclerosis.

Electron microscope studies showed that the progress of IHD is accompanied by lowering of the significance of the compensation and adaptive changes in the cardiac myocytes [386]. In connection with hypoxia of the myocardium, the compensating and adapting mechanisms are disturbed, and the restoration processes become unbalanced. The disruption of these mechanisms in the cardiac myocytes because of hypertrophy or atrophy, and also of reorganization of the archtectonis of the myocardium cells increases. Limitations of the transportation and trophic provision of the myocardium cells set in, and their main factor is the lack of energy. Suppression of tissue respiration and the reproduction of creatine phosphate in a contracting myocardium are combined with the activation of glycolysis [386, 387, 388, 389, 434]. A large number of inferior mitochondria with an unusual arrangement of the cristae and lowering of the activity of the redox enzymes in them appear.

These changes in the compensating and adaptive mechanisms in the cardiac myocytes lower the amount of myofibrills ensuring the contracting function of the myocardium. The plastic insufficiency of the compensation and adaptive mechanisms in the cardiac myocytes causes a progressive lowering of the adaptation possibilities of the myocardium cells and lowering of the tolerance of the myocardium to a physical load. It is the regenerating insufficiency of the contracting myocardium due to its hypoxia and ischemia that causes an increasing deficiency of energy and the development of cardiac insufficiency.

Disturbance of the compensation and adaptive mechanisms in the cardiac myocytes combined with energy deficiency and the accelerated wear of the organelles hyperfunctioning under conditions of inadequate plastic provision create a favorable basis for the development and progress of IHD. The inadequacy of the transportation and trophic provision of the myocardium tissues and the regenerative insufficiency of the contracting cells exhaust the adaptation possibilities of the myocardium and create conditions for the development of cardiac insufficiency.

The electrocardiographic changes in DM patients are associated with metabolic disorders: decomposition of glycogen and metabolism of protein cause the excretion from the myocardium cells of the potassium bound to it [179, 190, 198, 256, 312, 416, 619]. This is also facilitated by the acidosis observed in diabetes. The polyuria accompanying DM also increases the emergence of potassium from the myocardium cells.

The drop in the end diastolic volume of the left ventricle on the background of a growth in the end diastolic pressure is a result of the increased diastolic rigidity of the myocardium in DM.

Not only the relaxation processes, but also the contractions are disturbed in DM. Many investigators explain the drop in the contracting function of the myocardium by the improper supply of blood to it and by diabetic microangiopathy [4, 190, 227, 535]. Some authors divide the pathology of the myocardium in DM into three types: (i) microangiopathic disorders, (ii) metabolic disorders, and (iii) vegetoneuropathic changes.

The morphofunctional hypertrophy of the myocardium in DM ensures a growth in the function of the heart that activates the genetic apparatus of the myocardial cells. This increases the synthesis of nucleic acids and proteins. A result is the incidence of hypertrophy of the myocardium. In DM (especially combined with obesity), additional injuring factors act on the myocardium, in particular the lack of oxygen, and the separation of oxidation and phosphorylation [26, 42, 205, 465].

The numerous pathological processes in the myocardium in DM are based pathophysiologically on disturbance of the structure and function of the cell and intracellular membranes. This is why the restoration of the altered membrane interactions is an important condition when treating decompensated DM. Under DM decompensation conditions, the pathology of the membrane structures of the myocardium acquire a special significance because the main functions of the myocardium (excitation of the cells, the combination of excitation and contraction, neurohumoral and hormonal regulation of biochemical processes) are connected with the activity of the membrane structures of the heart.

We can determine the following main directions in studying the pathology of the myocardium (especially in the decompensation of DM) that, as indicated above, are associated with studying the pathology of the cell membranes:

(1) studying of the structure and function of the mitochondria and, accordingly, the incidence of ischemia of the myocardium;

(2) studying of the effect on membranes of cholesterol, free-radical peroxide oxidation of lipids, and the relation between disorders in the structure of the membranes and the progress of DM and development of IHD;

(3) studying the excitation of the myocardium cells associated with the functioning of the ion channels and movement of the sodium pump;

(4) studying of the electrical phenomena in the myocardium, the appearance of the disturbed relations of calcium metabolism, and, accordingly, the processes of relaxation of the myocardium and its contraction.

All these problems are solved at the boundary between biochemistry, biophysics, morphology, and clinical disciplines. A number of authors revealed that tissue hypoxia, and ischemia of the myocardium cells point to the great significance of the membrane phospholipases in the irreversible lesion of cells [386, 388, 389, 434]. Electrical instability of the lipid bilayer of the membranes is created and they lose their barrier functions.

Major significance in recent years in the development of IHD in DM patients is attached to the peroxide oxidation of lipids. The use of antioxidants inhibiting the peroxide oxidation of lipids was shown to affect the lipid composition and physical properties of all the membrane structures of the myocardium cells, and this may be the theoretical grounds for using antioxidants as an additional factor in the treatment of DM [434].

Lesion of the membranes of live cells under the influence of activation of the free-radical peroxide oxidation of lipids (FPOL) is a very important pathogenetic link in the action on an organism of factors such as stress, a physical load, and radiation. The preservation of homeostasis here in an organism is a directed systemic reaction of the organism that ensures the adaptation of various organs and systems. Processes of a change in the amount of functioning structures in cells, the adaptive reorganization of the enzyme systems, and the hormonal provision of the metabolic processes occur in an organism. Under the influence of stressor factors, activation of the reactions of FPOL occur and lesion of the lipid bilayer of the membranes appears with the relevant disturbance of their functioning [434]. In the cell membranes, the content of FPOL products lowers sharply, the level of cholesterol drops, and phospholipids accumulate, primarily at the expense of a growth in the fraction of the readily oxidized ones.

In this stress period, the rate of blood flow in the organs and tissues is disturbed, hypoxia appears, the content of biological antioxidants in the cells lowers, and the activity of superoxidedismutase (the main enzyme of antiradical protection) is inhibited. When analyzing the reaction of an organism under conditions of the action of stressor factors, most authors consider that a multilevel fluctuating process goes on with regularly changing parameters (hormonal interactions, activation of the enzyme systems, a change in the sensitivity of the receptor apparatus, etc.) [538, 539].

The dynamics of the changes in the FPOL processes in different organs and tissues under the influence of stress is not identical. It depends, on the one hand, on the stability of the FPOL systems, and on the other on the force of action of the stressor factor.

A great importance in lesion of the myocardium in DM is attached to the development of ischemia and the effect of catecholamines. Experiments showed that moderate ischemia of the myocardium (produced by dosed narrowing of the coronary arteries or by the introduction of large doses of noradrenaline) lowers the concentration of creatine phosphate and ATP, and then activates the synthesis of nucleic acids and proteins. This increases the mass of the myocardium by 30–80% [386, 389].

It was shown by experiments that catecholamines suppress the secretion of insulin and increase glycogenolysis in the liver and muscles. This results in the development of hyperglycemia [205, 301, 589, 592]. Adrenaline, glucagon, and corticotropin activate the adenylatecyclase system and form cAMP – a mediator of biochemical processes. In contrast to this, insulin suppresses adenylatecyclase and lowers the formation of cAMP, stimulating the formation of cGMP.

Physiological studies showed that the human myocardium has an increased adrenoreactivity, i.e. in response to the discharge of noradrenaline, the myocardium increases the inotropic effect (the contractility of the myocardium grows) [434]. This is due to the fact that under the influence of noradrenaline, the synthesis of cAMP grows in a human organism (including in the myocardium), and the membrane cell systems of the myocardium (responsible for processing of the controlling signal) ensure an increased efficiency of this signal from the adrenergic system.

Of major significance in the pathogenesis of IHD is the violation of the regulatory function of the catecholamines [14, 54, 85, 262, 368, 388, 539]. The primary stage in the action of catecholamines on organs and tissues is their reaction with adrenergic receptors. Their predominating form in the heart is beta-1, and to a smaller extent, alpha-1-adrenoreceptors. The high concentrations of catecholamines in stress exhibit a cardiotoxic effect that is realized chiefly via excess beta-adrenergic stimulation. In hypercatecholemia in the myocardium, a number of unfavorable metabolic changes appear in DM and IHD patients. Special significance here is attached to activation of FPOL and overloading of the cardiac myocytes with calcium.

One way of correcting the unfavorable action of stressor factors on the myocardium of a DM patient is the blocking of beta-adrenoreceptors leading to inhibition of the slow

Table 43. Parameters of Hemodynamics in Practically Healthy Persons and in Patients with Type II DM and IHD before and after Taking Anaprilin (Propranolol) (Data of Echocardiography).

Parameter	Healthy persons		Patients with type II DM and IHD	
	before taking drug	after taking drug	before taking drug	after taking drug
End systolic volume, ml	42.35±4.76	57.35±7.46	75.08±11.8	83.8±14.2
End diastolic volume, ml	147.21±5.72	151.83±8.98	178.33±13.82	176.14±11.8
Minute volume, l/min	6.88±0.48	4.93±0.31*	6.34±0.59	5.03±0.71
Cardiac index, l/min·m^2	3.562±0.26	2.55±0.17*	3.03±0.24	2.36±0.26
Stroke volume, ml	104.9±4.94	94.48±4.43	103.95±13.2	104.63±11.81
Stroke index, ml/m^2	54.4±2.77	49.0±2.77	49.6±4.28	44.42±4.647
Unit power, kgm/min·m^2	4.57±0.3	3.19±0.18*	4.29±0.34	3.23±0.32*
Peripheral resistance, dyn·s·cm^{-5}	1168.6±136.5	1558.3±124.3*	1379.0±107.0	1845.0±160.6*
Ejection fraction, %	70.94±2.81	62.88±2.57*	60.28±3.68	54.53±5.47
%ΔS	41.61±2.5	34.61±1.79*	33.53±2.37	29.6±3.45
V_{cf}, s^{-1}	1.46±0.08	1.05±0.05*	1.09±0.15	0.87±0.10
Contraction rate, cm/s	5.82±0.41	4.76±0.16*	3.77±0.35	3.46±0.29
Relaxation rate, cm/c	12.23±2.0	9.81±0.99	7.14±0.51	5.88±0.61

* The difference of the parameters is authentic (in comparison with the background level).

entry of a calcium flow into the cells and to diminishing of the intensity of the peroxide oxidation of the lipids.

To answer the question on the significance and effectiveness of using beta-blocking agents in type II DM and IHD patients, we studied 22 such patients. As a control group, we observed 15 healthy persons. Both the patients and healthy persons were subjected to echocardiography before taking anaprilin (propranolol) and in 1.5 hours after taking 40 mg of the drug.

In healthy persons, the reaction to the beta-blocking agent was quite definite (Table 43): the end systolic volume increased from 42.3 to 57.3 ml. The end diastolic volume did not virtually change, but the minute volume of the blood lowered from 6.8 to 4.9 l/min. The stroke volume of the blood diminished from 104.9 to 94.4 ml, and the ejection fraction diminished from 70.9 to 62.8%.

The degree of shortening of the front-rear distance of the left ventricle during systole lowered appreciably (from 41.6 to 34.6%), and the rate of circular shortening of the myocardium fibers (V_{cf}) decreased from 1.4 to 1.05 s^{-1}.

The lowering of the rate of relaxation under the influence of beta-blocking agents is a very important parameter. In healthy persons, it lowered from 12.2 to 9.8 cm/s after taking the drug.

When comparing the reaction of healthy persons and DM and IHD patients to the use of beta-blocking agents, we noted the following differences (Table 43). The end systolic volume grew from 75.0 to 83.8 ml, while the end diastolic volume did not alter (178.3 and 176.1 ml). The minute volume of the blood diminished insignificantly from 6.3 to 5.0 l/min. The stroke volume did not practically change (103.9 and 104.6 ml). The ejection fraction changed insignificantly – from 60.2 to 54.5%.

Similar insignificant changes were noted in the degree of shortening of the front-rear distance of the left ventricle during systole (33.5 and 29.6%). The rate of circular shortening of the myocardium fibers lowered insignificantly from 1.0 to 0.8 s^{-1}. The rate of relaxation changes more definitely. In DM patients, the rate of relaxation is lowered considerably. It is 12.2 cm/s in healthy persons, and 7.1 in type II DM patients (background). After the taking of beta blocking agents, it lowers to 5.8 cm/s.

Clinical observations with the use of beta blocking agents showed that in DM patients, the hormonal regulation of the myocardium is disturbed drastically, especially stimulation by catecholamines. Unlike healthy persons, in DM patients a hyperdynamic syndrome is observed with pronounced hypertrophy of the myocardium. Disturbance of the diastolic function of the myocardium (a growth in the end diastolic volume) is attended by lowering of its contractility (a reduction in the ejection factor, %ΔS, V_{cf}) and, which is very important, a drop in the rate of relaxation.

Clinical experimental studying of the correlation between the rate of relaxation of the left ventricle and the degree of lesion of the coronary vessels did not yield positive results. Pathogenesis of disorders of the diastolic relaxation of the left ventricle in IHD and type II DM patients has not been established finally [4, 102, 619]. The emergence of calcium from the myosin complex into the rapid phase of diastolic relaxation of the myocardium occurs. Sufficient energy is needed for normal proceeding of this process. Experiments on animals revealed that in ischemia of the myocardium, disorder of the diastolic relaxation sets in much earlier than when the systolic ejection lowers [388, 434]. Several factors probably participate in the process of the change in diastolic relaxation, a role being played not only by an acute disorder of the coronary blood flow, but also of chronic ischemia. Surgeons indicate that after an operation of aortocoronary shunting, the diastolic relaxation of the left ventricle improves [187].

In summarizing our analysis of the functional state of the myocardium in DM of type II, we can name the pathogenetic factors that lead to decompensation and development of cardiac insufficiency:

(1) incomplete ejection of the blood;

(2) a growth in the diastolic pressure;

(3) an initial drop in the minute volume;
(4) lowering of the main metabolism;
(5) mobilization of extracardiac factors;
(6) excitation of the renin-aldosterone mechanism;
(7) unloading of the heart, the loss of potassium.

The insufficiency of blood circulation in type II DM patients with a hypertrophied myocardium develops as a result of the action of the following factors:

(1) lowering of the influence of the adrenergic system;
(2) lowering of the power of the ion transportation system;
(3) lowering of the capacity of energy supply;
(4) lowering of the coronary reserve;
(5) lowering of the ATPase activity of the myosin of the myofibrills.

This results in diminishing of the maximum force and rate of contraction of the myocardium.

3.4 Conclusion

The results of our studies devoted to unrevealing the pathology of the myocardium in DM patients show that the progress of IHD in these patients is due to metabolic disorders of carbohydrate and lipid metabolism in the coronary vessels and cardiac myocytes.

We attach special significance in type II DM to lesion of the myocardium with sharply pronounced hypertrophy of the left ventricle. Activation of the genetic apparatus of the myocardial cells with an increase in the synthesis of nucleic acids and proteins is the pathophysiological basis of morphofunctional hypertrophy of the myocardium in DM. The function of the myocardium is disturbed especially with decompensation of DM, when the peroxide oxidation of lipids is activated with lesion of the membranes of the cardiac myocytes.

An important role in the progress of atherosclerotic changes in DM is played by arterial hypertension, which appears especially frequently in type II DM patients [18, 151, 156, 259]. Its pathogenesis is due to the presence of hyperinsulinemia and insulin resistance.

When DM is combined with obesity, the ischemic lesion of the myocardium is accelerated on the background of coronarogenic and metabolic disorders.

Hyperlipidemia increases the content of free fatty acids in the myocardium with a cardiotoxic effect [18, 80, 157, 179, 247]. Generally hyperkalemia is noted in DM patients. It is provoked by the development of electrical instability of the myocardium [102, 119, 152, 232, 416]. DM patients always have stable tachycardia whose pathogenesis is due to an increased amount of circulating catecholamines and the presence of diabetic myocardial neuropathy.

When studying the tolerance to a physical load in DM patients, we noted a sharp drop in the total and threshold power of the performed work. This is due, on the one hand, to disturbance of energy formation in the myocardium in DM, and on the other to morphofunctional lesion of the myocardium.

The energy deficiency in the myocardium in DM is due to metabolic disorders with the predomination of anaerobic metabolism and the development of hypokalemia. The insufficiency of the myocardium is especially manifest when DM is combined with obesity and the growth in the volume of the circulating blood is attended by a growth in the preload, a defect in perfusion of the myocardium, and lowering of the minute volume.

The hyperkinetic type of myocardium contraction (especially in type II DM patients) is characterized by lowering of the energy of isometric relaxation of the myocardium and diminishing of the end diastolic volume of the left ventricle on the background of a growth in the end diastolic pressure.

The hyperkinetic type of myocardium contraction in DM (especially of type II) is due, first, to a growth in the mass of the myocardium, and, second, to lowering of the total peripheral resistance.

An important role in the development of cardiac insufficiency in DM is played by decompensation of diabetes [4, 225, 227, 232, 416]. Here the diastolic function of the left ventricle and its filling are disturbed, relaxation of the myocardium is retarded, the rigidity of the wall and cavity of the left ventricle is increased, and its elastic properties are lowered.

All these functional disorders in the myocardium in DM patients are based on morphobiochemical lesions: microangiopathic, metabolic (a change in the structure of collagen under the influence of glycosylation, hyperkalemia, disturbance of calcium metabolism), and vegetoneuropathic [63, 80, 96, 112, 166, 247].

All the indicated pathological processes in the myocardium in DM patients require combined treatment of both metabolic disorders in various organs and systems of the organism, and of atherosclerotic changes in the myocardium.

Cardiac insufficiency in DM patients develops in several directions. First of all, the formation of energy in the myocardium lowers as a result of the predomination of anaerobic metabolism, sharp hypertrophy of the myocardium (especially in type II DM patients) with ischemia and hypokalemia.

The energy metabolism in an organism is a complicated thermodynamic system integrating the interaction of physiological and biochemical components, and also regulator and hormonal relations.

Cardiac vegetoneuropathy attended by tachycardia and disturbance of the rhythm is an important fragment of cardiac insufficiency in DM.

Clinicians noted long ago that in DM patients, myocardial infarction develops without pains [66, 190, 198, 201, 248, 416]. Generalized capillaropathy and lesion of the small intramural vessels with disturbance of the function of the neuroreceptor apparatus are considered to be the pathophysiological basis of this phenomenon.

The progressing development of atherosclerosis in DM is due, on the one hand, to the disturbance of lipid metabolism, and on the other to a morphofunctional change in the activity of the vessel wall. IHD develops especially rapidly in type II DM, when hyperinsulinemia and insulin resistance are clearly determined on the background of a sharp drop in the content of HDLP. Coronarographic studies in DM patients revealed the most frequent injury of the left and right coronary arteries with their complete occlusion.

Clinical investigations of the tolerance to a physical load in DM patients revealed numerous pathological changes, primarily a drop in the power of the threshold load, and considerable electrocardiographic changes: elongation of the electrical systole, and shifts of the relation $T_{V_1} - T_{V_6}$. In our studies, the repolarization of tooth T was disturbed especially frequently (from its lowering to the appearance of a negative tooth T).

When studying the fluctuations of the arterial pressure in DM patients, we noted its considerable increase under a physical load and emotional stress. Diabetic retinopathy and IHD progressed rapidly in these patients. We noted that arterial hypertension in type II DM (due to lesion of the kidneys and their vessels) has a number of features: (i) hyperinsulinemia and insulin resistance, (ii) hypertriglyceridemia and hypercholesterolemia, (iii) lowering of the HDLP content, (iv) a hyperkinetic type of hemodynamics on the background of an increase in the amount of circulating blood, and (v) lowering of the renin level in the blood plasma.

Our scintigraphical investigations of DM patients under conditions of a physical load revealed a considerable number of zones with reduced perfusion. We noted here that in type II DM patients, the myocardial blood flow was lowered more sharply than in type I patients. We consider that the standing out of transient defects of perfusion in the scintigrams for type II patients is related closely to the disturbance of central hemodynamics and the development of atherosclerosis in the coronary arteries of the myocardium.

Diabetic myocardiodystrophy is due to the disturbance of metabolism of the cardiac myocytes and the development of microangiopathy. In these patients, the coronarogenic and metabolic disorders mutually enhance the energy insufficiency of the myocardium.

Lowering of the rate of myocardium relaxation due to inhibition of oxidative phosphorylation, a drop in the level of ATP, and ischemia of the myocardium are an important pathological phenomenon in the activities of the myocardium in DM. Lowering of the relaxation rate is especially noticeable in type II DM patients when hypertrophy of the myocardium is manifest. It is attended by disturbance of the density of sympathetic innervation, lowering of the noradrenaline concentration and accordingly a drop in the inotropic function of the myocardium.

Hypertrophy of the myocardium (in type II DM patients) is attended by diminishing of the number of coronary capillaries per unit mass of the myocardium and, accordingly, a growth in the diffusion distance of oxygen occurs with phenomena of ischemia.

Cardiac insufficiency develops especially unfavorably when type II DM is combined with obesity. In these patients, lowering of the end diastolic volume of the left ventricle on the background of a growth in the end diastolic pressure occurs very early. Clinically, with this pathology we noted an increase in dyspnea, edema of the lower extremities, changes of an ischemic type in electrocardiography, and pronounced occlusion of the coronary arteries in coronarography.

We must note in concluding that atherosclerotic lesion of the myocardium in DM includes integrative units of carbohydrate and lipid metabolism, and the treatment must be complex and essentially individual.

CONCLUSION

Diabetes mellitus and atherosclerosis are integratively related diseases with a disorder of the lipid and carbohydrate metabolism attended by hypercholesterolemia, hypertriglyceridemia, and hypoalpha-cholesterolemia. A combination of risk factors such as hyperlipoproteinemia, arterial hypertension, and increased coagulation of the blood are rooted in the metabolic disorders in DM patients.

Insulin affects both the synthesis of triglycerides and their removal from circulation.

During recent years, clinicians and endocrinologists are devoting a lot of attention to studying type II DM. This is due not only to the large number of these patients, but also to the features of the development of microangiopathies and atherosclerotic changes in the vessels. In this type of DM, clinicians note a combination of diabetes and obesity. Many endocrinologists consider that IIDM appears as a result of disturbance of the balance between the sensitivity to insulin and the secretion of this hormone. We noted a drop in the number of insulin receptors in type II DM patients attended not only by hyperinsulinemia, but also by insulin resistance. Our clinical and biochemical investigations proved the drop in the activity of glucose transportation in the muscles and lipocytes in type II DM patients. The development of microangiopathies and atherosclerotic changes in the vessels increases the significance of IHD risk factors from four to five times. We attach special significance here to hypertension combined with obesity in type II DM patients.

The pathogenesis of diabetic microangiopathies is due, from our viewpoint, to a change in the basal membrane of vessels, hemodynamic disorders, and disorders in the blood coagulation and function of the thrombocytes. The changes in the basal membrane are due to the increase in the cells of sorbitol (a product of glucose transformation).

The hemodynamic disorders are attended by an increase in the permeability of the vessels of the myocardium, kidneys, and retina. These processes increase the transcapillary penetration of plasmoproteins and the development of diabetic complications. Diabetic nephropathy is one of serious complications of DM. It is manifest by albuminuria, hypertension, and azotemia.

When studying the clinical and biochemical characteristic of lipid metabolism In healthy persons and DM (with IHD) patients, we noted (both in a state of rest and under conditions of a lipid load) the great significance of lipoproteinlipase in the regulation of the interaction of lipoproteins and the smooth muscle cells of an artery wall. Our studies prove that an important role in the coordination of the interaction of the lipoproteins in the blood plasma and cholesterol in the cells is played by the LDLP receptors whose number and function affect the increase or decrease of the cholesterol level in the cells of an organism.

We found that the rate of the direct and reverse transportation of cholesterol from the blood into the cells of the cardiovascular system organs in DM patients also suffering from obesity is disturbed drastically. We attach major significance here to the peroxide oxidation of the lipids in the cells and tissues.

Hypertriglyceridemia is an important link in the disturbance of the metabolism of lipids in DM patients. A growth in the triglyceride level in the blood is an integrative parameter associated both with the nature of the course of DM (its type, state of compensation or decompensation) and with the nature of the patient's diet, his or her hormonal activity and level of physical activity. The pathogenesis of hypertriglyceridemia is due, on the one hand, to the disturbed excretion of the triglycerides from an organism because of the lowering in DM of the extrarenal LPL. On the other hand, here insulin resistance appears, and it results in disturbance of the lipolysis of VLDLP. We attach special significance to insulin resistance in type II DM, when equilibrium between the sensitivity to insulin and secretion of the latter is disturbed.

We consider that the lowering of the level of cholesterol in the HDLP is of major importance in activation of the atherosclerotic process in DM. The antiatherogenic func tion of the HDLP was known long ago, and their participation in inhibiting the atherosclerotic process in the coronary arteries was proved. We noted that in DM patients of both types the level of cholesterol in the HDLP is lowered, and the degree of adaptation of the cellular cholesterol by the high-density lipoproteins correlates clearly with the activation of the intracellular prostacyclin catalyzing the process.

When studying the significance of the insulin level in the blood of type I and II DM patients, we noted that in hyperinsulinemia the development of hypertriglyceridemia is stimulated, and the level of the cholesterol of the HDLP lowers. In our opinion, the combination in type II DM patients of hyperinsulinemia and insulin resistance is especially unfavorable. In these patients, the high level of insulin in the blood plasma stimulates the synthesis of fatty acids in the liver. The predominating synthesis of fatty acids in the liver is attended by their transportation in the composition of the VLDLP into the adipose tissue, where they accumulate in the form of triglycerides. When DM is combined with obesity, the amount of receptors to insulin in the adipose tissue, liver, and muscles is revealed very clearly.

We attach major importance to a lipid load in determining the disorders in fat and carbohydrate metabolism. It is the most physiological test for finding the functional possibilities of the metabolic activity of an organism (especially of lipid, carbohydrate metabolism and the hormonal regulation of these systems).

Our investigations with a lipid load showed that the activity in the development of an atherosclerotic process in the myocardium, kidneys, vessels of the fundus of the eye and of the lower extremities depends on the relation between two nonenzymatic processes: transportation of cholesterol esters into the liver for excreting the cholesterol from the organism, and the activity of incorporating the esters in the composition of the VLDLP and LDLP into the smooth muscle cells and endothelium of a vessel wall. We noted that in type II DM patients (especially also suffering from obesity) together with lowering of the activity of the enzyme LCAT and the level of the HDLP cholesterol, the development of atherosclerotic changes in the vessels accelerates drastically, and the appearance of microangiopathies progresses.

Attaching great significance to the disturbance of fat metabolism, we studied patients with obesity for determining the relation between the insulin level and the sensitivity to it of the fat cells. We noted that in obesity patients, the sensitivity of the fat cells to insulin lowers and it is due to improper activity of the LPL.

In continuing our clinical and biochemical analysis of this link of lipid metabolism, we noted the lowering of the cAMP level in the cells. In DM, insulin stimulates the increased synthesis of fatty acids and triglycerides from the products of carbohydrate metabolism.

The development of atheromatous plaques in the aorta and the coronary vessels is closely related to numerous hemodynamic, microcirculation, and rheological factors, especially with the intravascular aggregation of erythrocytes and thrombocytes.

Our investigations showed that disturbance of microcirculation and the rheological properties of the blood in DM patients is closely related to the pathology of lipid metabolism (an increased level of cholesterol, triglycerides, and free fatty acids). These changes occur not only in the blood plasma, but also in the erythrocyte membranes. In the latter, we noted a growth in the fraction of cholesterol, triglycerides, and free fatty acids, and a drop in the total percentage of phospholipids. This indicates an increase in the density of the lipid packing of a membrane and lowering of its deformability. These changes in the erythrocyte membranes are seen especially clearly in type II DM patients with the relevant change in the rheological parameters of the blood, namely, the Kesson viscosity and an increased cohesion of the erythrocytes. With a marked dysfunction of the plasma factors of blood coagulation in DM, the physicochemical properties of the erythrocyte membranes changed with the corresponding hindrance in the fluidity of the blood and the development of microthromboses. The thrombocytes play the important role of a binding link between the coagulation and thrombocyte-vascular components of hemostasis.

The microscopic changes in the vessels in DM in combination with atherosclerosis are based on pathological changes in the functional activity of the endothelial cells of the vessels, lowering of the activity of prostacyclin, an increase in the amount of thromboxan B_2, and in the Willebrand factor in the blood.

Our studies revealed an increased tendency of thrombocytes to aggregate in DM. This is due on the one hand to the greater formation of thromboxan B_2, and on the other to a higher sensitivity of the thrombocytes to ADP. Consequently, in DM patients, the interaction between the endotheliums of vessels and the thrombocytes is disturbed. We gave special attention to studying the interaction of disorders in lipid metabolism and the increased aggregation of thrombocytes and proved that in DM the increased sensitivity of thrombocytes to aggregation is due, among other factors, to hypertriglyceridemia.

The disorders in the rheological properties of the blood in DM are witnessed by changes in the retina: perivascular hemorrhages, the presence of an edema, a change in the size of the vessels, and the appearance of empty zones. The intravascular changes are registered in the form of a sludge phenomenon with a sharp retarding of the space velocity of the blood flow, the appearance of a counter flow, and stasis of the blood.

The rapid development of atherosclerosis in DM patients is also associated with an increase in the fibrinogen level in the blood with the relevant growth in the viscosity of the blood, the development of microthromboses, and disturbance of the diffusion of oxygen. Here the combination of a high level of fibrinogen with lowering of the level of antithrombin III is especially dangerous.

In DM (especially in combination with obesity), the DIC syndrome often develops, when a large amount of activators of blood coagulation enter the blood flow in a short time and thrombocytes aggregate with the formation of a large amount of thrombin and exhaustion of the plasma enzyme systems.

In our studies, we attach major significance to the growth in the amount of thromboxan B_2 and lowering of the prostacyclin level in DM patients. The predomination of the thromboxan-generating ability over the prostacyclin-synthesizing function of an endothelium in DM ensures progressing thrombus formation and the development of the DIC syndrome.

By using a cuff test in healthy persons and DM patients, we determined more accurately the functional significance of the endothelium. Simultaneously we determined the level of the Willebrand factor whose considerable increase was established in DM patients. A growth in the content of this factor in DM patients is a marker of lesion of the endothelium cells of the vessels and facilitates the adhesion of thrombocytes to the injured endothelium and subendothelium layer. These changes appeared especially frequently in type II DM patients on a background of hyperinsulinemia and insulin resistance. Capillaroscopic studying of the fundus of the eye in these patients showed the appearance of sleeves near the capillaries in the sections of diapedesis of the erythrocytes and of microthromboses.

In our opinion, a considerable effect was obtained when treating DM patients with the aid of plasmapheresis. We performed extracorporal therapy using a domestic fractionator of the blood. A course of plasmapheresis in type I and II DM patients was attended by lowering of the content of triglycerides, LDLP, VLDLP, and fibrinogen. The rheological parameters of the blood improved accordingly (the viscosity and the forces of cohesion of the erythrocytes lowered), diabetic retinopathy did not progress, but returned to its initial level.

BIBLIOGRAPHY

1. Abgel A, Roncari D. A. K. Medical complications of obesity. Can. Med. Assoc. J, 119, 1408-1411 (1978).
2. Acheson K, Jequier E, Nahren J. Influence of beta-adrenergic blockade on glucose-induced thermogenesis in man. J. Clin. Invest, 72, 3, 981-986 (1983).
3. Adamyan K.G., Sidorenko B.A., Nranyan N.V. Determining the Functional State of the Myocardium in IHD. Kardiologiya, 8, 114-118 (1984)[*].
4. Ahmed S. S, Jaferi C. A, Narang R. M. Preclinical a bnormality of left ventricular function in diabetes mellitus. Am. Heart J, 89, 153-158 (1975).
5. Alaupovic P. Atherosclerosis, 13, 141 (1971).
6. Alekseeva N.V. Central Hemodynamics, Hemorheology and Peroxide Oxidation of Lipids in Arrhythmias in IHD Patients. Synopsis of Candidate's Thesis. Moscow (1985)[*].
7. Alhenc-Galas F, Tsai S. J, Callahan K. S. Stimulation of prostaglandin formation by vasoactive mediators in cultured human endothelial cells. Prostaglandins, 4, 723-742 (1982).
8. Aliev T.A., Glezer M.G., Moskalenko N.P. Blood Circulation and Its Regulation in DM. Baku (1984)[*].
9. Alizadze I.G., Mamedova E.M., Azimova R.L. Influence of Plasmapheresis on the Parameters of Lipid Metabolism and Hemorheology in Hypertension Patients. Azerb. Med. Zh., 5, 33-37 (1991)[*].
10. Allo S. N, Schaffer S. W. Defective sarcolemmal phosphorylation associated with non-insulin-dependent diabetes. Biochem. Biophys. Acta, 1023, 206-212 (1990)
11. Alzarka A, Haft J. I, Bassil H. Correlation of cross-linked fibrin degradation products (d-dimer) with coronary angiografic lesion morphology. Int. J. Cardiol, 10, 1, 77-87 (1991).
12. Amosov N.M., Bendet Ya.A., Physical Activity and the Heart. Kiev (1984)[*].
13. Anderson D. N, Schaefter E. J, Bronzert T. J. J. Clin. Invest, 663, 857-866 (1981).
14. Anestiadi V.V. The Hormonal and Metabolic Aspects of a Disposition to Atherosclerosis in DM. Synopsis of Candidate's Thesis. Kiev (1985)[*].
15. Arner P. Role of antilipolytic mechanisms in adipose tissue distribution and function in 30. man. Acta Med. Scand, 723 (suppl.), 147-152 (1988).
16. Aronov D.M., Guliev A.A., Samoilenko L.E. The Change in Hemodynamics and Oxygen Consumption in Dosed Psychoemotional Stress in IHD Patients. Kardiologiya, 10, 55-58 (1990)[*]
17. Arustamian K.K., Eloian A.Ya., Arutyunova E.G. The State of the Vegetative Nervous System in Women Suffering from the Neuro-Metabolic-Endocrine Form of the Hypothalamic Syndrome. In: Modern Problems of Reproductology: Abstracts of Reports, pp. 232-233. Tbilisi (1988)[*].
18. Assmann G, Schutle H. The Prospective Cardiovascular Munster (PROCAM) Study: prevalence of hyperlipidemia in persons with hypertension and/or diabetes mellitus and the relationship to coronary heart disease. Am. Heart J, 116, 1713-1724 (1988).
19. Astrup A, Buemann B, Christensen N. J. 24-Hour energy expenditure and sympathetic activity in postobese women consuming a high-carbohydrate diet. Am. J. Physiol, 262, 3, E282-E288 (1992).

[*] All publications marked with an asterisk are in Russian.

20. Atherton B. T, Hynes R. O. A difference between plasma and cellular fibronectins located with monoclonal antibodies. Cell, 25, 133-141 (1981).
21. Atroshchenko E.S., Shagisultanov E.R. Influence of Hemosorption on State of Microcirculation in Patients with Stable Stenocardia according to Data of Radioindication. Kardiologiya, 10, 25-27 (1991)*.
22. Attali J. R, Valensi P. Diabetes and hemorheology. Diabetes Metab, 16, 1, 1-6 (1990).
23. Austin M. A, Goto Y, Lenfant C. Epidemiology. Am. J. Cardiol, 68, 3, 22-25 (1991).
24. Babsky E.B., Berdyaev S.Yu. Automatics of Heart. In: Koradi G.P. (ed.) Physiology of Blood Circulation. Physiology of Heart, pp. 63-87. Leningrad: Nauka (1980)*.
25. Baenziger N. L, Force L. E, Becherer P. R. Histamine stimulates prostacyclin synthesis in cultered human umbilical vein endothelial cells. Biochemical and Biophysical Research Communications, 92, 1435-1440 (1980).
26. Baidalin S.D., Tritek V.S., Yugai L.A. Synthesis of Fatty Acids in the Liver and Adipose Tissue in Insulin-Independent DM and Heriditary Obesity in Mice. In: Questions of Endocrinology: Collected Papers, pp. 56-57. Alma-Ata (1983)*.
27. Bakhsalieva S.B., Aliev T.A., Rakhmani S.A. The Nature of the Changes in Lipid Metabolism in DM Patients Combined with IHD. Kardiologiya, 9, 61-64 (1990)*.
28. Balashov V.A., Dikker V.E., Galenok V.A. Contractive Function of the Myocardium and Oxygen Metabolism in Diabetes Mellitus. Terapevt. Arkhiv, 10, 139-142 (1985)*.
29. Baluda V.P., Deyanov I.I. Significance of Determining the Antithrombogenic Properties of the Wall of Vessels in the Prophylaxis of Thrombosis. Kardiologiya, 5, 103-105 (1988)*.
30. Bang H. O, Dyeberg J, Hjorne N. The composition of food consumed by Greenland Eskimos. Acta Med. Scand, 200, 69-73 (1976).
31. Bang H. O, Dyeberg J. Plasma lipids and lipoproteins in Greenlandic West Coast Eskimos. Acta Med. Scand, 192, 85-94 (1972).
32. Barakat H. A, Carpenter J. W, McLendon V. D. Influence of obesity, impaired glucose tolerance and NIDDM on LDL structure and composition. Possible link between hyperinsulinemia and atherosclerosis. Diabetes, 39, 12, 1527-1533 (1990).
33. Barbir M, Wile D, Trayner I. High prevalence of hypertriglyceridemia and apolipoprotein abnormalities in coronary artery disease. Br. Heart J, 60, 397-403 (1988).
34. Barkagan Z.S. Hemorrhagic Diseases and Syndromes. Moscow: Meditsina (1988)*.
35. Baumgartner H. R. The subendothelial surface and thrombosis. Thrombosis: Pathogenesis and clinical trials. Ed.: Deutsch E, Brinkhous K. et al, Stuttgart: F. K. Schattauer-Verlag, 91-105 (1974).
36. Becker R. C. Seminars in thrombosis, thrombolysis and vascular biology: Part 5: Cel lular rheology and plasma viscosity. Cardiology, 79, 4, 265-270 (1991).
37. Bel'gov A.Yu. The Role of the Hypothalamus in the Development of Endocrine Pathology in Teenagers. In: Pathology of Internal Organs in Neurohormonal and Metabolic Disorders: Features of Hemodynamics in Teenagers with Obesity, pp. 77-86. Leningrad (1988)*.
38. Belinsky V.P. State of Adipose and Muscle Tissue in Women with Alimentary Obesity. Vrach. Delo, 9, 75-77 (1988)*.
39. Beloshitskaya N.I. State of the Thrombocyte and Vascular Link of the Hemostasis System in DM Patients with IHD. Candidate's Thesis. Moscow (1990)*.
40. Belousov Yu.B. The Clinico-Pathogenetic Aspects of Intravascular Coagulation and Rheological Disorders in IHD Patients. Doctor's Thesis. Moscow (1982)*.
41. Belova A.K. Clinico-Genealogical and Morphofunctional Characteristic of Obesity in Prepuberty and Puberty Periods and Features of Lipid Metabolism. Candidate's Thesis. Orenburg (1988)*.

42. Belovalova I.M. Studying the Qualitative Composition of IRI in DM Patients with a Normal Body Weight and with Obesity. Candidate's Thesis. Moscow (1988)*.
43. Benjamin C. W, Hopkins N. K, Oglesby J. D. Agonist specific desensitization of leukotriene C_4-stimulated PGI_2 biosynthesis in human cells. Biochemical and Biophysical Research Communications, 117, 780-787 (1983).
44. Bergstrom R. W, Newell-Morris L, Leonetti D. L. Association of elevated fasting, C-peptide level and increased intra-abdominal fat distribution with development of NIDDM in Japanese-American men. Diabetes, 39, 104-111 (1990).
45. Beridze N.G. The Level of Insulin, Glucose, Free Fatty Acids and Triglycerides in the Blood of Patients with Various Types of Obesity. Candidate's Thesis. Tbilisi (1985)*.
46. Bertman B, Wolf S, Kaufhald F. Rheo logic findings in patients with diabetic retinopathy. Fortshr. Ophthalmol, 88, 4, S.321-325 (1991).
47. Bevilacque M. P., Pober J. S, Majean G. R. Interle ukin I induced biosynthesis and cell surface expression of procoagulant activity in human vascular endothelial cells. J. Ex. Med., 160, 618-623 (1984).
48. Beyul E.A., Oleneva V.A., Shaternikov V.A. Obesity. Moscow: Meditsina (1986)*.
49. Bieger W. P., Michel G, Barwich D. Diminished insulin receptors on monocytes and erythrocytes in hypertriglyceridemia. Metabolism, 33, 982-987 (1984).
50. Bihari-Varga M, Gruber E, Rotheneder M. Interaction of lipoprotein(a) and low density lipoprotein with glycosoaminoglycans from human aorta. Atherosclerosis, 8, 851-857 (1988).
51. Bjorntorp P. Metabolic implications of body fat distribution. Diabetes Care, 14, 1132-1143 (1991).
52. Blair I. A, Barrow S. E, Waddell K. A. Prostacyclin is not a circulating hormone in man. Prostaglandins, 23, 579-583 (1982).
53. Bodnar P.N., Zaume S., Pristupnyuk A.M. Deformability of Erythrocytes and Thiol Compounds in DM Patients. Vrach. Delo, 10, 87-89 (1984)*.
54. Boev S.S. Analysis of the Features of the Sympathico-Adrenal Sysytem Activity in Patients with Chronic IHD with Obesity. In: Systemic Analysis of Functional Manifestations of Protective Specific and Nonspecific Reactions of an Organism, pp. 12-13. Moscow (1980)*.
55. Boeynaems J. M, Galand N. Stimulation of vascular prostacyclin synthesis by extracellular ATP and AD Biochemical and Biophysical Research Communications, 112, 290-295 (1983).
56. Bogardus C, Lillioja S, Howard B. Relationships between insulin secretion, insulin action and fasting plasma glucose concentration in non-diabetic and non-insulin-dependent diabetic subjects. J. Clin. Invest, 74, 1238-1246 (1984).
57. Bokarev I.N. Thrombophilic States and Their Clinical Aspects. Klinich. Meditsina, 8, 11-17 (1991)*.
58. Bonora E, Zenere M, Branzi P. Influence of body fat and regional localization on risk factors for atherosclerosis in young men. Am. J. Epidemiol, 135, 1, 1271-1278 (1992).
59. Bortone A. S, Hess O. M, Eberli F. R. Abnormal coronary vasomotion during exercise in patients with normal coronary arteries and reduced coronary flow reserve. Circulation, 79, 516-527 (1989).
60. Borzova L.V., Zaletsky L.L. Morphofunctional Properties of Blood Cells and Rheological Parameters of Blood after Plasmapheresis in Patients with Unstable Stenocardia. Gemat. i Transfuz., 9, 11-15 (1991)*.
61. Broekmans A. W, Veltcamp L. L, Bertina R. Congenital protein C deficiency and venous thromboembolism. N. Engl. J. Med., 309, 340-344 (1983).

62. Brook G, Aviram M, Winterstein G. Platelet function and lipoprotein levels after plasma exchange in patients with familial hypercholesterolemia. Clin. Sci., 54, 7, 637-672 (1983).
63. Brownlee M, Cerami A. The biochemistry of the complications of diabetes mellitus. Ann. Rev. Biochem, 50, 385-432 (1981).
64. Bubnova M.G. The Influence of Food Fat and Physical Loads on the Lipid Transportation System in IHD. Synopsis of Candidate's Thesis. Moscow (1993)*.
65. Bugiardini R. Epicardial coronary artery reactivity in patients with syndrome X: the role of increased adrenergic tone. Cor. Art. Dis., 3, 547-554 (1992).
66. Bukharov S.I., Sergienko V.B., Maiorov N.I. The Use of Myocardium Infarction Scintigraphy for Patients with Typical Stenocardia and Atypical Pain Syndrome. Kardiologiya, 10, 100-103 (1985)*.
67. Busch C, Owen W. G. Identification in vitro of an endothelial cell surface cofactor for antithrombin III. Parallel studies with isolated perfused rat heart and microcarrier cultures of bovine endothelium. J. Clin. Invest, 69, 725-729 (1982).
68. Buslenko N.S., Belyavsky V.I., Sarygin D.B. Relation between the Level of Lipids in Blood Plasma and Lesion of the Coronary Channel in IHD Patients. Kardiologiya, 11, 28-32 (1984)*.
69. Candrina R Double-filtration plasmapheresis in the ter ozygous familial hypercholesterolemia: our experience over 25 treatments. Beitr. Infusiother, Bd.23, S.189-19 (1988).
70. Cannon R. O, Camici P. G, Epstein S. E. Pathophysiological dilemma of syndrome X. Circulation, 85, 883-892 (1992).
71. Cannon R. O, Leon M. B, Watson R. M. Chest pain and "normal" coronary arteries. Am. J. Cardiol, 55, 3, 50B-60B (1985).
72. Carlson L. A, Bottiger L. E. Risk factors for ischemic heart disease in men and women. Results of the 19-year follow-up of the Stockholm Prospective Study. Acta Med. Scand., 218, 207-211 (1985).
73. Carlson L. A, Hamsten A, Asplund A. Pronounced lowering of serum levels of lipoprotein Lp(a) in hyperlipidemic subjects treated with nicotinic acid. J. Int. Med., 226, 271-276 (1989).
74. Carlson L. A, Wallidus G. Fatty acid incorporation into human adipose tissue in hypertriglyceridemia. Eur. J. Clin. Invest., 6, 195-211 (1976).
75. Caro J. F. Insulin resistance in obese and nonobese men. J. Clin. Endocrinol. Metab, 73, 691-695 (1991).
76. Castelli W. P., Garrison R. J, Wilson P. W. F. Incidence of coronary heart disease and lipoprotein cholesterol levels. JAMA, 256, 2835-2838 (1986).
77. Castro C. R, Fielding C. J. Effects of postprandial lipemia on plasma cholesterol metabolism. J. Clin. Invest, 75, 874-882 (1985).
78. Chan T. K, Chan V. Antithrombin III, the major modulator of intra-vascular coagulation, is synthesized by human endothelial cells. Thrombosis and Haemostasis, 46, 504-506 (1981).
79. Cheishvili L.D. Gegenava L.S., Gvaladze G.P. Hemomorphological Equivalents of Blood in Health Persons and IHD Patients. Sovrem. Vopr. Kardiologii, 137-143 (1987)*.
80. Chen Y.-D. I, Jeng C.-V, Reaven G. M. HDL metabolism in Diabetes. Diabet es Metab. Rev, 3, 653-668 (1987).
81. Chernukh A.M., Aleksandrov P.N., Alekseev O.V. Microcirculation. Moscow: Meditsina (1975)*.
82. Chernyshov V.A., Tseluiko V.I. A Clinico-Genealogical Characteristic of IHD Patients with Various Disorders of Lipid Metabolism. Kardiologiya, 3, 13-17 (1992)*.

83. Chiappe de Cingolani G. E. Beta-adrenergic receptors and adenylate cyclase activity in diabetic rat fat cells. Diabetes, 35, 1229-1232 (1986).

84. Cho N. H, Becker D. I, Ellis D. Spontaneous whole blood platelet aggregation, hematological variables and complications in insulin-dependent diabetes mellitus: the Pittsburg Epidemiology of Diabetes Complications study. J. Diabetes Complicat., 1, 12-18 (1992).

85. Christensen N. J, Gundersen H. J, Hegedus L. Acute effects of insulin on plasma noradrenaline and the cardiovascular system. Metabolism, 29, 11, 1138-1145 (1980).

86. Christensen N. J. Plasma norepinephrine and epinephrine in untreated diabetics, during fasting and after insulin administration. Diabetes, 23, 1-8 (1974).

87. Ciaraldi T. P., Molina J. M, Olefsky J. M. Insulin action kinetics in adipocytes from obese and noninsulin-dependent diabetes mellitus subjects: identification of multiple cellular defects in glucose transport. J. Clin. Endocrinol. Metab, 72, 4, 876-882 (1991).

88. Cohen J. C, Chiesa J, Hobbs H. Sequence polymorphisms in the apolipoprotein(a) gene: evidence of dissociation between apolipoprotein(a) size and plasma lipoprotein(a) levels. J. Clin. Invest, 91, 1630-1636 (1993).

89. Cohn J. S, McNamara J. R, Cohn S. D. Postprandial plasma lipoprotein changes in human subjects of different ages. J. Lipid Res, 29, 469-479 (1988).

90. Cohn J. S, McNamara J. R, Schaefer E. J. Lipoprotein cholesterol concentrat ion in the plasma of human subjects as measured in the fed and fasted states. Clin. Chem., 34, 12, 2456-2459 (1988).

91. Collier G. R, Greenberg G. R, Wolever T. M. S. The acute effect of fat on insulin secretion. J. Clin. Endocrinol. Metab, 66, 2, 323-326 (1988).

92. Colwell J. A, Winocour P. D, Lopes-Virella M. New concepts about the pathgenesis of athrosclerosis in diabetes mellitus. Am. J. Med., 75, 67-80 (1983).

93. Comp P. C, Esmos C. T. Generation of fibrinolytic activity by infuson of activated protein C into dogs. J. Clin. Invest., 68, 1221-1228 (1981).

94. Comp P. C, Esmos C. T. Recurrent venous thromboembolism in patients with a partial deficiency of protein S. N. Engl. J. Med., 311, 1525-1528 (1984).

95. Comp P. C, Jacocks R. M, Ferell G. G. Activation of protein C in vivo. J. Clin. Invest., 70, 127-134 (1982).

96. Coneoli A, Nurhyan N, Reilly J. J. Mechanism of increased glyconeogenesis in non-insulin-dependent diabetes mellitus: role of alterations in systemic, hepatic, and muscle lactate and alanin metabolism. J. Clin. Invest, 86, 2038-2045 (1990).

97. Copley A. L. The endothelial fibrin lining, fibrinogen gel clotting, and the endothelium, blood interface. Surface phenomena in hemorheology: their theoretical, experimental and clinical aspects. Eds.: Copley A. L, Seaman G. V. F, New York, 377-396 (1983).

98. Copley A. L. The future of the science of biorheology. Biorheology, 19, 47-69 (1982).

99. Cowan L. D, Wilcosky T, Criqui M. H. Demographic, behavioural, biochemical and dietary correlates of plasma triglycerides. Lipid Research Clinics Program study. Arteriosclerosis, 5, 466-480 (1985).

100. Cralnik H. R, Williams S. B, Coller B. S. Fibrinogen competes with von Willebrand factor binding to the glycoprotein IIbIIIa complex when platelets are stimulated with thrombin. Blood, 64, 797-801 (1984).

101. Crawford M. A. Background to Essential Fatty Acids and their prostanoid derivatives. British Medical Bulletin, 39, 210-213 (1983).

102. Crepaldi G, Nosadini R. Diabetic cardiopathy: Is it a real entity? Diabetes Metab. Rev, 4, 273-288 (1988).

103. Csaszar A, Dieplinger H, Sandholzer C. Plasma lipoprotein(a) concentration and phenotypes in diabetes mellitus. Diabetologia, 36, 47-51 (1993).

104. Curwen K. D, Gimbrone M. A, Handin R. I. In vivo studies of thromboresistance: the role of prostacyclin (PGI_2) in platelet adhesion to cultured normal and virally transformed human vascular endothelial cells. Lab. Invest, 42, 366-374 (1980).

105. D'Angelo V, Villa S, Mysliewiec M. Defective fibrinolytic and prostacyclin-like activity in human atheromatous plaques. Thrombosis and Hemostasis, 39, 535-536 (1978).

106. Dabels J., Preussner S., Braher P. Plasmaaustansch als neues Behandlungsverfahren der homozygoten famillaren Hypercholesterolamie. Dtsch. Gesundh. Wesen, 37, 4, 165-168 (1982).

107. Dairou F. Plasma exchange treatment for severe familial hypercholesterolemia: a comparison of two different techniques. Beitr. Infusionther, 23, S.152-159 (1988).

108. Daubresse J. C. Les triglycerides en Rev. Med. Liege, 1991, 46, 245-255 (1991).

109. Davidenko D.N., Mozzhukhin A.S., Telegin V.V. Mobilization of Physiological Reserves in Strenuous Muscular Activity. Fiziol. Chel., 1, 127-132 (1987)*.

110. Dean J. D, Jones C. J, Hutchison S. J. Hyperinsulinemia and microvascular angina ("syndrome X"). Lancet, 337, 456-457 (1991).

111. DeFronzo R. A, Ferrannini E. Insulin resistance. A multifaceted syndrome respon sible for NIDDM, obesity, hypertension, dyslipidemia and atherosclerotic cardiovascular disease. Diabetes Care, 14, 173-194 (1991).

112. DeFronzo R. A, Ferrannini E. The pathogenesis of non-insulin-dependent diabetes: an update. Medicine, 61, 125-140 (1982).

113. DeFronzo R. A. The triumvirate: beta-cell, muscle, liver: A collusion responsible for NIDDM. Diabetes, 37, 667-687 (1988).

114. D'Elia J. A, Weinrauch L. A, Healy R. W. Myocardial dysfunction without coronary artery disease in diabetic renal failure. Am. J. Cardiol., 43, 193-199 (1979).

115. Demant T, Carlson L. A, Holmquist L. Lipoprotein metabolism in hepatic lipase deficiency: studies on the turnover of apolipoprotein B and on the effect of hepatic lipase on high density lipoprotein. J. Lipid Res., 29, 1603-1611 (1988).

116. Dobordzhginidze L.M. Association of Apolipoproteins B and A-1 of Blood Plasma with IHD and Nutrition in Population. Synopsis of Candidate's Thesis. Moscow (1987)*.

117. Dobashina A.N. Change in the Aggregate State of the Blood and Microcirculation in Stenocardia Patients under the Influence of Gravitational Plasma- and Thrombocytopheresis. Synopsis of Candidate's Thesis. Moscow (1988)*.

118. Donish R.M. The Functional State of the Heart in DM Patients. Synopsis of Candidate's Thesis. Kiev (1977)*.

119. Ducimetiere P., Eschwege E, Papoz L. Relationship of plasma insulin levels to the incidence of myocardial infarction and coronary heart disease mortality in a middle-aged population. Diabetologia, 19, 3, 205-210 (1980).

120. Dudaev V.A., Gorin V.V. Factors Facilitating the Development of Hyperinsulinemia and Its Role in Atherogenesis. Kardiologiya, 26, 4, 106-113 (1986)*.

121. Dudaev V.A., Parfenov A.S., Torkhovskaya T.I. Influence of Disorders in Lipid Metabolism on the Rheological Properties of the Blood in IHD Patients. Kardiologiya, 3, 37 (1983)*.

122. Duerden J. M, Barlett S. M, Gibbons G. F. Regulation of very-low-density lipoprotein lipid secretion in hepatocyte cultures derived from diabetic animals. Biochem. J., 262, 313-319 (1989).

123. Dufaux J. Agregation erythrocytaire et resistance a l'ecoulement. Evolution dans le resean circulatoire. Arteres Veines, 6, 4, 315-318 (1987).

124. Duguid J. B. Thrombosis as factor in the pathogenesis of coronary atherosclerosis. J. of Pathology and Bacteriology, 58, 207-212 (1946).
125. Duman V.L., Rozhdestvenskaya E.D., Oransky I.E. Determining the Effect of B-Adrenergic Blocking in IHD Patients according to the Parameters of Veloergometry and Cardiointervalography. Kardiologiya, 2, 112-113 (1984)*.
126. Durrington P. N, Ishola M, Hunt L. Apolipoproteins(a), AI and B and parental history in men with early onset heart disease. Lancet, 1, 1070-1073 (1988).
127. Dutta T. K, Das A. K. Haemorheology in diabetes. J. Assoc. Physic, 38, 2, 919-920 (1990).
128. Dyeberg J, Bang H. O, Hjorne N. Fatty lipid composition of the plasma lipids in Greenland Eskimos. Am. J. Clin. Nutr, 28, 958-966 (1975).
129. Dynnik V.V. Some Aspects of the Energy Metabolism of the Myocardium. In: Regulation of the Contractive Function and Metabolism of the Myocardium, pp. 113-135. Moscow (1987)*.
130. Dzhanashiya P.Kh. The Clinical Significance of Blood Rheology and the Possibility of Correcting It in Patients with Congenital and Acquired Defects and Ischemic Heart Disease. Synopsis of Doctor's Thesis. Moscow (1988)*.
131. Eckel J, Asskamp B, Reinauer H. Induction of insulin resistance in primary cultured adult cardiac myocytes. Endocrinology, 129, 345-352 (1991).
132. Efendic S, Luft R, Wajngot A. Aspects of the pathogenesis of type 2 diabetes. Endocrine Rev,, 5, 395-410 (1984).
133. Efimov A.S. Diabetic Angiopathies. Meditsina (1989)*.
134. Efimov A.S. Some Problems of Clinical Diabetology. Probl. Endokrinol., 4, 52-57 (1990)*.
135. Eldor A, Fridman R, Vlodavsky I. Interferon enhances prostacyclin production by cultured vascular endothelial cells. J. Clin. Invest., 73, 251-257 (1984)
136. Encholm C, Kuusi T. Preparation, characterization and measurement of hepatic lipase. Methods in Enzymology. Eds.: Albers J. J, Segrest J. P., London: Academic Press, V.129, 716-738 (1986).
137. Enikeev R.G. A Glucose-Tolerant Test and the Dynamics of the Level of NEFA, Cholesterol in Children with Exogeno-Constitutional Obesity. In: The Early Identification of Endocrine Diseases and New Treatment Methods, pp. 97-98. Moscow (1984)*.
138. Episkoposyan N.G., Sisakaya S.A. Microcirculatory and Microhemorheological Mechanisms of Cardiodamaging Effect of Endogenic Monoaminas. Eksperim. i Klinich. Meditsina, 31, 1, 61-65 (1991)*.
139. Epshtein E.V. Hormonal Mechanisms of Obesity Development. Doctor's Thesis. Kiev (1980)*.
140. Erickson H. P., Carrell N. P. Fibronectin in extended and compact conformations. J. Biol. Chem., 258, 14539-14544 (1983).
141. Eriksson J, Franssila-Kalunki A, Ekstrand A. Early metabolic defects in persons at increased risk for non-insulin dependent diabetes mellitus. N. Engl. J. Med., 321, 337-343 (1989).
142. Ermolayev V.V. The Coagulation Activity and Electrophoretic Mobility of Erythrocytes in IHD. Synopsis of Candidate's Thesis. Novosibirsk (1980)*.
143. Ernst E, Kollar L, Matrai A. Placebo-controlled, double-blind study of hemodilution in peripheral arterial disease. Lancet, 1 (8548), 1449-1451 (1987).
144. Eschwege E, Richard J. I, Thibult N. Coronary heart disease mortality in relation with diabetes, blood glucose and plasma insulin levels. The Paris Prospective Study, ten years later. Horm. Metab. Res., 15 (suppl.), 41-46 (1985).

145. Eschwege E. Hyperinsulinemia and hypertriglyceridemia as risk factors for coronary heart disease in diabetic and non-diabetic patients: Presented at the Symposium on Insulin Resistance, Hyperinsulinemia and Dyslipidemias in the Etiology of Coronary Heart Disease, Atlanta, Georgia, 19 Jun (1990).
146. Esmos C. T, Owen W. G. Identification of an endothelial cell cofactor for thrombin-catalyzed activation of protein C. Proc. Natl. Acad. Sci. USA, 78, 2249-2252 (1981).
147. Esmos N. B, Owen W. G, Esmos C. T. Isolation of a membrane-bound cofactor for thrombin-catalyzed activation of protein C. J. Biol. Chem, 257, 859-864 (1982).
148. European Atherosclerosis Society Study Group. The recognition and management of hyperlipidemia in adults: a policy statement of the European Atherosclerosis Society. Europ. Heart J., 9, 571-600 (1987).
149. Expert Panel. Report of the National Cholesterol Education Program Expert Panel on detection, evaluation and treatment of high blood cholesterol in adults. Arch. Intern. Med., 148, 36-69 (1988).
150. Fauvel F, Grant H. E, Legrand Y. J. Interaction of blood platelets with a microfibrillar extract from adult bovine aorta. Requirement for von Willebrand factor. Proc. Natl. Acad. Sci. USA, 80, 551-554 (1983).
151. Feher M. D, Rampling M. W, Sever P. S. Diabetic hypertension, the importance of fibrinogen and blood viscosity. J. Hum Hypertens, 2, 117-122 (1988).
152. Fein F. S, Sonnenblick E. H. Diabetic cardiomyopathy. Prog. Cardiovasc. Res, 27, 255-270 (1985).
153. Felley C. P., Kleiber H, van Melle G. D. Resistance to insulin mediated glucose disposal in obese subjects: respective effect of lipid metabolism and glycemia. Int. J. Obes., 1992, 16, 3, 185-191.
154. Ferland M, Despres J. P., Nadeau A. Contribution of glucose tolerance and plasma insulin levels to the relationships between body fat distribution and plasma lipoprotein levels in women. Int. J. Obes., 15, 10, 677-688 (1991).
155. Ferrannini E, Barrett E. J, Bevilacqua S. Effect of fatty acids on glucose production and utilization in man. J. Clin. Invest, 72, 5, 1737-1747 (1983).
156. Ferrannini E, Bazzigoli G, Bonnadonna R. Insulin resistance in essential hypertension. N. Engl. J. Med., 317, 350-357 (1987).
157. Ferrannini E, Haffner S. M, Mitchell B. D. Hyperinsulinemia: the key feature of a cardiovascular and metabolic syndrome. Diabetologia, 34, 6, 416-422 (1991).
158. Ferstrate M., Fermilen Zh. Thromboses. Moscow: Meditsina (1986)*.
159. Fields L. E, Daugherty A, Bergmann S. R. Effect of fatty acid on performance and lipid content of hearts from diabetic rabbits. Am. J. Physiol., 250, 1079-1085 (1986).
160. Firsov N.N. Macro- and Microrheology of Blood Normally and in Pathology. Synopsis of Doctor's Thesis. Kupavna (1983)*.
161. Firsov N.N., Sinitsin A.A., Lavrent'ev V.V. Rheological Features of the Blood and Erythrocyte Suspensions. Probl. Gematol. i Perelivaniya Krovy, 9, 32-34 (1975)*.
162. Firsov N.N., Zhvaniya G.M. On the Mechanisms of Regulating the Macrorheological Properties of the Blood. Fiziol. Zh. SSSR im. Sechenova, 71, 11, 1362-1366 (1985)*.
163. Fitilev S.B., Saprygin D.B., Migalina L.A. The Basal Level of the Somatotropic Hormone, Cortisol and Insulin in Patients with Chronic Forms of IHD. Klinich. Meditsina, 63, 10, 106-110 (1985)*.
164. Flack J. M, Sowers J. R. Epidemiologic and Clinical Aspects of Insulin Resistance and Hyperinsulinemia. Am. J. Med., 91, 1A, 11S-21S (1991).

165. Fontbonne A, Charles M. A, Thibult N. Hyperinsulinemia as a predictor of coronary heart disease mortality in a healthy population: the Paris Prospective Study, 15-years follow-up. Diabetologia, 34, 5, 356-361 (1991).
166. Fontbonne A, Eschwege E. Diabetes, hyperglycemia, hyperinsulinemia and atherosclerosis: epidemiological data. Diabetes Metab., 13, 3, 350-353 (1987).
167. Fontbonne A, Eschwege E. M, Cambien V. Hypertriglyceridemia as a risk factor of coronary heart disease mortality in subjects with impaired glucose tolerance with impaired glucose tolerance or diabetes: results from the 11-year follow-up of the Paris Prospective Study. Diabetologia, 32, 300-304 (1989).
168. Fontbonne A. Relationship between diabetic dyslipoproteinemia and coronary heart disease risk in subjects with non-insulin-dependent diabetes mellitus. Diabetes Metab. Rev., 7, 179-189 (1991).
169. Forte T. High Density Lipoproteins in Atherosclerosis. Moscow (1983)*.
170. Foster D. W. Insulin resistance, a secret killer? N. Engl. J. Med., 320, 733-734 (1989).
171. Frayn K. N, Coppack S. W. Insulin resistance, adipose tissue and coronary heart disease: Editorial review. Clin. Sci., 82, 1-8 (1992).
172. Fregatova L.M. The Use of Plasmacytopheresis for Treating Thrombus-Forming Diseases of the Peripheral Extremity Arteries. Synopsis of Candidate's Thesis. Moscow (1986)*.
173. Fried S. K, Turkenkopf I. J, Goldberg I. J. Mechanism of increased lipoprotein lipase in fat cells of Obese Zucker rats. Am. J. Physiol., 261, 5, E653-E660 (1991).
174. Friedman J. J. Vascular sensitivity and reactivity to norepinephrine in diabetes mellitus. Am. J. Physiol., 256, 1134-1138 (1989).
175. Frolova I.A., Oleneva V.A. Features of the Metabolism of Lipids and Catecholamines in Complication of Alimentary Obesity by Diabetes Mellitus in the Latent Form. Voprosy Pitaniya, 4, 7-10 (1981)*.
176. Felig P., Baxter J.D., Broadus A.E. (eds.). Endocrinology and Metabolism. New-York: McGraw-Hill (1982).
177. Fujimoto T, Hawiger J. Adenosine diphosphate induced binding of von Willebrand factor to human platelets. Nature, 297, 154-156 (1982).
178. Furkalo N.K., Bol'shakova R.M., Dukhina M.A. Hemostasis in IHD Patients with Microthrombosis in the Microcirculation System. Kardiologiya, 3, 85-89 (1982)*.
179. Galderisi M, Anderson K. M, Wilson P. W. F. Echocardiographic evidence for the existence of a distinct diabetic cardiomyopathy (The Framingham Heart Study). Am. J. Cardiol., 68, 85-89 (1991).
180. Galenok V.A., Dikker V.E., Gostinskaya E.V. Spectrum of Free Fatty Acids and Rheological Properties of Erythrocytes in Persons with Disturbed Tolerance to Glucose and in DM Patients. Probl. Endokrinol., 2, 17-20 (1991)*.
181. Galenok V.A., Gostinskaya E.V., Dikker V.E. Hemorheology in Disorder of Carbohydrate Metabolism. Nauka (1987)*.
182. Galenok V.A., Khanykina I.V. Glycosylized Proteins and Rheological Properties of Erythrocytes in Disorders of Carbohydrate Metabolism. Terapevt. Arkhiv, 12, 113-116 (1991)*.
183. Gandzha I.M., Myagkaya I.P., Bobrik M.V. Immune Reactions in Healthy Persons under Conditions of Emotional Stress and a Single Food Load with Lipids. Terapevt. Arkhiv. 51, 5, 50-52 (1979)*.
184. Ganguly P. K, Dhalla K. S, Innes I. R. Altered norepinephrine turnover and metabolism in diabetic cardiomyopathy. Circ. Res., 59, 684-693 (1986).
185. Ganguly P. K, Thliveris J. A, Mehta A. Evidence against the involvement of non-enzymatic glycosylation in diabetic cardiomyopathy. Metabolism, 39, 769-773 (1990).

186. Gard A, Helderman J. H, Kaffler M. Relationship between lipoprotein levels and in vivo insulin action in normal young white men. Metabolism, 37, 10, 982-987 (1988).
187. Gatsura V.V. Pharmacological Correction of Energy Metabolism in Ischemized Myocardium. Moscow (1993)*.
188. Gavrilov A.O. Effects of Gravitational Plasmapheresis. Gemat. i Transfuz., 9, 6-8 (1991)*.
189. Geller G., Ganefeld M., Yaross V. Lesions of Lipid Metabolism: Diagnosis, Clinic, Therapy. Moscow: Meditsina (1979)*.
190. Genee V.S. Diabetes Mellitus and the Heart. Terapevt. Arkhiv, 80, pp. 142-146 (1980)*.
191. Genest J, Jenner J. L, McNamara J. R. Prevalence of lipoprotein(a) (Lp(a)) excess in coronary artery disease. Am. J. Cardiol., 67, 1039-1045 (1991).
192. Genest J, Sniderman A, Cianflone K. Hyperapobetalipoproteinemia plasma lipoprotein responses to oral fat load. Arteriosclerosis, 6, 1, 297-304 (1986).
193. Gerasimova E.N., Zhukovsky G.S., Nikul'cheva N.G. Dyslipidemia in Populations of Men and Women and Its Relation to IHD. In: Epidemiology and Risk Factors of IHD, pp. 57-71. Leningrad: Meditsina (1989)*.
194. Gerrard J. M, Stuart M. J, Rao G. H. R. Alteration in the balance of prostaglandin and synthesis in diabetic rats. J. Lab. Clin. Med., 95, 950-958 (1980).
195. Gibbons G. F. Assembly and secretion of hepatic very-low-density lipoprotein. Biochem. J, 268, 1-13 (1990).
196. Giorgadze M.L., Panchenko E.P., Gratsiansky N.A. Thromboxan and Prostacyclin to the Blood from the Coronary Sinus in Spontaneous Ischemia of the Myocardium. Kardiologiya, 8, 7-11 (1991)*.
197. Girma J. P., Pietu G, Chopek M. W. Localization of distinct von Willebrand factor domains reacting with ristocetin or thrombin in stimulated platelets using proteolysis and monoclonal antibodies. Circulation, 70 (II), 209 (abstract) (1984).
198. Glezer M.G., Moskalenko N.P. Changes in the Blood Circulation System in DM, a Scientific Review. Moscow (1983)*.
199. Goldstein J. L, Brown M. S, Anderson R. G. W. Receptor-mediated endocytosis: concepts emerging from the LDL receptor system. Ann. Rev. Biol., 1, 1-39 (1985).
200. Goldstein J. L, Brown M. S. Regulation of low density lipoprotein receptors: implications for pathogenesis and therapy of hypercholesterolemia and atherosclerosis. Circulation, 76, 504-507 (1987).
201. Golubyatnikova G.A. Ischemic Heart Disease in Diabetes Mellitus. Probl. Endokrinol., 6, 47-51 (1988)*.
202. Gordon J. L, Pearson J. D. Biology of the vascular endothelium. In: Hemostasis and Thrombosis. Eds.: Bloom A. L, Thomas D. P., Churchill Livingstone, 303-311 (1987).
203. Gotto A. M, Pownall H. J, Havel R. J. Introduction to the plasma lipoproteins. In: Methods in Enzymology. Eds.: Segrest J. P., Albers J. J, London: Academic Press, 3-41 (1986).
204. Grachev S.P. Change in the Hemodynamics and Metabolism of the Myocardium with Acute Complications of Myocardial Infarction under the Influence of Plasmapheresis. Synopsis of Candidate's Thesis. Moscow (1988)*.
205. Granovskaya-Tsvetkova A.M., Ibragimova G.V., Vinogradova I.M. Immunoreactive Insulin and Hormonal Antagonists of Insulin in Obesity. In: Materials of Scientific Conference of Young Georgian Medical Workers, pp. 866-867 (1981)*.
206. Greenberg M. A, Grose R. M, Neuberger N. Impaired coronary vasodilator responsiveness as a cause of lactate production during pacing-induced ischemia in patients with angina pectoris and normal coronary arteries. J. Am. Coll. Cardiol., 9, 743-751 (1987).

207. Griffin J. N, Evan B, Zimmerman T. S. Deficiency of protein C in congenial thrombotic disease. J. Clin. Invest, 68, 1370-1373 (1981).
208. Griffin J. N, Mosher D. F, Zimmerman T. S. Protein C, an antithrombotic protein, is reduced in hospitalized patients with intravascular coagulation. Blood, 60, 261-264 (1982).
209. Griffin J. N. Clinical studies of protein C. Sem. Thromb. Hemost, 10, 162-166 (1984).
210. Grigor'ev P.Ya., Isakov V.A., Yakovenko E.P. Pathogenetic Fundamentals of Alimentary Obesity. Terapevt. Arkhiv, 61, 9, 120-125 (1989)*.
211. Gritsyuk A.I., Angelutsa P.A. The Coagulating and Fibrinolytic Activity of Erythrocytes with a Various Course of the Acute Period of Myocardial Infarction. Kardiologiya, 2, 89-91 (1987)*.
212. Groop L. C, Bonadonna R. C., Simonson D. C. Effect of insulin on oxidative and nonoxidative pathways of free fatty acid metabolism in human obesity. Am. J. Physiol., 263, 1, E79-E84 (1992).
213. Groop L. The glucose/fatty acid cycle. Implications for pathogenesis and treatment of NIDDM. Europ. J. Endocrinol., 130 (suppl. 2), 61 (1994).
214. Groot P. H. E, Scheek L. M. Effects of fat ingestion on high density lipoprotein profiles in human sera. J. Lipid Res., 25, 684-692 (1984).
215. Gryglewski R. J, Bunting S, Moncada S. Arterial walls are protected against deposition of platelets thrombi by a substance (prostaglandin X) which is made from prostaglandin endoperoxides. Prostaglandins, 12, 685-713 (1976).
216. Gryglewski R. J, Dembinska-Kies A, Zmuda Z. Prostacyclin and thromboxan A_2 biosynthesis capacities of the heart, arteries and platelets at various stages of experimental atherosclerosis in rabbits. Atherosclerosis, 31, 385-394 (1978).
217. Gryglewski R. J, Korbut R, Ocetkiewicz A. Generation of prostacyclin by lungs in vivo and its release into the arterial circulation. Nature, 273, 765-767 (1978).
218. Gryglewski R. J, Korbut R, Ocetkiewicz A. Lungs as a generator of prostacyclin - hypothesis on physiological significance. Naunyn-Schmiedberg's Arch. Pharmacol., 304, 45-50 (1978).
219. Gryglewski R. J, Szczeklik A. Prostacyclin and atherosclerosis. Clinical Pharmacology of prostacyclin. Eds.: Lewis P. J, O'Grandy J, New York: Raven Press, 89-95 (1981).
220. Gryglewski R. J. Prostacyclin and atherosclerosis. IX World Congress of Cardiology, 1, 190 (1982).
221. Gudbjarnason S, El-Hage A. N, Whitehurst V. E. Reduced arachidonic acid levels in major phospholipids of heart muscle in the diabetic rat. J. Mol. Cell. Cardiol., 19, 1141-1146 (1987).
222. Haffner S. M, Fong D, Hazuda H. P. Hyperinsulinemia, upper body adiposity, and cardiovascular risk factors in non-diabetics. Metabolism, 37, 4, 338-345 (1988).
223. Haffner S. M, Stern M. P., Hazuda H. Hyperinsulinemia in a population at high risk for non-insulin dependent diabetes mellitus. N. Engl. J. Med., 315, 22-224 (1986).
224. Haffner S. M, Stern M. P., Mitchell B. D. Incidence of type II diabetes in Mexican Americans predicted by fasting insulin and glucose levels, obesity and body fat distribution. Diabetes, 39, 283-288 (1990).
225. Haffner S. M, Valdez R. A, Hazuda H. Prospective analysis of the insulin-resistance syndrome (syndrome X). Diabetes, 41, 6, 715-722 (1992).
226. Hahn R, Muller-Seydlitz P. M, Jockel K. H. Viscoelasticity and red blood cell aggregation in patients with coronary heart disease. Angiology, 40, 10, (1988).
227. Hamby R. I, Zoneraich S, Sherman L. Diabetic cardiomyopathy. JAMA, 229, 1749-1754 (1974).

228. Harpel P. C, Rosenberg R. A_2-macroglobulin and antithrombin-heparin cofactor: modulators of hemostasis and inflammatory reaction. Prog. Haemost. Thromb., 3, 145-189 (1976).
229. Havel R. J, Kane J. P., Kashyap M. L. J. Clin. Invest., 52, 32-35 (1973).
230. Hayashi M, Yamada K. Differences in domain structures between plasma and cellular fibronectins. J. Biol. Chem., 256, 11292-11300 (1981).
231. Heller F. G, Jamart J, Honore P. Serum lipoprotein(a) in patients with diabetes mellitus. Diabetes Care, 16, 819-823 (1993).
232. Heyliger C. E, Prakash A, McNeill J. H. Alterations in cardiac sarcolemmal Ca^{2+} pump activity during diabetes mellitus. Am. J. Physiol., 252, H540-H544 (1987).
233. Higgs E. A, Moncada S, Vane J. P. Effects of prostacyclin (PGI_2) on platelet adhesion to rabbit arterial subendothelium. Prostaglandins, 16, 17-22 (1978).
234. High Density Lipoproteins and Atherosclerosis: Materials of 1st Soviet-American Symposium. Moscow: Meditsina (1983)*.
235. Hiramatsu K, Nozaki H, Ariori S. Reduction of platelet aggregation induced by englycaemic insulin clamp. Diabetologia, 30, 5, 310-313 (1987).
236. Hirsh P. D, Hillis L. D, Campbell W. B. Release of prostaglandin and thromboxan into the coronary circulation in patients with ischemic heart disease. N. Engl. J. Med., 304, 685-696 (1981).
237. Hodgkin D. D, Boucek R. J, Purdy R. E. Dietary lipids modify receptor- and non-receptor-dependent components of a alpha-adrenoreceptor-mediated contraction. Am. J. Physiol., 261, 1465-1469 (1991).
238. Hollenbeck C. B, Chen N, Chen Y. D. Relationship between the plasma insulin response to oral glucose and insulin-stimulated glucose utilization in normal subjects. Diabetes, 33, 5, 460-463 (1984).
239. Hombach V, Boberg H, Godzkowski A. Regression der Koronarsklerose bei Samiliarer Hypercholesterinamie Iia durch sperifische LDL Apherese. Dtsch. Med. Wschr., 111, 45, S.1709-1715 (1986).
240. Hong S. L. Effect of bradikinin and thrombin on prostacyclin synthesis in endothelial cells from calf and pig aorta and human umbilical cord vein. Thromb. Res., 18, 787-795 (1980).
241. Hormann H. Fibronectin, mediator between cells and connective tissue. Klinische Wochenschrift, 60, S.1265-1277 (1982).
242. Hornstra G, Haddeman E, Ten Hoor F. Fish oils, prostaglandins and arterial thrombosis. Lancet, 2, 1080 (1979).
243. Hornstra G, Haddeman E. Effects of dietary fats on the role of platelets in arterial thromboembolism. Platelets and Thrombosis, London: Academic Press, 115 (1977).
244. Horrobin D. F. The roles of essential fatty acids in the development of diabetic neuropathy and other complications of diabetes mellitus. Prostaglandins, 31, 181-197 (1988).
245. Houdijk W. P. M, De Groot P. N. G, Nievelstein P. Von Willebrand factor and fibronectin but not thrombospondin are involved in platelet adhesion to the extracellular matrix of human vascular endothelial cells. Arteriosclerosis, 6, 24-33 (1986).
246. Houdijk W. P. M, Sakariassen K. S, Nievelstein P. Role of factor VIII, von Willebrand factor and fibronectin in the interaction of platelets in flowing blood with monomeric and fibrillar human collagen types I and III. J. Clin. Invest., 75, 531-540 (1985).
247. Howard B. V. Lipoprotein metabolism in diabetes mellitus. J. Lipid Res., 28, 613-628 (1987).

248. Hulper B, Wilms B. Investigations of autonomic diabetic neuropathy of the cardiovascular system. Eds.: Gries F. A, Freund F, Rabe F, Berger H. Aspects of autonomic neuropathy in diabetes, Stuttgart: Georg Thieme Verlag, 77 (1980).
249. Ibragimova T.K. Spectrum of Blood Serum Lipids in DM. Probl. Endokrinol., 5, 33-34 (1990)*.
250. Ikonnikova E.I., Eremin A.K. Influence of a Fat Load on the Lipid Metabolism Parameters in Clinically Healthy Persons and IHD Patients with Initial Normolipidemia. Dep. At NPO Soyuzmedinform, Nov. 15, 1988. No. 16554 (1988)*.
251. International Collaborative Group: Asymptomatic hyperglycemia and coronary heart disease. J. Chron. Dis., 32, 829-837 (1979).
252. Ionov B.V. The Morphofunctional Features of Erythrocytes under Conditions of Disturbance of Microcirculation. Synopsis of Candidate's Thesis. Moscow (1984)*.
253. Ip C, Tepperman H. M, Holohan P. Insulin binding and insulin response of adipocytes from rats adapted to fat feeding. J. Lipid. Res., 17, 6, 588-599 (1976).
254. Ivanov V.V. Hypoinsulinemia and Peroxide Oxidation of Lipids in Emotional-Pain Stress. Probl. Endokrinol., 2, 77-80 (1990)*.
255. Ivashchenko T.I. Functional Properties of Thrombocytes and Their Role in Hemorheological and Microcirculatory Disorders in IHD Patients. Synopsis of Candidate's Thesis. Kiev (1982)*.
256. Iverieli V.P., Metreveli D.S., Tsintsadze G.I. On the Pathogenesis of the Diabetic Lesion of the Myocardium. Kardiologiya, 8, 30-34 (1984)*.
257. Jackson R. L, Gotto A. M. Phospholipids in biology and medicine. N. Engl. J. Med., 280, 24-29, 87-93 (1974).
258. Jarrett R. J, Keen H, McCartney M. Glucose tolerance and blood pressure in two population samples. Their relation to diabetes mellitus and hypertension. Int. J. Epidemiol., 7, 15-24 (1978).
259. Jarrett R. J. Cardiovascular disease and hypertension in diabetes mellitus. Diabetes Metab. Rev, 5, 547-558 (1989).
260. Jarrett R. J. Is insulin atherogenic? Diabetologia, 31, 71-75 (1988).
261. Jarrett R. J. Type 2 (non-insulin-dependent) diabetes mellitus and coronary heart disease chicken, egg or neither? Diabetologia, 26, 99-102 (1984).
262. Jeanrenaud B. Neuroendocrine and metabolic basis of type II diabetes as studied in animal models. Diabetes Metab. Rev, 4, 603-614 (1988).
263. Jennings A. M., Ford I., Murdoch The effects of diet and insulin therapy on coagulation factor VII, blood viscosity and platelet release proteins in diabetic patients with secondary sulphonylurea failure. Diabetic Med., 8, 4, 346-353 (1991).
264. Jensen M, Haymond M, Rizza R. Influence of body fat distribution on free fatty acid metabolism in obesity. J. Clin. Invest., 83, 1168-1173 (1989).
265. Johnson P. Peripheral Blood Circulation. Moscow: Meditsina (1982)*.
266. Joost H. G, Weber T. M. The regulation of glucose transport in insulin-sensitive cells. Diabetologia, 32, 831-838 (1989).
267. Kacharova A.G., Tertov V.V., Metel'skaya V.A. Relation between Level of Lipoprotein-Containing Immune Complexes and Lipid Level in Blood Serum of IHD Patients. Kardilogiya, 2, 39-42 (1991)*.
268. Kadyrova R.Kh., Gendrikson L.N., Mukhameszhanova S.U. Intrasecretory Function of the Pancreas in Alimentary Obesity Patients. In: Pathology of the Endocrine System. Pp. 143-144. Alma-Ata (1984)*.

269. Kahn C. R. Role of insulin receptors in insulin-resistant states. Metabolism, 29, 5, 455-466 (1980).
270. Kalmykov S.A. Some Features of Lipid Metabolism Disorders in Obesity. Klinich. Meditsina, 62, 8, 107-109 (1984)[*].
271. Kandror V.I. The Role of Insulin in the Regulation of Glycemia in Hypermetabolism. Uspekhi Sovrem. Biologii, 96, 2, 280-295 (1983)[*].
272. Kane J. P. Lipid metabolism in mammals. Ed.: Snyder F, New York, V.1, 209-257 (1977).
273. Kannel W. B, Gordon T, Gastelli W. P. Role of lipids and lipoprotein fractions in assessing atherogenesis. The Framingham Study. Prog. Lipid Res., 20, 339-348 (1981).
274. Kannel W. B, McGee D. L. Diabetes and cardiovascular disease: the Framingham study. JAMA, 241, 2035-2038 (1979).
275. Kannel W. B, McGee D. L. Diabetes and cardiovascular risk factors. The Framingham study. Circulation, 59, 8-13 (1979).
276. Karimov E.A., Yunusov M.Yu., Trankin A.V. Use of Plasmapheresis in Combined Treatment of Patients Suffering from Critical Ischemia of the Lower Extremities. Med. Zh. Uzbekis., 11, 26-28 (1991)[*].
277. Karpman V.L., Kaidinova G.A., Lyubina B.G. Hemodynamic Mechanisms of Ensuring the Maximum Transport of Oxygen in an Organism. Fiziol. Chel., 3, 456-462 (1978)[*].
278. Kartelishev A.V. Modern Problems of Obesity in Children. In: Collected Works of 2nd MOLGMI, 167 (issue 35), pp. 45-50 (Series Pediatry) (1981)[*].
279. Kasanicki M. A, Pilch P. F. Regulation of glucose-transporter function. Diabetes Care, 13, 219-227 (1990).
280. Kazimov R.G. Insulin and Contrainsular Hormones (Glucagon, Cortisol, STH) in IHD Patients Depending on the State of the Carbohydrate Metabolism. Synopsis of Candidate's Thesis. Moscow (1982)[*].
281. Kasuga M, Inoue S, Akanuma Y. Insulin receptor function and insulin effects on glucose metabolism in adipocytes from ventromedial hypothalamus-lesioned rats. Endocrinol., 107, 5, 1549-1555 (1980).
282. Katz A. M, Messineo F. C. Lipid-membrane interactions and the pathogenesis of ischemic damage in the myocardium. Circ. Res, 48, 1-16 (1981).
283. Kavtaradze N.P., Kurashvili R.B., Shatirashvili E.G. Lipid Metabolism in the Hypothalamic Syndrome. Sov. Meditsina, 10, 112-114 (1986)[*].
284. Kendysh I.N. Regulation of Carbohydrate Metabolism. Moscow (1985)[*].
285. Kennedy L, Baynes J. W. Non-enzymatic glycosylation and the chronic complications of diabetes: An overview. Diabetologia, 26, 93-98 (1984).
286. Kesaniemi Y. A, Grundy S. M. Increased low density lipoprotein production associated with obesity. Atherosclerosis, 3, 170-177 (1983).
287. Kiesewetter H, Erlenwein S, Jung E. Isovolemic hemodilution in patients with coronary heart disease. Klin. Wochenschr, 66 (suppl. 15), S.8-14 (1988).
288. Kiortsis D, Bruckert E, de Gennes J. L. Atherosclerotic role of lipoprotein little A. Europ. J. Endocrinol., 130 (suppl. 2), 98 (1994).
289. Kipshidze N.N., Varsaladze D.K. On the Possible Pathogenetic Role of Hyperinsulinemia in the Development of Obesity, DM and Atherosclerosis. In: Urgent Problems of Therapy, pp. 449-455. Tbilisi (1980)[*].
290. Kirichenko L.L., Lekokhmakher S.S., Porunova A.K. The State of Lipid Metabolism, Aggregation of Thrombocytes and Microcirculation in Stenocardia Patients during Treatment by Plasmapheresis. Kardiologiya, 8, 10-14 (1990)[*].
291. Klimov A.N. Atherosclerosis. In: Preventive Cardiology, pp. 239-316. Moscow (1987)[*].

292. Klimov A.N., Perova N.V., Tristanov V.F. Lipids and Lipoproteins of Blood Plasma in Male and Female Populations from an Age Aspect. In: Epidemiology and Risk Factors of IHD, pp. 36-56. Leningrad: Meditsina (1989)*.

293. Kliorin A.I. Obesity at a Children's Age. Leningrad: Meditsina (1989)*.

294. Klip A, Paquet M. R. Glucose transport and glucose transporters in muscle and their metabolic regulation. Diabetes Care, 13, 228-243 (1990).

295. Knyazev Yu.A., Turkina T.I, Lebed'kova S.E. Dyslipidemia in Children. Pediatriya, 6, 60-67 (1985)*.

296. Kobaladze A.S., Medulashvili G.V. Prospects of Using Cod Liver Oil in Prophylaxis and Treatment of Dyslipoproteinemia and Atherosclerosis. Terapevt. Arkhiv, 11, 151-154 (1990)*.

297. Koga S., Jamanaga J., Fujii S. Effects of oral and introvenous fat administration on the levels of apoproteins A-1, A-11 and C-111 in human subjects. Atherosclerosis, 41, 115-124 (1982).

298. Kogan B.M. Kuz'mishin L.E., Sokolova D.A. The Functional State of the Cardiovascular System in DM. Sov. Meditsina, 4, 76-79 (1985)*.

299. Kohler N, Lipton A. Platelets as a source of fibroblast growth-promoting activity. Experim. Cell Res., 87, 297-301 (1974).

300. Kokarev A.N., Kardakov Yu.I., Kas'yanov L.S. Pathogenetic Role of Disorders of Metabolism in Erythrocyte Membranes in Patients with Stenocardia of Stress. Kardiologiya, 2, 45-47 (1991)*.

301. Kolodin V.A. The Role of Surplus Body Weight, Insulin, Testosterone, Cortisol in the Development of Hyperlipoproteinemia and Its Treatment in Men with IHD Risk Factors in a Polyclinic. Synopsis of Candidate's Thesis. Moscow (1989)*.

302. Kolodin V.A., Gasilin V.S., Baranov V.A. Relation between the Surplus Body Weight and the Concentration of Cholesterol, Triglycerides and the Level of the Basic Hormones Regulating Lipid Metabolism in Healthy Men 30-50 Years Old. Terapevt. Arkhiv, 58, 12, 24-27 (1986)*.

303. Kolterman O. G, Saekow M, Olefsky J. M. The effects of acute and chronic starvation on insulin binding to isolated human adipocytes. J. Clin. Endocrinol. Metab., 48, 5, 836-842 (1979).

304. Konovalov G.A., Pokrovsky S.N., Trakht I.N. Plasmoimmunosorption of Low-Density Lipoproteins in Treatment of Patients with Hereditary Hypercholesterolemia. Terapect. Arkhiv, 7, 65-70 (1991)*.

305. Koplatadze M.G., Chrkviani N.L. Influence of Growth in Body Weight on Selected Parameters of Lipid Metabolism in Women with Normal Tolerance to Glucose. In: Materials of 8th Congress of Therapeutists of Georgia, pp. 614-616. Kutaisi (1988)*.

306. Korolev S.V., Eventov A.Z., Khadzhidis P.K. A Semiquantitative Appraisal of Scintigrams of the Myocardium with ^{201}Tl. Med. Radiologiya, 4, 24-27 (1983)*.

307. Koroleva T.V. The State of the Blood Viscosity and Its Correction with Drugs in DM Patients. Synopsis of Candidate's Thesis. Moscow (1983)*.

308. Korsakova N.N. Appraising the Efficiency of Hemosorption and Plasmapheresis in Progressing Stenocardia. Synopsis of Candidate's Thesis. Sverdlovsk (1990)*.

309. Kosaka K, Kuzuya T, Akanoma Y. Increase in insulin response after treatment of overt maturity-onset diabetes is independent of the mode of treatment. Diabetologia, 18, 23-28 (1980).

310. Kostareva T.A. The Clinical and Diagnostic Significance of Studying the Lipid Spectrum of Blood in Obesity in Children as a Risk Factor of the Early Development of Atherosclerosis. Candidate's Thesis. Omsk (1989)*.
311. Koudo T, Ogawa K, Satake T. Plasma-free eicosapentaenic acid/arachidonic acid ratio: a possible new coronary risk factor. Clin. Cardiol., 9, 413-416 (1986).
312. Kovsh O.Ya., Dundure B.L., Pirogova T.F. A Fat Load as a Method of Determining Latent Disorders of Lipid Metabolism. In: Specialized Medical Service in the Latvian SSR, pp. 173-175. Riga (1980)*.
313. Kozlov M.P. Pathogenetic Factors of Setting in of Relative Insulin Insufficiency in Hypothalamic Obesity in an Experiment. In: Questions of Endocrinology: Collected Articles, pp. 55-56. Alma-Ata (1989)*.
314. Kozovsky M.I., Gulyamov T.D., Khusainova F.A. Studying the Frequency of Lowering the Sensitivity to Insulin of Obesity Patients. Probl. Endocrinol., 30, 2, 14-18 (1984)*.
315. Kravets E.B. Features of Lipid Metabolism in Children in Risk Groups with Respect to Obesity and with Obesity. Pediatriya, 5, 48-49 (1985)*.
316. Krol J. V, Lundholm K, Bjorntorp P. Hepatic lipid metabolism in severe human obesity. Metabolism, 26, 9, 1025-1031 (1977).
317. Krol V.A. Some Methods of Determining Latent Cardiac Insufficiency by Studying Gas Metabolism. In: Modern Methods of Diagnosis of Cardiac Insufficiency: A Scientific Review, pp. 35-38. Moscow (1974)*.
318. Kucherova K.I., Maksimova E.A., Gubanova Yu.K. Metabolic Syndromes Due to Disturbance of the Functions of the Endocrine Glands in Obesity in Children. In: Coll. Scientific Works of Saratov Med. Inst., pp. 35-37 (1983)*.
319. Kukharchuk V.V., Arabidze G.G. The Use of Extracorporal Methods in Cardiology. Kardiologiya, 10, 7-11 (1986)*.
320. Kuo T. H, Moore K. H, Giacomelli F. Defective oxidative metabolism of heart mitochondria from genetically diabetic mice. Diabetes, 32, 781-787 (1983).
321. Kurmanova A.K. Hypothalamic Syndrome of the Puberty Period: Etiology, Pathogenesis, Clinic, Therapy, Prognosis. Synopsis of Candidate's Thesis. Moscow (1973)*.
322. Kuznik B.I., Morozov V.G., Khavinson V.Kh. A Vessel Wall as an Efferent Regulator of Immunogenesis, Hemostasis, Kallikrein-kinin System and Regenerator Processes. Fiziol. Zh. SSSR, 4, 499-505 (1987)*.
323. Kuznik B.I., Rusyaev V.D. On the Role of the Formed Elements of the Blood and Vascular System in Blood Coagulation: a Review of the Literature and Own Information. Probl. Gemat. i Perelivaniya Krovy, 3, 50-56 (1974)*.
324. Laakso M, Sarlund C, Pyorala K. Prevalence of insulin deficiency among initially non-insulin-dependent middle-aged diabetic individuals. Diabetes Care, 9, 228-231 (1986).
325. Landgraf-Leurs M. M. C, Landgraf R, Loy A. Aggregation and thromboxan formation by platelets and vascular prostacyclin production from BB rats. An animal model for type I diabetes. Prostaglandins, 24, 35-46 (1982).
326. Landin K, Krotkiewski M, Smith U. Importance of obesity for the metabolic abnormalities associated with an abdominal fat distribution. Metabolism, 38, 572-576 (1989).
327. Landin K, Lonnroth M, Krotkiewski M. Increased insulin resistance and fat cell lipolysis in obese but not lean women with a high waist/hip ratio. Euro J. Clin. Invest, 20, 530-535 (1990).
328. Landsberg L. Insulin resistance, energy balance and sympathetic nervous system activity. Clin. and Exper. Hyper, Theory and Practice, A12, 5, 817-830 (1990).

329. Laskin S.A., Methodical Features of Myocardium Scintigraphy with ^{201}Tl Chloride. Med. Radiologiya, 3, 79-81 (1989)[*].
330. Lavau M, Fried S. K, Susini C. Mechanism of insulin resistance in adipocytes of rats fed a high-fat diet. J. Lipid. Res., 20, 1, 8-16 (1979).
331. Lebed'kova S.E., Maleeva N.P., Belova O.K. Functional State of the Cardiovascular System, Vegetative System and Lipoprotein System of Blood in Obesity Patients. Pediatrya, 2, 20-22 (1985)[*].
332. Ledet B, Neubauer B, Christensen N. J. Diabetic cardiopathy. Diabetologia, 16, 207-209 (1979).
333. Leites S.M. Problems of Metabolism Regulation Normally and in Pathology: Selected Works. Moscow: Meditsina (1978)[*].
334. Lekokhmakher S.S. Influence of Gravitational Surgery of the Blood on the Metabolism of Lipids in IHD Patients. Gemat. i Transfuz., 9, 8-12 (1991)[*].
335. Leonova M.V., Razumov V.B. The Role of Erythrocytes in the Pathogenesis of Disorder of the Functional Activity of Thrombocytes in IHD Patients and the Possibility of Correction by Chemotherapy. Kardiologiya, 4, 107-111 (1990)[*].
336. Leschke M, Saddig C, Kleophas W. Rheology changes in the postprandial phase. Klin. Wochenschr, 68 (suppl. 22), S.89-94 (1990).
337. Levin G.Ya., Sheremet'ev Yu.A., Yakhno V.G. A New Approach to Studying the Aggregation of Erythrocytes. Byul. Eksperim. Biol. I Med., 93, 3, 94-96 (1982)[*].
338. Levina L.I. The Heart in Endocrine Diseases. Leningrad: Meditsina (1989)[*].
339. Levine P. H. An acute effect of cigarette smoking on platelet function. A possible link between smoking and arterial thrombosis. Circulation, 48, 619-623 (1973).
340. Levtov V.A., Regirer S.A., Shadrin N.Kh. Rheology of Blood. Nauka (1982)[*]
341. Lillioja S, Mott D. M, Howard B. V. Impaired glucose tolerance as a disorder of insulin action: longitudinal and cross-sectional studies in Pima Indians. N. Engl. J. Med, 318, 1217-1224 (1988).
342. Lindgren F. T. Analysis of lipids and lipoproteins. Ed.: Perkins E. C, Champaigne, 204-225 (1975).
343. Lineitseva I.E. A Fat Load Test in Determining the State of Lipid Metabolism in Healthy Persons, IHD Patients and Their Children. Synopsis of Candidate's Thesis. Novosibirsk (1982)[*].
344. Lipovetsky B.M. Functional Appraisal of the Coronary Blood Flow in Humans. Leningrad (1985)[*].
345. Lipovetsky B.M., Shlimovich P.G., Tryufanov V.F. Tolerance to Carbohydrates Containing Immunoreactive Insulin and Somatotropic Hormone in the Blood Serum in IHD Patients with a Normal and Increased Level of Lipids in the Blood. Terapevt. Arkhiv, 52, 5, 26-30 (1980)[*].
346. Lobzeva V.I., Rylova A.K., Zhemerikina E.V. Adaptation Reserves of Myocardium in IHD Patients. Kardiologiya, 9, 22-24 (1990)[*].
347. Loiko V.I., Kolmakov V.N. Content of Cholesterol in Blood Plasma of Metabolic and Alimentary Obesity. Klinich. Meditsina, 56, 11, 118-121 (1978)[*].
348. Lollar P., MacIntosh S. C, Owen W. G. Reaction of antithrombin III with thrombin bound to the vascular endothelium. Analysis in a recirculating perfused rabbit heart preparation. J. Biol. Chem., 259, 4335-4338 (1984).
349. Lollar P., Owen W. G. Active-site dependent thrombin-induced release of nucleotides from cultured human endothelial cells. Ann. of the New-York Academy of Science, 370, 51-56 (1991).

350. Lopaschuk G. D, Tahiliani A. G, Vadlamudi R. V. S. V. Cardiac sarcoplasmic reticulum function in insulin- or carnitine-treated diabetic rats. Am. J. Physiol., 245, H969-H976 (1983).
351. Lopukhin Yu.M., Archakov A.I., Vladimirov Yu.A. Cholesterolosis. Moscow: Meditsina (1983)*.
352. Loskutoff D. J, Levin E. Properties of plasminogen activators produced by endothelial cells. Biology of endothelial cells. Ed.: Jaffe E. A, Boston, 200-208 (1984).
353. Lowe G. D. O. Blood rheology in arterial disease. Clin. Sci., 71, 2, 137-146 (1986).
354. Lubawy W. C, Valentovic M. A, Atkinson J. E. Chronic cigarette exposure adversely alters ^{14}C-arachidonic acid metabolism in rat lungs, aortas and platelets. Life Sciences, 33, 577-584 (1983).
355. Luk'yanchikov V.S. Controversial Matters of Etiology, Pathogenesis and Treatment of Diabetic Microangiopathy. Kardiologiya, 11, 88-94 (1991)*.
356. Lupanov V.P Tests under Physical Loads in IHD: Criteria, Achievements and Prospects. Kardiologiya, 4, 110-124 (1984)*.
357. Lutai M.I., Ivashchenko T.I. The Clinical Course of the Disease, Functional State of the Myocardium and Rheological Properties of the Blood in Patients with Postinfarction Cardiosclerosis and Unchanged Coronary Arteries according to Data of Coronarography. Kardiologiya, 5, 38-40 (1992)*.
358. Lyubarskaya L.G., Gracheva N.K., Knyazeva A.P. Insulin-Receptor Interaction on the Plasma Membrane Cells of the Liver and Isolated Adipocytes of Rats in Experimental Obesity. Probl. Endokrinol., 27, 1, 66-71 (1981)*.
359. Lyusov V.A., Redchits E.G. The Role of Cholesterol of the Erythrocyte Membranes in Disturbing Their Structural and Functional State in IHD Patients. Kardiologiya, 1, 86-88 (1987)*.
360. Lyusov V.A., Soboleva V.N., Katyshkina N.I. Changes in the Hemostasis System in IHD Patients with Isovolemic Hemodilution. Kardiologiya, 9, 63-67 (1985)*.
361. Mac Intyre D. E, Pearson J. B, Gordon J. L. Localization and stimulation of prostacyclin production in vascular cells. Nature, 271, 549-551 (1978).
362. MacRuzy S. M, Small M, MacCuish A. C. Association of hypertension with blood viscosity in diabetes. Diabetic Med., 5 (9), 839-834 (1988).
363. Mahley R. W. Apolipoprotein E: cholesterol transport protein with expanding role in cell biology. Science, 240, 622-630 (1988).
364. Maichuk E.Yu. Features of Oxygen Metabolism in IHD Patients. Candidate's Thesis. Moscow (1982)*.
365. Makhmudov E.A. Some Aspects of Carbohydrate, Lipid Metabolism and Hormone Level in Obesity and Its Treatment. Candidate's Thesis. Tashkent (1985)*.
366. Malov G.A., Oblovatskaya O.G. Perfusion Scintigraphy of Myocardium with ^{201}Tl in Patients with Anomalous Withdrawal of Left Coronary Artery from Lung Stem: A Clinico-Anatomical Comparison. Kardiologiya, 12, 68-73 (1989)*.
367. Malysheva A.A., Maslennikov O.V. Studying Blood Viscosity for Appraising Peripheral Blood Circulation in IHD Patients. Kardiologiya, 5, 36-40 (1977)*.
368. Malyukov V.M. The Functional State of the Hypothalamo-Hypophyseal-Adrenal System in IHD Patients According to Data of Insulin Test. In: Stress and Its Pathogenetic Mechanisms, pp. 350-351. Kishinev (1973)*.
369. Manson J. E, Colditz G. A, Stampfer M. J. A prospective study of obesity and risk of coronary heart disease in women. N. Engl. J. Med., 322, 882-889 (1990).

370. Mant M. J, Kappagoda C. T, Taylor R. F. Transcoronary platelet activation and consumption in coronary artery disease: studies at rest. Thromb. Res, 50, 1, 202-211 (1988).
371. Manual of Laboratory Operations: Lipid Research Clinical Program, Washington, V.1 (1974).
372. Marcus R. J, Weksler B. B, Jaffe R. A. Enzymatic conversion of prostaglandin endoperoxide H_2 and arachidonic acid to prostacyclin by cultured human endothelial cells. J. Biol. Chem., 253, 7138-7141 (1978).
373. Marin P., Anderson B, Ottosson M. The morphology and metabolism of intra-abdominal adipose tissue in men. Metabolism, 41, 1242-1248 (1992).
374. Markov Kh.M. Pharmacological Correction of Disorders of the Prostacyclin-Thromboxan Balance in IHD: Problems and Prospects. Kardiologiya, 4, 111-118 (1990)*.
375. Markov Kh.M. The Prostacyclin-Thromboxan Balance and IHD Risk Factors. Kardiologiya, 9, 5-9 (1989)*.
376. Martin M. J, Hulley S. B, Browner W. S. Serum cholesterol, blood pressure and mortality: implications from a cohort of 361662 men. Lancet, 2, 933-936 (1986).
377. Martynyuk I.O. Blood Cholesterol in Patients with Alimentary-Metabolic Obesity. Vrach. Delo, 3, 88-90 (1984)*.
378. Masurov A. V. Transitory decrease of platelet count, aggregation and adhesion after plasmapheresis and immunosorption procedure. Thromb. Res., 48, 4, 505-510 (1987).
379. Matkovics B, Varga S. I, Szabo L. The effect of diabetes on the activities of the peroxide metabolism enzymes. Horm. Metabol. Res., 14, 77-79 (1982).
380. Mazhul L.M., Belozerova L.R., Gavrilov V.B. Diagnosis and Prophylaxis of Diabetic Angiopathies. Klin. Meditsina, 11, 76-79 (1991)*.
381. Mazovetsky A.G., Velikov V.K. Diabetes Mellitus. Moscow: Meditsina (1987)*
382. Mazovetsky A.G., Suntsov Yu.I. Commonplace Insulinemia in Healthy Men and Ones Suffering from IHD from 20 to 59 Years Old: Population Investigation. Sov. Meditsina, 1, 7-11 (1983)*.
383. Mc Donald S. W.D, Dupre J, Rodger N. W. Comparison of platelet thromboxan synthesis in diabetic patients on conventional insulin therapy and continuous insulin infusions. Thromb. Res., 28, 705-712 (1982).
384. McGarry J. D, Foster D. W. Regulation of hepatic fatty acid oxidation and ketone body production. Ann. Rev. Biochem, 49, 395-420 (1980).
385. McLean L. R, Demee R. A, Socorro L. Mechanism of action of lipoprotein lipase. Methods in Enzymology. Eds.: Albers J. J, Segrest J. P., London: Academic Press, V.129, 738-763 (1986).
386. Meerson F.Z. Pathogenesis and Prevention of Stressor Ischemic Lesions of the Heart. Moscow (1984)*.
387. Meerson F.Z. The Concept of Prolonged Adaptation. Moscow (1993)*.
388. Meerson F.Z., Kapel'ko V.I. Contraction and Relaxation of the Cardiac Muscle in Compensatory Hypertrophy and Training. In: Metabolism of the Myocardium, pp. 124-144. Moscow (1977)*.
389. Meerson F.Z., Pshennikova M.G. Adaptation to Stressor Situations and Physical Loads. Moscow: Meditsina (1988)*.
390. Memeh C. U, Reid H. L. Plasma and serum viscosity in Nigerian diabetics. Acta Diabetol. Lat., 25, 2, 101-108 (1988).
391. Memeh C. U. Relation between blood plasma viscosity and presence of vascular and neurological complications in diabetic Africans. Horm. Metab. Res., 23, 6, 278-280 (1991).

392. Metreveli D.S., Megreladze I.I., Ordzhonikidze S.P. Influence of Compensation of Diabetes Mellitus on Contracting Function and Mass of Myocardium of Left Ventricle according to Data of Echocardiography. Terapevt. Arkhiv, 10, 18-21 (1982)*.

393. Miettinen T, Mancini M. Management of hypertriglyceridemic patient. Other non-pharmacological approaches. Am. J. Cardiol., 68, 3, 38-39 (1991).

394. Mikaelyan N.P., Maksina A.G., Knyasev Yu.A. State of Erythrocyte Membranes in DM. Lab. Delo, 9, 41-44 (1991)*.

395. Miller T. B. Jr. Phosphorylase activation hypersensitivity in hearts of diabetic rats. Am. J. Physiol., 246, E134-E140 (1984).

396. Minaker K. L, Rowe J. W, Young J. B. Effect of age on insulin stimulation of sympathetic nervous system activity in man. Metabolism, 31, 12, 1181-1184 (1982).

397. Mitchell B. D, Haffner S. M, Hazuda H. P. The relation between serum insulin levels and 8-year changes in lipid, lipoprotein and blood pressure levels. Am. J. Epidemiol., 136, 1, 12-22 (1992).

398. Modan M, Halkin H, Almog S. Hyperinsulinemia, a link between hypertension, obesity, and glucose intolerance. J. Clin. Invest., 75, 809-817 (1985).

399. Modan M, Halkin H, Lusky A. Hyperinsulinemia is characterized by jointly disturbed plasma VLDL, LDL and HDL levels. A population-based study. Arteriosclerosis, 8, 3, 227-236 (1988).

400. Modan M, Halkin H. Hyperinsulinemia or increased sympathetic drive as links for obesity and hypertension. Diabetes Care, 14, 6, 470-487 (1991).

401. Model S.V., Kubantseva I.V., Postnikov A.A. Influence of Plasmapheresis with Substitution Using Dextrane Solutions on Hemostasis Parameters in IHD Patients. Kardiologiya, 9, 37-40 (1991)*.

402. Mogosh G. Thromboses and Embolias in Cardiovascular Diseases. Bukharest (1979)*.

403. Moller L, Kristensen T. S. Plasma fibrinogen and ischemic heart disease risk factors. Ateroscler. Thromb., 11, 2, 344-350 (1991).

404. Molnar D. Insulin secretion and carbohydrate tolerance in childhood obesity. Klin. Pediatr, 202, S.131-135 (1990).

405. Moncada S, Gryglewski R. J, Banting S. A lipid peroxide inhibits the enzyme in blood vessel microsomes that generates from prostaglandin endoperoxides the substance (prostaglandin X) which prevents platelet aggregation. Prostaglandins, 12, 715-733 (1976).

406. Montorsi P., Manfredi M, Loaldi A. Comparison of coronary vasomotor responses to nifedipine in syndrome X and in Prinzmetal's angina pectoris. Am. J. Cardiol., 63, 1198-1202 (1989).

407. Morita I, Kanayasu T, Murota S. Kallikrein stimulates prostacyclin production in bovine vascular endothelial cells. Biochimica et Biophysica Acta, 792, 304-309 (1984).

408. Mosher D. E. Fibronectin. Progress in Haemostasis and Thrombosis. Ed.: Spaet T. H, Grune u Stratton inc, V.5, 11-151 (1980).

409. Mosseri M, Yarom R, Gotsman M. S. Histologic evidence for small-vessel coronary disease in patients with angina pectoris and patent large-coronary arteries. Circulation, 74, 964-972 (1986).

410. Motz W, Vogt M, Rabenay O. Evidence of endothelial dysfunction in coronary resistance vessels in patients with angina pectoris and normal coronary angiograms. Am. J. Cardiol., 68, 996-1003 (1991).

411. Mrochek A.G. Thrombocytes, Lipids and Hormones in Atherosclerosis and IHD. Kardiologiya, 9, 102-104 (1989)*.

412. Murthy K. V, Shipp J. C. Heart triglyceride synthesis in diabetes: selective increase in activity of enzymes of phosphatidate synthesis. J. Mol. Cell. Cardiol., 12, 299-309 (1980).
413. Nabel E. G, Selwyn A. P., Ganz P. Large coronary arteries in humans are responsive to changing blood flow: an endothelium-dependent mechanism that fails in patients with atherosclerosis. J. Am. Coll. Cardiol., 16, 349-356.
414. Nadler J. L, Velasco J. S, Horton R. Cigarette smoking inhibits prostacyclin formation. Lancet, 1, 1248-1250 (1983).
415. Neely J. R, Morgan H. E. Relationship between carbohydrate and lipid metabolism and the energy balance of heart muscle. Ann. Rev. Physiol., 36, 413-459 (1974).
416. Nesto R. W, Philipps R. T. Asymptomatic myocardial ischemia in diabetic patients. Am. J. Med., 80 (suppl. 4C), 40-47 (1986).
417. Neumann F. J, Waas W, Zimmermann R. Hemorheological studies in patients with reduced coronary vasodilatator capacity but normal coronary angiogram (syndrome X). Europ. Heart J, 10, 6, 509-513 (1988).
418. Nicholl T. A, Lopaschuk G. D, McNeill J. H. Effects of free fatty acids and dichloroacetate on isolated working diabetic rat heart. Am. J. Physiol, 261, 1053-1059 (1991).
419. Nichols A. V, Blanche P. J, Krouss R. M. High density lipoprotein methodology. Workshop Report, Bethesda, 303-309 (1979).
420. Nielsen F. S, Voldsgaard A. I, Gall M. A. Apolipoprotein(a) and cardiovascular disease in type 2 (non-insulin-dependent) diabetic patients with and without diabetic nephropathy. Diabetologia, 36, 438-444 (1993).
421. Nikkila E. A, Miettinen T. A, Vesenne M. R. Plasma insulin in coronary heart disease. Lancet, 2, 508-511 (1965).
422. Nikkila E. A. Plasma triglycerides in human diabetes. Proc. R. Soc. Med., 67, 18-21 (1974).
423. Nikul'cheva N.G. Lipoproteins, Dyslipoproteinemia and Atherosclerosis. Leningrad: Meditsina (1984)*.
424. Nikul'cheva N.G., Soliternova I.B., Krivoruchenko I.V. The Activity of Lipoproteinlipase in IHD Patients with Different Types of Hyperlipoproteinemia on the Background of a Fat Load. Vopr. Med. Khimii, 6, 808-812 (1980)*.
425. Niskanen L, Mykkanen L, Karonen S. L. Apoprotein(a) levels in relation to coronary heart disease and risk factors in type 2 (non-insulin-dependent) diabetes. Cardiovascular Risk Factors, 3, 76-81 (1993).
426. Niyazmukhamedov M.B. The Insulin Effect in the Adaptation of an Organism to Muscular Activity. Zdravookhr. Tadzhikistana, 3, 77-79 (1984)*.
427. Nizheradze I.Sh. Morphofunctional Characteristics of Erythrocytes and Rheological Blood Properties in Chronic IHD Patients with Stagnant Blood Circulation Insufficiency. Synopsis of Candidate's Thesis. Tbilisi (1984)*.
428. Nordoy A. Lipids as triggering factors in thrombosis. Thromb. Haemost., 35, 32-48 (1976).
429. Oberly L. W. Free radicals and diabetes. Free Rad. Biol. Med., 5, 113-124 (1988).
430. O'Dea K, Esler M, Leonard P. Noradrenaline turnover during under-and over-eating in normal weight subjects. Metabolism, 31, 9, 896-899 (1982).
431. Oganov R.G., Aliev T.A. Disorder of Carbohydrate Metabolism and Ischemic Heart Disease. Baku (1983)*.
432. Ohlson L. O, Larsson B, Svardsudel K. The influence of body fat distribution on the incidence of diabetes mellitus: 13.5 years of follow-up of the participants in the study of men born in 1913. Diabetes, 34, 1055-1058 (1985).
433. Oivin I.A., Chekalina S.I. A Functional Method of Studying the Ability of the Fibrinolytic System of the Blood to Be Activated. Haemostase, 4, 387-389 (1964)*.

434. Ol'binskaya L.I., Litvitskaya P.F. Coronary and Myocardial Insufficiency. Moscow (1986)*.
435. Olefsky J. M, Saekow M. The effects of dietary carbohydrate content on insulin binding and glucose metabolism by isolated rat adipocytes. Endocrinol., 103, 6, 2252-2263 (1978).
436. Opherk D, Zede H, Weihe E. Reduced coronary dilatory capacity and ultrastructural changes of the myocardium in patients with angina pectoris, but normal coronary arteriograms. Circulation, 63, 817-825 (1981).
437. O'Rahilly S. P., Nugent Z, Rudensky A. S. Beta-cell dysfunction, rather than insulin insensitivity, is the primary defect in familial type 2 diabetes. Lancet, 2, 360-364 (1984).
438. O'Rahilly S. P., Turner R. C, Matthews D. R. Impaired pulsatile secretion of insulin in relatives of patients with non-insulin-dependent diabetes. N. Engl. J. Med., 318, 1225-1230 (1988).
439. Orchard T. J, Becker D. J, Bates M. Plasma insulin and lipoprotein concentrations: an atherogenic association? Am. J. Epidem, 118, 326-337 (1983).
440. Orlov L.L., Shilov A.M., Roitberg G.E. The Contractive Function and Ischemia of the Myocardium. Moscow (1987)*.
441. Orlov V.N., Vasil'eva E.Yu, Shpektor A.V. Features of Disorder of Hemostasis and Lipid Metabolism in Patients with Various Pathogenetic Forms of Stenocardia. Kardiologiya, 8, 14-17 (1990)*.
442. Orlov V.N., Vasil'eva E.Yu., Baspal'ko I.A. The Content of the Willebrand Factor in the Blood Plasma of IHD Patients. Terapevt. Arkhiv, 3, 57-60 (1986)*.
443. Owen W. G, Esmos C. T. Functional properties of an endothelial cell cofactor for thrombin-catalyzed activation of protein C. J. Biol. Chem., 256, 5532-5535 (1981).
444. Packard C. J, Munro A, Lorimer A. R. Metabolism of apolipoprotein B in large triglyceride-rich very low density lipoproteins of normal and hypertriglyceridemic subjects. J. Clin. Invest, 74, 2178-2192 (1984).
445. Panchenko A.V. Clinico-Diagnostic Significance of a Change in the Internal Viscosity of Erythrocytes in IHD and Ways of Correcting It. Synopsis of Candidate's Thesis. Moscow (1986)*.
446. Panina E.B. The Role of the Vegetative Dysfunction in the Pathogenesis of Lipid Metabolism in Patients with Metabolic-Alimentary Obesity. Zh. Nevrologii i Psikhiatrii, 92, 5-12, 22-24 (1992)*.
447. Panze J. A, Quyyumi A. A, Brush J. E. Abnormal endothelium-dependent vascular relaxation in patients with essential hypertension. N. Engl. J. Med., 323, 22-26 (1990).
448. Parfenov A.S., Belousov Yu.B. Relation between Disorders of the Rheological Properties of the Blood and the Hemostasis System in IHD Patients. Kardiologiya, 10, 47-50 (1989)*.
449. Patsch J. R, Karlin J. B, Scott L. W. Inverse relationship between blood levels of high density lipoprotein subfraction 2 and magnitude of postprandial lipemia. Proc. Natl. Acad. Sci. USA, 80, 5, 1449-1453 (1983).
450. Paulson D. J, Crass M. F. Endogenous triacylglycerol metabolism in diabetic heart. Am. J. Physiol., 242, H1084-H1094 (1982).
451. Pavlovsky Yu.M., Regirer S.A., Skobeleva I.M. The Hydrodynamics of Blood. Moscow University Press (1969)*.
452. Peiris A. N, Sothman M. S, Hoffmann R. G. Adiposity, fat distribution and cardiovascular risk. Ann. Intern. Med., 110, 867-872 (1989).
453. Perova N.V. Apolipoproteins in Dyslipoproteinemias and Atherosclerosis. Doctor's Thesis. Moscow (1982)*.

454. Perova N.V. Prospective Directions of Studying the Role of Lipoproteins in the Atherogenesis of Humans: Influence on the Progress of the Disease, Significance in Diagnosis, Possibilities of Correcting Interferences. Kardiologiya, 6, 5-9 (1989)*.
455. Perova N.V., Shcherbakova I.A., Nechaev A.S. Kardiologiya, 10, 91-95 (1985)*.
456. Pieper G. M. Arachidonic acid causes postischemic dysfunction in control but not diabetic hearts. Am. J. Physiol., 258, H923-H930 (1990).
457. Pierce G. N, Dhalla N. S. Mitochondrial abnormalities in diabetic cardiomyopathy. Can. J. Cardiol., 1, 48-54 (1985).
458. Pierce G. N, Dhalla N. S. Sarcolemmal Na^+–K^+ ATPase activity in diabetic rat heart. Am. J. Physiol., 245, 241-247 (1983).
459. Pietu G, Cherel G, Marguerrie G. A new role for fibrinogen: inhibition of von Willebrand factor, platelet interaction. Nature, 309, 6488-6490 (1984).
460. Pimenov L.T., Leshchinsky L.A., Trusov V.V. The Secretion of Insulin and C-Peptide and the Insulin Sensitivity in IHD Patients. Klin. Med., 7, 86-91 (1986)*.
461. Pittilo R. M, Mackie I. J, Rowles P. M. Effects of cigarette smoking on the ultrastructure of rat thoracic aorta and its ability to produce prostacyclin. Thromb. Haemost., 48, 173-176 (1982).
462. Popova Yu.P., Fedortsova L.P., Shevyakova L.V. Surplus Body Weight in Young Persons – a Risk Factor of Incidence of Metabolism Diseases. Vopr. Pitaniya, 13-15 (1989)*.
463. Porte D. Jr, Kahn S. E. Hyperinsulinemia and amyloid in NIDDM: clues to etiology of beta-cell dysfunction? Diabetes, 38, 1333-1336 (1989).
464. Postiglione A, Rubba P., Scarpato N. Increased blood flow to lower after plasma exchange in two patients with familial hypercholesterolemia. Atherosclerosis, 41, 2-3, 421-425 (1982)
465. Potemkin V.V., Zenovko E.I., Gudukina G.N. On the Problem "Insulin and Obesity". In: Questions of Endocrinology: Republican Collection of Scientific Articles, pp. 85-88. Moscow (1982)*.
466. Preobrazhensky D.V., Makhmutkhoszhaev S.A. Influence of DM on the Development of Atherosclerosis. Kardiologiya, 3, 116-121 (1987)*.
467. Pyorala K, Savolainen E, Kaukola S. Plasma insulin as coronary heart disease risk factor: relationship to other risk factors and predictive value during 9 12-year follow-up of the Helsinki Policemen Study population. Acta Med. Scand., 701 (suppl. 1), 38-52 (1985).
468. Pyorala K, Uusitupa M, Laakso M. Macrovascular complications in relation to hyperinsulinemia in non-insulindependent diabetes mellitus. Diabetes Metabol., 13, 3bis, 345-349 (1987).
469. Pyorala K. Hyperinsulinemia as predictor of atherosclerotic vascular disease: epidemiological evidence. NIDDM Today. Glucophage Symposium. Eds.: Maul A. L, Lane E, Schnall D, Wood J. P., Paris: Time Code, G.L.CC. Partners, 11 (1990).
470. Pyorala K. Relationship of glucose tolerance and plasma insulin to the incidence of coronary heart disease: results from two population studies in Finland. Diabetes Care, 2, 2, 131-141 (1979).
471. Quyyumi A. A, Cannon R. O, Panze J. A. Endothelial dysfunction in patients with chest pain and normal coronary arteries. Circulation, 86, 1864-1871 (1992).
472. Raison J, Bonithon-Kopp C, Egloff M. Hormonal influences on the relationships between body fatness, body fat distribution, lipids, lipoproteins, glucose and blood pressure in French working women. Atherosclerosis, 85 (2-3), 185-192 (1990).
473. Randle P. J, Kerbey A. L, Espinal J. Mechanisms decreasing glucose oxidation in diabetes and starvation: Role of lipid fuels and hormones. Diabetes Metab. Rev, 4, 623-638 (1988).

474. Reaven G. M. Abnormal lipoprotein metabolism in non-insulin-dependent diabetes mellitus. Am. J. Med., 83 (suppl. 3A), 31-40 (1987).
475. Reaven G. M. Insulin and hypertension. Clin. and Exper. Hyper. Theory and Practice, A12, 5, 803-816 (1990).
476. Reaven G. M. Insulin resistance and compensatory hyperinsulinemia: role in hypertension, dyslipidemia, and coronary heart disease. Am. Heart J, 121, 4, 1283-1288 (1991).
477. Reaven G. M. Role of insulin resistance in human disease. Diabetes, 37, 1595-1607 (1988).
478. Redchits E.G. The Role of Cholesterol of the Membranes of Erythrocytes and Thrombocytes in the Disturbance of Their Functional State in IHD Patients. Synopsis of Candidate's Thesis. Moscow (1984)*.
479. Regan T. J, Lyons M. M, Ahmed S. S. Evidence for cardiomyopathy in familial diabetes mellitus. J. Clin. Invest, 60, 885-889 (1977).
480. Rillaerts E, Van-Camp K, Wandewoude M. Blood viscosity parameters in coronary heart disease: effect of fish oil supplementation. Acta Clin. Belg., 44, 1, 17-23 (1989).
481. Rimmer T, Fleming J, Kohner E. M. Hypoxic viscosity and diabetic retinopathy (see comments). Br. J. Ophthalmol, 74, 7, 400-404 (1990).
482. Robertson D. A, Sing B. M, Natrass M. Effects of obesity on circulating intermediary metabolite concentrations in the absence of impared glucose tolerance. Int. J. Obes., 1991, 15, 10, 635-645.
483. Rodgers G. M, Greenberg C. S, Shuman M. A. Characterization of the effects of cultured vascular cells on the activation of blood coagulation. Blood, 61, 1155-1162 (1983).
484. Rodrigues B, McNeill J. H. Cardiac function in spontaneously hypertensive diabetic rats. Am. J. Physiol., 251, 571-80 (1986).
485. Romanenko A.I. Rheological Properties of Blood in IHD Patients. Synopsis of Candidate's Thesis. Kiev (1982)*.
486. Romashkin A.V. Features of Reaction of Cardiovascular System under a Physical Load in IHD Patients. Synopsis of Candidate's Thesis. Moscow (1982)*.
487. Rosen P., Buddle T, Reinauer H. Triglyceride lipase activity in the diabetic heart. J. Mol. Cell. Cardiol, 13, 539-550 (1981).
488. Rosenstock J, Strowig S, Cercone S. Reduction in cardiovascular risk factors with intensive diabetes treatment in insulin-dependent diabetes mellitus. Diabetes Care, 10, 729-734 (1987).
489. Ross R, Glomset J. A. The pathogenesis of atherosclerosis, Part 1. N. Eng. J. Med., 295, 369-377.
490. Ross R, Harker L. Hyperlipidemia and atherosclerosis. Science, 193, 1094-1100 (1976).
491. Ross R, Wight T. N, Strandness E. Human Atherosclerosis. Cell constitution and characteristics of advanced lesions of the superficial femoral artery. Am. J. Pathology, 114, 79-93 (1984).
492. Ross R. Platelet-derived growth factor. Ann. Rev. Med., 38, 71-79 (1987).
493. Ross R. The pathogenesis of atherosclerosis, an update. N. Engl. J. Med., 314, 489-500 (1986).
494. Ross R. The smooth muscle cell in connective tissue metabolism and atherosclerosis. Proceedings of the Sigrid Euselius Foundation Symposium in Turku, Finland. The biology of the fibroblast. Ed.: Kulonen E, London: Academic Press, 623-636 (1973).
495. Rowe J. W, Young J. B, Minaker K. L. Effect of insulin and glucose infusions on sympathetic nervous system activity in normal man. Diabetes, 30, 3, 219-225 (1981).
496. Rubba P., Postiglione A, Scarpato N. Improved reactive hyperemia test after plasma exchange in familial hypercholesterolemia. Atherosclerosis, 56, 2, 237-242 (1985).

497. Ryabov G.A. Hypoxia of Critical States. Moscow: Meditsina (1988)*.
498. Ryan J. W, Ryan U. S. Endothelial surface enzymes and the dynamic processing of plasma substrates. Int. Rev. Ex Pathol, 26, 1-43 (1984).
499. Sakariassen K. S, Ottenhof-Rovers M, Sixma J. J. Factor VIII, von Willebrand factor requires calcium for facilitation of platelet adherence. Blood, 63, 996-1003 (1984).
500. Sako Y, Grill V. E. A 48-hour lipid infusion in the rat time-dependently inhibits glucose-induced insulin secretion and B cell oxidation through a process likely coupled to fatty acid oxidation. Endocrinol., 127, 4, 1580-1589 (1990).
501. Sale G. J, Randle P. J. Occupancy of phosphorylation sites in pyruvate dehydrogenase phosphate complex in rat heart in vivo. Biochem J., 206, 221-229 (1982).
502. Salkhanov B.A., Kadyrov R.Kh. Metabolism of Carbohydrates and Lipids in Obesity Patients. In: Matters of the Pathology of the Endocrine System, pp. 145-147. Alma-Ata (1984)*.
503. Saprunova M.M. Investigation of Daily Excretion of Catecholamines in Children Suffering from Constitutional-Exogenic Obesity. Voprosy Pitaniya, 6, 18-20 (1980)*.
504. Savel'eva G.M., Dzhiveletova G.D., Shalina R.I. Hemorheology in Obstetrics. Moscow (1986)*.
505. Sax F. L, Cannon R. O, Watson R. M. Forearm flow in patients with syndrome X: evidence of a generalized disorder of vascular tone? N. Engl. J. Med., 317, 1366-1370 (1987).
506. Saynor R, Verel D, Gillot T. The long-term effect of dietary supplementation with fish lipid concentrate on serum lipids, bleeding time, platelets and angina. Atherosclerosis, 50, 3-10 (1984).
507. Scharf R. E, Harker L. A. Thrombosis and atherosclerosis: regulatory role of interactions among blood components and endothelium. Blood, 55, 131-144 (1987).
508. Schliere G. Discussion-High-density lipoproteins: epidemiology and physiologic influences. Am. J. Cardiol., 52, 4, 208-209 (1983).
509. Schmid-Schonbein H, Skalak R, Vsami S. Cell distribution in capillary networks. Microvasc. Res., 19, 18-44 (1980).
510. Schmid-Schonbein H. Antithixotrope Therapy bei arteriellen VerschBerkrankungen: Was leisten "rheologische MaBnahmen" bei unvermindert bestahen der obliteration? Wien. Med. Wschr, 135, 15-16, S.368-378 (1985).
511. Schmid-Schonbein H. Hemorheology and the experimental basis of classical humoral pathology. Clin. Hemorheol., 1, 179-195 (1981).
512. Schmid-Schonbein H. Microrheology of erythrocytes and thrombocytes, blood viscosity, and the distribution of blood flow in the microcirculation. Handbuch der allgemeinen Pathologie. Ed.: Messen H, Berlin, 289-384 (1977).
513. Schmid-Schonbein H. Microrheology of erythrocytes, blood viscosity, and the distribution of blood flow in the microcirculation. International review of physiology: cardiovascular physiology. Eds.: Guyton A. C, Cowley A. W, Baltimore, V.9, 1-62 (1976).
514. Schwarz H. P., Fischer M, Hopmeier P. Plasma protein S deficiency in familial thrombotic disease. Blood, P., 64, 1297-1300 (1984).
515. Seinberg D. Lipoproteins and the pathogenesis of atherosclerosis. Circulation, 76, 508-514 (1987).
516. Sergeev P.V., Shimanovsky N.L. Receptors of Pathophysiologically Active Substances. Moscow (1987)*.
517. Shabashova N.I. Rheological Properties of Blood in Atherosclerosis Risk Factors and Ways of Their Correction. Synopsis of Candidate's Thesis. Moscow (1984)*.

518. Sharandak A.P., Postnikov A.A., Model S.V. Tolerance to a Physical Load in Patients with Stress Stenocardia after Therapeutic Plasmapheresis. Kardiologiya, 8, 58-60 (1990)*.
519. Shchukina G.N. Tolerance to Fat Load in IHD Patients. Candidate's Thesis. Moscow (1990)*.
520. Shebits I, Krosh G. On the Role of Blood Viscosity in Myocardial Infarction. Klinich. Meditsina, 59, 2, 63-66 (1981)*.
521. Shorin Yu.P., Selyatskaya V.G., Pal'chikova N.A. Features of Hormonal Regulation Mechanisms and Their Correction: All-Union Congress of Pathophysiologists, Abstracts and Reports, p.223. Moscow (1989)*.
522. Sicree R. A, Zimmet P. Z, King H. O. M. Plasma insulin response among Nauruans: protection of deterioration in glucose tolerance over 6 years. Diabetes, 36, 179-186 (1987).
523. Simpson H. C. R, Mann J. I, Meade T. W. Hypertriglyceridemia and hypercoagulability. Lancet, 1, 786-790 (1983).
524. Simpson H. S, Williamson C. M, Olivecrona T. Postprandial lipemia, fenofibrate and coronary artery disease. Atherosclerosis, 85, 2-3, 193-202 (1990).
525. Simpson L. O. Blood viscosity and diabetes mellitus (letter). N. Z. J. Med., 101 (852), 550 (1988).
526. Sinzinger H, Feigl W, Silberbauer K. Prostacyclin generation in atherosclerotic arteries. Lancet, 2, 100-102 (1981).
527. Sinzinger H, Kefalides A. Passive smoking decreases severely platelet sensitivity to antiagreggatory prostaglandins. Lancet, 2, 392-393 (1982).
528. Sinzinger H, Silberbauer K. P., Feigl W. Diminished prostacyclin formation by human atherosclerotic arteries. Lancet, 1, 469-471 (1979).
529. Sinzinger H, Silberbauer K. Winter M. Altered prostacyclin generation of vascular tissue: effect of smooth muscle cells? A review. Atherogenesis, 4, 14-16 (1979).
530. Sixma J. J, Sakariassen K. S, Beeser-Viser N. H. Adhesion of platelets to human artery subendothelium. Effect of factor VIII, von Willebrand factor of various multimeric composition. Blood, 63, 128-139 (1984).
531. Sixma J. J, Sakariassen K. S, Stel H. V. Functional domains in von Willebrand factor. Recognition of discrete factor with platelets and with collagen. J. Clin. Incest, 74, 736-744 (1984).
532. Sixma J. J. Role of blood platelets, plasma proteins and the vessel wall in hemostasis. Hemostasis and Thrombosis. Eds.: Bloom A. L, Thomas D. P., Churchill Livingstone, 283-302 (1987).
533. Skidgel R. A, Printz M. P. PGI_2 production by rat blood vessels, diminished prostacyclin formation in veins compared to arteries. Prostaglandins, 16, 1, 1-16 (1978).
534. Skulachev V.P. Bioenergetics. Membrane Energy Converters. Moscow: Vyssh. Shkola (1989)*.
535. Slavina E.S. Lesions of Heart and Vessels in Diseases of Endocrine Glands. In: Manual on Cardiology, vol. 4, pp. 417-451. Moscow (1982)*.
536. Smith E. B. Relationship between lipids and atherosclerosis. Hemostasis and Thrombosis. Eds.: Bloom A. L, Thomas D. P., Churchill Livingstone, 679-696 (1987).
537. Sokhor K. Ischemic Heart Disease and Diabetes Mellitus. Voenno-Med. Zhurn., 11, 58-59 (1982)*.
538. Sokolov E.I. Emotions and Atherosclerosis. Moscow: Nauka (1987)*.
539. Sokolov E.I. Emotions, Hormones and Atherosclerosis. Utrecht: VSP (1994).
540. Sokolov E.I., Zaev A.P., Sergienko V.B. Physical Workability and Perfusion of the Myocardium in DM Patients. Probl. Endokrinol., 6, 26-28 (1991)*.

541. Solomon S. S, Chiang T. M, Palazzolo M. R. Activation of cyclic AMP phosphodiesterase by phorbol and protein kinase C pathway: differences in normal and diabetic tissue. Endocrine Res., 16, 391-402 (1990).
542. Solov'ev A.N., Kartelishev A.V., Turkina T.I. The "Lipolysis-Lipogenesis" System and Features of Its Regulation in Obesity. In: Obesity and Dyslipidemia in Children, pp. 34-40. Moscow (1980)*.
543. Solov'ev V.V., Bukharov S.I., Sergienko V.B. The Use of Scintigraphy of the Myocardium with ^{201}Tl Combined with a Load Test for Early Diagnosis of IHD. In: 3rd All-Russian Congress of Cardiologists: Abstracts and Reports, pp. 466-467. Sverdlovsk (1985)*.
544. Solun M.N., Kalashnikov A.I, Goch A.N. On the Mechanisms of Lowering the Tolerance to a Dosed Physical Load in Young DM Patients. Probl. Endokrinol., 1, 20-24 (1990)*.
545. Sparke H.W. et al. Regulation of Coronary Blood Circulation. In: Sperelaksis N. (ed.) Physiology and Pathophysiology of the Heart (in two volumes), Trans. from English, vol. 2, ch. 39. Moscow: Meditsina (1990)*.
546. Srivastava K. S, Awasthi K. K. Unsaturated Fatty Acids and Platelet. Clin. Med., 14, 465-472 (1983).
547. Stalder M, Pometta D, Suenram A. Relationship between plasma insulin levels and high density lipoprotein cholesterol levels in healthy men. Diabetologia, 21, 544-548 (1981).
548. Starkova N.T. (ed.) Clinical Endocrinology: a Guide. Moscow: Meditsina (1991)*.
549. Starkova N.T. Clinical Endocrinology. Moscow: Meditsina (1983)*.
550. Starkova N.T., Kandror V.I. Causes and Mechanisms of Development of Endocrine Diseases. In: Clinical Endocrinology, pp. 5-12. Moscow (1991)*.
551. Steiner G. Diabetes and atherosclerosis: an overview. Diabetes, 30 (suppl. 2), 1-7 (1981).
552. Stel H. V, Sakariassen K. S, De Groot P. Von Willebrand factor in the vessel wall mediates platelet adherence. Blood, 65, 85-90 (1985).
553. Stern D, Brett J, Hams K. Participation of endothelial cells in protein C-protein S anticoagulation pathway: the synthesis and release of protein S. J. Cell Biol., 102, 1971-1978 (1986).
554. Stone P. H, Muller J. E, Hartwell T. The effect of diabetes mellitus on prognosis and serial left ventricular function after acute myocardial infarction. Contribution of both coronary disease and diastolic left ventricular dysfunction to the adverse prognosis. J. Am. Coll. Cardiol., 14, 49-57 (1989).
555. Stout R. W, Vallance-Owen J. Insulin and atheroma. Lancet, 1, 1078-1080 (1979).
556. Stout R. W. Insulin and atheroma: 20-year perspective. Diabetes Care, 13, 631-654 (1990).
557. Su X. L, Xin D. I, Xiao F. X. Analysis of hemorheological manifestation and serum lipids in patients with diabetes mellitus. Ien Ko Hsuch Pao, 5, 3-4, 98-100 (1989).
558. Suntsov Yu.I., Zhukovsky G.S., Mazovetsky A.G. The Role of the Insulinemia Level in the Dissemination of Dyslipoproteinemia and IHD in Men 20-69 Years Old. Kardiologiya, 24, 11, 36-39 (1984)*.
559. Svedberd J, Bjorntorp P., Smith U. Insulin binding, degradation and action in rat hepatocytes is reduced in a dose-dependent way by free fatty acids. Int. J. Obes., 13, 556A (1989).
560. Szatkowski N. S, Miller C. M, Endres-Brooks J. L. Clinical studies of human protein S and identification of a patient with protein S deficiency and thrombotic complication. Circulation, 70 (II), 204-206 (1984).
561. Taskienen M. R, Kuusi T. High density lipoproteins in postrandial lipemia. Relation to sex and lipoprotein lipase activity. Atherosclerosis, 59, 121-130 (1986).

562. Tassava T. M, Okuda T, Romsos D. R. Insulin secretion from ob/ob mouse pancreatic islets: effects of neurotransmitters. Am. J. Physiol., 262, 3, E338-E343 (1992).
563. Taton Ya. Obesity: Pathophysiology, Diagnosis, Treatment. Trans. From Polish. Warsaw: Polish Med. Publ. (1981)*.
564. Teien A, Abilgaard U, Hook M. The anticoagulant effect of heparin sulphate and dermatan sulphate. Thromb. Res., 8, 859-867 (1976).
565. Temple R. C, Carrington C. A, Luzio S. D. Insulin deficiency in non-insulin-dependent diabetes. Lancet, 1, 293-295 (1989).
566. Teplov S.I. Hormonal Regulation Factors. In: Physiology of Blood Circulation. Regulation of Blood Circulation, pp. 94-110. Leningrad (1986)*.
567. Tepperman J., Tepperman H.M. Metabolic and Endocrine Physiology, 5th ed. Chicago: Year Book Med. Publ., Inc. (1987).
568. Teuscher A, Egger M, Herman J. B. Diabetes and hypertension. Blood pressure in clinical diabetic patients and a control population. Arch. Int. Med., 149, 1942-1945 (1989).
569. The Clinical Significance of Studying Disorders of the Rheological Properties of the Blood in Patients with Hypertension. Terapevt. Arkhiv, 5, 89-94 (1990)*.
570. The University Group Diabetes Program. A study of the effects of hypoglycemic agents on vascular complications in patients with adult-onset diabetes. V. Evaluation of pheniformin therapy. Diabetes, 24 (suppl.), 165-184 (1975).
571. Thiebaud D, De Fronzo R. A, Jacot E. Effect of long chain triglyceride infusion on glucose metabolism in men. Metabolism, 31, 12, 1128-1136 (1982).
572. Thompson G. R. Plasma lipids and lipoproteins, and the hyperlipoproteinemias. Biochemical Aspects of Human Disease. Eds.:Elkeles R. S, Tavill A. S, Oxford: Blackwell, 85-123 (1983).
573. Thorburn A. W, Gumbiner B, Brechtel G. Effect of hyperinsulinemia and hyperglycemia on intracellular glucose and fat metabolism in healthy subjects. Diabetes, 1990, 9, 1, 22-30.
574. Thorens B, Charron M. J, Lodish H. F. Molecular physiology of glucose transporters. Diabetes Care, 13, 209-218 (1990).
575. Thorngren M, Shafi S, Born G. V. R. Delay in primary hemostasis produced by a fish diet without change in local thromboxane A_2. Br. J. Hemat., 58, 567 (1984).
576. Tishenina R.S. Hormones, Lipids, Lipoperoxides and Antioxidant Activity of Blood in Puberty-Adolescent Dyspituitarism. In: Problems of Endocrinology: Republican Collection of Scientific Works, pp. 19-22. Moscow (1983)*.
577. Todd A. S. Fibrinolysis autographs. Nature, 181 (4607), 495-496 (1958).
578. Trovati M, AnfossiG, Cavalot F. Insulin directly reduced platelet sensitivity to aggregation agents. Diabetes, 37, 6, 780-786 (1988).
579. Tuck M. L, Sowers J. R, Dornfeld L. Reductions in plasma catecholamines and blood pressure during weight loss in obese subjects. Acta Endocrinol., 102, 2, 252-257 (1983).
580. Tulukov A.N., Tatulian S.A., Andreeva E.V. On the Role of Plasma and Cell Factors in Erythrocyte Aggregation Processes. Gemat. i Transfuz., 31, 6, 12-15 (1986)*.
581. Turitto V. T, Weiss H. J, Baumgartner H. K. Decreased platelet adhesion on vessel segments in von Willebrand's disease: defect in initial platelet attachment. J. Lab. Clin. Med., 102, 551-564 (1983).
582. Uglanova T.M. The Pathogenesis, Classification and Therapy of Obesity. In: Pathology of Metabolism: Collected Scientific Works, pp. 83-91. Nizhny Novgorod (1992)*.
583. Uusitupa M, Siitonen O, Pyorala K. Left ventricular function in newly diagnosed non-insulin-dependent (Type 2) diabetics evaluated by systolic time intervals and echocardiography. Acta Med. Scand., 217, 379-388 (1985).

584. Van de Werve G, Zaninetti D, Lang U. Identification of a major defect in insulin-resistant tissues of genetically obese (fa/fa) rats: impaired protein kinase C. Diabetes, 36, 310-314 (1987).

585. Van Gaal L. F, Nobels F. R, Rilaerts E. G. Hypertension in obese and non-obese non-insulin dependent diabetics: a matter of regional adiposity? Diabetes Metab., 14, 289-293 (1988).

586. Van Gaal L. F, Rilaerts E, Creten W. Relationship of body fat distribution pattern to atherogenic risk factors in NIDDM. Preliminary results. Diabetes Care, 11, 103-106 (1988).

587. Vane J. R, Dunting S, Moncasa S. Prostacyclin in physiology and pathophysiology. Int. Rev. Ex Pathol., 23, 161-207 (1982).

588. Vargaftig B. B, Benveniste J. Platelet activating factor today. Trends Pharmacol. Sci., 4, 341-343 (1983).

589. Vashchenko E.A., Doloman L.B. Features of Hemodynamics in Patients with Hypothalamic Syndrome. Fiziol. Zh., 5, 688-691 (1981)*.

590. Vasil'ev A.P. Determining the Index of Erythrocyte Deformability. Lab. Delo, 9, 44-46 (1991)*.

591. Vasil'ev V.N. Sympathico-Adrenal Activity in Various Functional States of a Human. Synopsis of Doctor's Thesis. Moscow (1981)*.

592. Vasil'ev V.N., Chugunov B.S. Sympathico-Adrenal Activity in Various Functional States of a Person. Moscow: Meditsina (1985)*.

593. Vasil'eva E.Yu., Shpektor A.V., Raskurazhev A.B. Thrombocyte Hemostasis in Patients with Various Forms of Stenocardia. In: Physiology and Pathology of Hemostasis, pp. 69-70. Poltava (1991)*.

594. Vertkin A.L., Svetova Yu.B., Vasyagin A.I. The Use of Cod Liver Oil for Correcting the Atherogenic Dyslipoproteinemias. Kardiologiya, 3, 88-92 (1992)*.

595. Virsaladze D.K., Beridze N.G., Asatiani N.G. The Parameters of the Sensitivity to Insulin in Obesity. In: Modern Methods of Diagnosis and Treatment in Diabetology, Cardiology and Hepatology: Materials of Scientific Conference, pp. 14-17. Tbilisi (1984)*.

596. Virsaladze D.K., Beridze N.G., Pachuliya L.S. Level of IRI in Blood and Parameter of Physical Workability in Patients with Various Types of Obesity. In: Scientific Session of Research Inst. of Experimental and Clinical Therapy: Abstracts of Reports, pp. 91-92. Tbilisi (1985)*.

597. Volzhin A.I., Subbotin Yu.K Adaptation and Compensation – a Universal Biological Mechanism of Adaptation. Moscow (1987)*.

598. Vorob'eva E.N. The Role of Erythrocytes in Regulating the Level of Blood Cholesterol in Experimental Hypercholesterolemia and in Constant Donors. Synopsis of Candidate's Thesis. Novosibirsk (1983)*.

599. Vouillamoz D, Temler E, Jeduier E. Importance of substrate competition in the mechanism of insulin resistance in man. Metabolism, 36 (8), 715-720 (1987).

600. Voznesenskaya T.G. Cerebral-Vegetative Relations in Obesity. Candidate's Thesis. Moscow (1980)*.

601. Wall S. R, Lopaschuk G. D. Glucose oxidation rates in fatty acid-perfused isolated working hearts from diabetic rats. Biochim. Biophys. Acta, 1006, 97-103 (1989).

602. Ward W. K, Beard J. E, Halter J. B. Pathophysiology of insulin secretion in non-insulin-dependent diabetes mellitus. Diabetes Care, 7, 491-502 (1984).

603. Watanabe K, Ischida T, Yoshitomi F. Fibrinogen degradation peptides influence PGI_2 synthesis by cultured porcine aortic endothelial and smooth muscle cells. Atherosclerosis, 51, 151-161 (1984).

604. Wattigney W. A, Harsha D. W, Srinivasan S. R. Increasing impact of obesity on serum lipids and lipoproteins in young adults. The Bogalusa Heart Study. Arch. Int. Med., 151, 10, 2017-2022 (1991).
605. Weed R. L, La 'Cella P. L. ATP dependence of erythrocyte membrane structure and function, Philadelphia, 310-336 (1970).
606. Weksler B. B, Marcus A. J, Jaffe E. A. Synthesis of prostaglandin I_2 (prostacyclin) by cultured human and bovine endothelial cells. Proc. Natl. Acad. Sci. USA, 74, 3922-3926 (1977).
607. Weksler B. B, Pett S. B, Alonso D. Differential inhibition by aspirin of vascular and platelet prostaglandin synthesis in atherosclerotic patients. N. Engl. J. Med., 308, 14, 800-805 (1983).
608. Welborn T. A, Wearne K. Coronary heart disease and cardiovascular mortality in Busselton with reference to glucose and insulin concentrations. Diabetes Care, 2, 154-160 (1979).
609. Wennmalm A, Alster P. Nicotine inhibits vascular prostacyclin but not platelet thromboxan formation. General Pharmacol, 14, 189-192 (1983).
610. Werns S. W, Walton J. A, Hsia H. H. Evidence of endothelial dysfunction in angiographically normal coronary arteries of patients with coronary artery disease. Circulation, 79, 287-291 (1989).
611. West K. M, Ahuja M. M. S, Bennett P. H. The role of circulating glucose and triglyceride concentrations and their interactions with other "risk factors" as determinants of arterial disease in nine diabetic population samples from the WHO Multi National study. Diabetes Care, 6, 361-369 (1983).
612. Woolf N. P., Pittilo R. M, Machin S. J. Cigarette smoking and platelet adhesion. Lancet, 2, 1091 (1983).
613. Woolf N. P. Thrombosis and atherosclerosis. Hemostasis and Thrombosis. Eds.: Bloom A. L, Thomas D. P., Churchill Livingstone, 651-678 (1987).
614. Yamamoto A, Yamamura T, Kawaguchi A. Triglyceride and glucose intolerance as a risk factor for coronary heart disease. Cardiology, 78 (3), 185-193 (1991).
615. Yareshchenko V.T., Zelikman A.M., Lyagkuti O.T. Clinical Features of Myocardial Infarction in DM. Vrach. Delo, 8, 70-73 (1983)*.
616. Yoshimatsu H, Niijima A, Oomura V. Effects of hypothalamic lesion on pancreatic autonomic nerve activity in the rat. Brain Res., 303, 1, 147-152 (1984).
617. Yui Y, Aoyama T, Morishita H. Serum prostacyclin stabilizing factor is identical to apolipoprotein A-1 (apo-A-1). J. Clin. Invest., 82, 803-807 (1988).
618. Zadionchenko V.S., Besprozvanny A.B., Dzhanashiya P.Kh. Influence of Therapeutic Plasmapheresis on Rheological Properties of Blood in IHD Patients. Sov. Meditsina, 10, 3-6 (1991)*.
619. Zaev A.P., Ol'kha R.P., Khar'kov S.A. Role of Changes in Central and Intracardial Hemodynamics in the Lowering of the Ability to Work in IHD and DM Patients. Terapevt. Arkhiv, 9, 53-56 (1989)*.
620. Zatulina M.V. State of the Hormonal and Metabolic Parameters in Patients Suffering from Puberty and Adolescent Dyspituitarism. Candidate's Thesis (1989)*.
621. Zavaroni I, Bonora E, Bagliara M. Risk factors for coronary artery disease in healthy persons with hyperinsulinemia and normal glucose tolerance. N. Engl. J. Med., 320, 702-706 (1989).
622. Zavaroni I, Chen Y-D. I, Mondon C. E. Ability of exercise to inhibit carbohydrate-induced hypertriglyceridemia in rats. Metabolism, 1981, 30, 5, 476-480.
623. Zelinsky B.A. The Functional State of the Cardiovascular System in DM. Kiev (1984)*.

624. Zick Y. The insulin receptor: structure and function. Critical Rev. Biochem. Mol. Biol., 24, 217-269 (1969).
625. Zingerman L.S., Bayurova N.V. Relation between Level of Lipids in Blood Serum and State of Coronary Arteries in IHD Patients. Kardiologiya, 2, 42-44 (1984)*.
626. Zubairov D.M. Coagulation of Blood. Kazan University Press (1966)*.
627. Zucker M. B, Mosesson M. W, Brockman M. J. Release of platelet fibronectin (cold-insoluble) from alpha granules induced by thrombin or collagen: Lack of requirement for plasma fibronectin in ADP-induced platelet aggregation. Blood, 54, 8-12 (1979).
628. Zuniga-Guarjardo S, Steiner G, Zinman B. Insulin resistance and action in hypertriglyceridemia. Diabetes Res. Clin. Pract., 14, 55-62 (1991).

INDEX